THE **BOREAL HERBAL**

INTERNATIONAL AWARDS:

Next Generation Indie Book Awards,
GOLD GRAND PRIZE, Best Design,
GOLD, Overall Design. Finalist in Science/ Nature/ Environment

GOLD Nautilus Award,
Health & Healing / Wellness / Prevention / Vitality

SILVER Independent Publishers Book Award (IPPY),
Health / Medicine and Nutrition

SILVER Benjamin Franklin Award,
Independent Book Publishers Association (IBPA),
Nature & Environment

Living Now Book Award
GOLD Book of the Year, Earth

"The North's Favourite Cookbook,"
2012 CBC Cross Country Cookbook Shelf

Important notice to readers: Plants have distinct effects on different people, so it is vital that you consult your health care provider before considering any self-directed treatment of a medical condition, including the use of wild food and medicine plants. Even a healthy person may experience an individual adverse reaction to a plant that is not harmful to other people. Be vigilant when consuming wild plants: if you notice any negative effects, discontinue use and consult your health care provider immediately.

The beneficial plants that are discussed in this book might closely resemble other plants that are toxic to human health. Before you consume any wild plant, be certain that you have correctly identified it. This book is intended for a general audience. It is not intended to diagnose or treat conditions in individual readers.

Great care has been taken in the preparation of this book, but there is still much to be learned about the use of northern wild plants for food and medicine. The information in this book is made available to readers on the basis that the publisher and those involved in the preparation and sale of this publication do not guarantee the accuracy or completeness of the information and will not be held responsible for errors or omissions of any kind.

Cover photographs: Spruce tips, BG; fireweed forest, FM; hands, BG; jars, BG; author photo, CA; map, Tanya Handley.

Previous page: After a forest fire, fireweed blazes across the charred landscape. FM

COPYRIGHT © 2011 BY BEVERLEY GRAY
03 04 05 06 07 — 16 15 14 13 12

All rights reserved. No part of this publication may be reproduced, stored in a retrieval system or transmitted, in any form or by any means, without prior permission of the publisher or, in the case of photocopying or other reprographic copying, a licence from Access Copyright, the Canadian Copyright Licensing Agency, 1 Yonge Street, Suite 1900, Toronto, Ontario, M5E 1E5, www.accesscopyright.ca, 1-800-893-5777, info@accesscopyright.ca.

Published by:
Aroma Borealis Press
Box 10056, Whitehorse, Yukon, Canada, Y1A 7A1
www.aromaborealis.com, www.borealherbal.com
ISBN 978-0-9868271-0-5

Editorial development and production services provided by:
    Black Swan Services,
    www.blackswanservices.com
Editing: Nadine Pedersen and Tamara Letkeman
Copy editing: Lily Gontard
Proofreading: Patricia Wolfe
Indexing: Elph Text Services
Cover and page design: Swanlight Design
Photographs by: Cathie Archbould (CA), Michelle Clusian (MC), Robert Frish (RF), Bev Gray (BG), Randi Hausken (RH), Birch Kuch (BK), Berwyn Larson (BL), Peter Long (PL), Fritz Mueller (FM), and Robert Rogers (RR). For more information on the main contributors, see page 440.

All botanical illustrations, as well as the glossary of botanical terms and the botanical descriptions (which were adapted from William J. Cody's book, *Flora of the Yukon Territory*, 2nd ed., NRC Research Press, Ottawa, Ontario) were provided with permission of Canadian Science Publishing, Ottawa.

Agriculture and Agri-Food Canada
Agriculture et Agroalimentaire Canada

Aroma Borealis Press and the author gratefully acknowledge financial assistance from the Government of Canada through the Agriculture and Agri-Food Canada, Advancing Canadian Agriculture and Agri-Food (ACAAF) Program.

Library and Archives Canada Cataloguing in Publication

Gray, Beverley, 1966-
    The boreal herbal : wild food and medicine plants of the North / Beverley Gray.

Includes bibliographical references and index.
ISBN 978-0-9868271-0-5 — 978-1-8964455-6-4 (CCIP)

    1. Medicinal plants—Canada, Northern. 2. Medicinal plants—Canada, Northern—Identification. 3. Materia medica, Vegetable—Canada, Northern. 4. Herbals—Canada, Northern.
I. Canadian Circumpolar Institute II. Title. III. Series: Occasional publication series (Canadian Circumpolar Institute) ; no. 70

QK99.C3G73 2011   581.6'340971   C2011-900489-5

To the people and the plants of the Northern boreal forest.

For my children, Ceilidh-Anne and Markie-May Gray Bailie, and my husband, Mike Bailie, for "living in the clover" with me.

My parents, Deanna and Jack Gray for giving me encouragement, support and most of all strong roots that gave me the foundation to grow and flower. Anne and Glenn Bailie for your constant love and support. And in memory of Colm Keane (1931–2009) for sharing his stories of the Gaelic ways.

I would like to add a special dedication to my dad, who passed away as I was nearing the completion of this book. John Stephen (Jack) Gray (1935–2011). I love you forever, Dad.

TABLE *of*

Foreword 10
Preface and Acknowledgements 11
Introduction 13

Part I: Getting Started

How to Use This Book 19
Before You Go Harvesting 21
Healing Plant Basics 25
Sacred-Spirit Plant Healing 28
Preserving Wild Plants 32
The Kitchen Apothecary 36

Part II: Plant Profiles

Herbs

Arnica 47

Wild Chamomile 59

Coltsfoot 73

Bearberry 51

Chickweed 63

Dandelion 77

Bedstraw 55

Red Clover 67

Devil's Club 83

CONTENTS

Dock 87

Fireweed 91

Gentian 95

Goldenrod 98

Horsetail 102

Labrador Tea .. 106

Lamb's Quarters 111

Lungwort 115

Mint 118

Stinging Nettle 123

Wild Onion ... 129

Plantain 132

Rhodiola 137

River Beauty .. 141

Rose 144

Wild Sage 149

Shepherd's Purse 154

Mountain Sorrel 157

Sheep Sorrel .. 160

Strawberry Blite 164

Sweetgrass ... 167

Twinflower 171

Valerian 174

Yarrow 178

Berries

- Blueberry 187
- Cloudberry 191
- Cranberry 195
- Highbush Cranberry ......200 
- Currants 204
- Juniper 208
- Mossberry 214
- Raspberry 217
- Saskatoon Berry 221
- Soapberry 225
- Strawberry 229

Trees

- Alder 235
- Trembling Aspen ....238 
- Birch 241
- Subalpine Fir 247
- Lodgepole Pine 251
- Balsam Poplar ...256 
- Spruce 259
- Tamarack 265
- Willow 269

Part III: Plant Preparations and Recipes

Medicinal Preparations and Recipes

Herbal Teas 279
Tinctures, Vinegars, Elixirs, and Bitters 291
Syrups 298
Capsules 301
Plant Essences 302
Topical Treatments 304
Essential Oils and Hydrosols 321
Boreal Baths and Steams 324

Getting Wild in the Kitchen

Cooking with Wild Foods 333
Raw Juices and Smoothies 334
Muffins, Biscuits, and Pancakes 342
Jam, Jellies, Chutney, and Topping Syrups 348
Sauces and Dressings 356
Soups and Salad 363
Main Courses and Sides 366
Desserts and Sweet Treats 371

Part IV: Economics

Starting Up a Non-timber Forest-products Business 381

Part V: For Reference

Herbalists through the Ages 389
The Celtic Tree Alphabet 392
Botany Basics 393
Botanical Terms 395
Understanding Herbal Constituents and Phytochemistry 401
Medicinal Actions of Plants 403
Charts: Herbal Usage .... 406, Medicinal Actions .... 416, Vitamins 420
Bibliography 424
Index 428
Contributors 440

FOREWORD

If we were to go back as far as we could go in time, we would find that man has always depended on plants of all kinds to be used for medicine, food, and shelter.

Today, we are fortunate to be able to bring this knowledge back into the open again, as it was not too many years ago that a lot of people were told that the medicine they made, and used, would not work. They were forced to continue using it in secret.

We say "thank you" to all of our grandmothers and grandfathers and parents for continuing to teach their younger generations all that they know about using these plants, our most precious gifts.

In daring so, we are taught to love and respect our land, water, air, and all the animals, so that we take heed—as we are losing these too quickly.

It is through people like Bev Gray, who have learned to live close to the land and learned from the people who live and use plants, that we will have a record of how to help ourselves stay healthy and strong. After all, it is these very same plants that are used to make the medicines that we get today from the drugstores.

And yes, we listen to and talk to the plants.

Have you ever hugged a tree?

Ruth Welsh
Gwich'in Elder

Above: Plump cloudberries can be stored for months in a cold cache. RF

Below: Tender young spruce tips. BG

Right: As its name suggests, Labrador tea is delicious when steeped in hot water. CA

PREFACE AND ACKNOWLEDGEMENTS

Fringing the northern part of the planet, the boreal forest is a vast and busy place. Rich in plant and animal life, its ecology has been essential to the dietary, medicinal, and spiritual health of indigenous people around the circumpolar north for thousands of years.

In many places, traditional knowledge has been passed down orally from one generation to the next but largely remains unrecorded. While I have included references to indigenous plant names and uses, where this information is already publicly available, I did not set out to write a book on traditional boreal ecological knowledge and it never became the intention of this book.

That said, I was granted permission by the Vuntut Gwitchin First Nation (VGFN), of Old Crow, Yukon, to include some Gwich'in Athapaskan plant names and uses in this book. Much of this information comes from *Plant Use in Vuntut Gwitchin Territory*, a wonderful book published by the VGFN, which also recently published *People of the Lakes: Stories of our Van Tat Gwich'in Elders* by the VGFN and anthropologist Shirleen Smith.

It is important to note that in the Yukon alone there are fourteen different First Nations, with eight language groups stemming from two major language families, Athapaskan and Tlingit—so the Gwich'in names and uses are likely different from those of other First Nations, even within neighbouring territories. As well, the information contained in these pages on traditional Gwich'in plant use is only a small fraction of their plant knowledge and that of other indigenous people around the circumpolar north.

While writing this book, I was invited by the VGFN heritage committee to be involved in a plant-medicine workshop at Rampart House in the northern Yukon. The workshop was both inspiring and enriching. Archeological evidence suggests the Van Tat Gwich'in have been occupying the area around their community of Old Crow for at least 15,000 years. During the plant-medicine workshop I saw just how much the harvesting of wild plants—along with fishing, trapping, and hunting—continues to be an integral part of their livelihood and culture. I feel very honoured that the Van Tat Gwich'in were willing to share some of their traditional plant knowledge with me and with the readers of this book, and would like to thank them for their contribution to *The Boreal Herbal*. I would like to extend a special thank you to Megan Williams, Mary Jane Moses, and all the people at the Rampart House camp.

I am also immensely grateful to Gwich'in Elder Ruth Welsh, originally of Dootat Gwitshik (Husky River), Northwest Territories, for sharing her wisdom and her plant teachings with me over the years, and for writing the foreword to this book. Ruth is well known in ethnobotanical circles for her vast knowledge of northern plants and their uses, and I feel blessed to know her as both a teacher and a friend.

While attending an agricultural conference in northern Norway, I was very fortunate to meet Laila Spik, a Sami Elder from northern Sweden, and am thankful to her for sharing some of her traditional Sami plant knowledge and some of her recipes with me.

This book was made possible thanks to funding from the Advancing Canadian Agriculture and Agri-Food Program (ACAAF). For overseeing the program, and for sharing my vision, I'd like to extend special thanks to Rick Tone and the Yukon Agriculture Association (YAA), and thanks to Valerie Whelan, of Agriculture and Agri-Food Canada (AAFC), for her encouragement, and to the ACAAF committee.

The plant identification sections of the book were made possible thanks to botanical illustrations and descriptions provided from the National Research Council (NRC).

Biologist Bruce Bennett shares with me a love of plants, and was very helpful at clarifying and helping me with the botany sections. Botanist Jennifer Line also provided input in this area. Herbalist Robert Dale Rogers is a walking encyclopedia of knowledge about the plants of the boreal forest and beyond, and I am indebted to him for sharing his knowledge with me and editing the medicinal information in this book, and for being a great friend throughout this whole project.

I am also thankful to Steve Johnson, of the Alaskan Flower Essence Project, for allowing me to use his Alaskan flower essence descriptions in this book.

Miche Genest, author of *The Boreal Gourmet: Adventures in Northern Cooking*, provided moral support throughout the writing process.

Photographer Cathie Archbould, who has captured images of Aroma Borealis and my family through the years, romped with me through the northern bush to snap photos for the book.

I am also thankful to Peter Long, Fritz Mueller, Teresa Earle, Michelle Clusiau, Birch Kuch, Berwyn Larson, Julie Frisch, Sylvia Frisch, and the late Robert Frisch, for their support and contributions of amazing plant photos. Thanks also to Tanya Handley for her illustrations.

I am thankful to Nadine Pedersen for her editorial support and direction, attention to detail, and for her dedication to this project, and to her son, Hudson, for sharing his mom with me throughout the project. Thanks also to editors Tamara Letkeman, Lily Gontard, and Patricia Wolfe, for fixing my mistakes, and to Swanlight Design for bringing the words alive with a beautiful layout and design.

I am also grateful to Elaine Maloney and CCI Press, at the University of Alberta, for co-publishing this book, and to the anonymous reviewer who provided feedback before it went to press.

Many thanks to Marie-France Campagna, Wendy Cooper, the late Rose Barlow, Elissa Miskey, Vanora Millar, Laila Spik, Susun Weed, and Steven Badhwar for sharing their recipes.

Aroma Borealis provided financial and health support during the writing process, and I am grateful to the Aroma Borealis staff, especially Cathy Gignac, for their support during this project and to Aroma Borealis's customers for their many years of support and encouragement.

Thanks also to Terry Willard, Wade Davis, Rosemary Gladstar, Janice Schofield-Eaton, Carl Eric, Jennifer Skinner, the Yukon Conservation Society (YCS), Lori Schroder, Nicole Edwards, Helen O'Connor, Lee Close, Gord Smith, Katie Delau, John McDougall, Tory Russell, Zola Dore, Lois Moorcroft, Al Pope, Lillian Strauss, Suzanne Catty, Lisa Thiel, Tim Brigham, Irwin Brodo, Anodea Judith, Jenni Holloway, and Kerrie, for their encouragement.

My sisters—Brenda, Barb, Kerry, and Deidre

—and my cousin, Sandy, and their families, provided lots of encouragement. I'm also grateful to all the members of my East Coast and Prairie families for teaching me about fishing and farming when I was young.

I am also grateful to my husband, Mike Bailie, and to my daughters, Ceilidh-Anne and Markie-May, for their love and patience during this project.

Lastly, I feel abundant gratitude to all the plant and healing teachers who have shared their knowledge with me throughout the years.

INTRODUCTION

NATURE HEALS! This is my intrinsic belief. It's why I wrote this book and it's why I live in the northern boreal forest with my family. Every day I am grateful for what nature provides us: shelter, heat, food, medicine, clean air, water, and energy.

Up North the summer months are intense. The profusion of light and warmth brings food and medicine in abundance. Nutritious and healing plants can be enjoyed, gathered, and preserved throughout the season for use during our long, dark, and cold winters.

In early spring, when snow still rests on the forest floor, we begin gathering medicine plants. The buds of the balsam poplar can be harvested to make healing ointments to soothe dry and cracked skin or help ease respiratory irritations, and the running sap of the birch tree provides a nourishing spring tonic full of micronutrients. We celebrate the first crocuses of the season that bravely sprout from the cool soils to colour meadows and hillsides bright purple. Before you know it, tasty vitamin C- and A-rich fireweed shoots emerge, usually in stride with juicy, light-green spruce tips, also full of vitamin C. Almost overnight, tender, nutrient-rich dandelion leaves start to unfurl throughout our garden beds and lawns. These can be steamed, enjoyed in salads, or added to stir-fries, soups, and sandwiches.

We know summer has truly arrived when the sweet scent of wild roses wafts all around us and their bright pink flowers draw us to their beauty. Later in the season, the magenta blossoms of fireweed blaze across the northern landscape and can be gathered and dried for winter teas, jelly making, and cake decorating. Over the course of the summer, plants such as horsetail, bedstraw, coltsfoot, sweetgrass, Labrador tea, wild onion, wild strawberries, and various edible mushrooms come into their full expression, and are ready to be made into foods, drinks, or medicinal remedies.

Crocuses mark the beginning of spring. BG

Late summer is sweetened by raspberries and blueberries. And with the early autumn frosts come ripened cranberries, mossberries, and rosehips to eat, dry for teas, can as jams, and freeze for later use in muffins, breads, and smoothies.

The Boreal Herbal is about understanding

Introduction

Top: In autumn the boreal forest explodes with colour. FM

Above: Cranberries ripen after the first frost. FM

our interconnection with the natural world through the exploration of wild food and medicine plants in the northern landscape. By its definition, a "herbal" is a book that combines information on botany, medicine, and traditional lore. What I present to you in this book is just one tiny piece of a broad picture of knowledge and information that is out there to be gathered, preserved, and shared. It's the synthesis of the seed that has been planted in me, and it's drawn from my experiences as a mother, organic gardener, wild harvester, cook, herbalist, aromatherapist, reflexologist, student, teacher, energy-healing practitioner, environmentalist, writer, and medicine maker.

The Boreal Herbal is part plant-identification guide, and part medicine-and food-making guide. It includes recipes for use at home and for use while traversing the wild lands of the boreal forest. The book features plants found in remote stretches of the wilderness and common "weeds" that grow around our homes, gardens, and recreational areas.

My passion is wild plants. When I am out foraging and collecting in wild and sacred places, my insights and perceptions flow. My thoughts are clear and quiet, allowing me to listen deeply; my spirit is nourished and in alignment with the knowledge that each plant gathered will be energetically embodied within those who use it as food and medicine.

I believe plants have inherent wisdom: when we are open to what they have to teach us, there is so much we can learn from them.

For this reason, in addition to scientific information about the plants' nutritional profiles and practical information on how to gather and preserve plants for medicinal and food purposes, I have also included material on how to connect energetically with plants through meditation and attunement exercises, as well as how to prepare energetic remedies such as plant essences. With the diverse ecosystems of the boreal forest, it's important to honour and respect the land. As you use this book, take care to "leave no trace," for nature provides shelter and food to a vast number of animals, birds, insects, and microorganisms. Aboriginal people intrinsically know this as they have been living off the land and using plants for medicine and food for countless generations.

Choosing the plants for this book was difficult and many were left out, including the moss, lichens, and mushrooms that cover the boreal forest floor. The plants in this book are ones I have used for years as food and medicine, for bush crafts and cosmetics, both at home and in my store, the Aroma Borealis Herb Shop, in Whitehorse, Yukon.

Every plant's life cycle is a journey, much like our own. We are both filled with the energy of spirit that links us to the generations who preceded us and carries with it the knowledge of our ancestors. We root in Mother Earth; our flowers are the expressions of

Introduction

our joy, and our seeds blow in the wind when we share knowledge and ideas.

The spirit of the Earth is calling us together to simplify, return to our roots, and live in a symbiotic relationship with nature. Now is the time—nature heals!

I humbly present to you *The Boreal Herbal*. I hope this book will act as a portal for you to discover and connect with wild plants, their sacred communities, and form relationships with the myriad ecosystems of the boreal forest.

Respectfully,

Beverley Gray

Top: After a brisk winter walk in the forest a steamy cup of wild herbal tea will warm you up. CA

Above: Fireweed seed. BG

PART I

GETTING STARTED

"Once we have...'fallen in love outwards,' once we have experienced the fierce joy of life that attends extending our identity into nature, once we realize that the nature within and the nature without are continuous, then we too may share and manifest the exquisite beauty and effortless grace associated with the natural world."

—John Seed, Deep Ecology advocate, teacher, and co-creator of the Council of All Beings

HOW TO USE THIS BOOK

THE BOREAL HERBAL is both a plant-identification guide and food- and medicine-making handbook.

In this section, Part I: Getting Started, I hope to provide you with an understanding of how you can use this book both at home and in the field.

Before you begin gathering wild plants, I suggest you read Before You Go Out Harvesting (page 21) because it will help you learn both the best times to harvest different plant parts and about sustainable harvesting practices.

In Healing Plant Basics (page 25), I provide a foundation for you to understand what a herb is, what the benefits are of using herbs, as well as some of the more esoteric aspects of herb use—for example the Doctrine of Signatures, which states that the shape of a plant part gives hints to its usage. I also explain why plants so often have so many varied applications.

If you want to enrich your experience by learning how to develop a spiritual connection to plants, I encourage you to read and explore the exercises in Sacred-Spirit Plant Healing (page 28).

In Part II: Plant Profiles, you'll find profiles of the fifty-five plants featured in this guide. The profiles are organized into three sections: herbs, berries, and trees. In each section, the plants are organized alphabetically by their common name. When the plant's common name begins with a descriptive adjective—like "stinging" nettle or "wild" chamomile—the plant appears under the main part of the name: nettle or chamomile.

Each plant profile contains a botanical description, photographs, and a botanical illustration so you can easily and correctly identify it.

Information is also given on the plant's family, where you can expect to find the plant (habitat and range), which parts of the plants are used, and when the wisest time is to cultivate the various parts of the plant for food and medicine (harvest time). I also explore the various ways the plant can be used for medicine, food, cosmetic, therapeutic, and spiritual uses.

Each plant profile includes at least one recipe featuring the plant, and lists the other recipes in the book where the plant appears.

In Part III: Plant Preparations and Recipes, I introduce the different ways of drying, storing, and preserving plants, how to determine dosage, and some general guidelines for taking herbs medicinally.

> "Boreal," meaning "of the North," comes from the Greek word *boreas*, meaning "god of the north wind." Personally, I like to think of *borealis* as meaning "goddess of the North."

Previous spread: The boreal forest is the world's largest terrestrial ecosystem. FM

Opposite: Alongside conifers grows an abundance of other boreal plants. BG

Valerian root can be made into a decoction, infusion, or tincture. BG

In this section of the book, you will also find dozens of medicine and food preparations that can be easily made at home or, in many cases, out in the bush. These recipes include herbal teas, tinctures, medicinal vinegars, elixirs, bitters, syrups, capsules, pills, powders, plant essences, creams, salves, poultices, fomentations, various baths and steams, juices and smoothies, jams, jellies, chutneys, and everything from appetizers and dips to entrées and desserts.

In Part IV: Economics, you will find information on how to turn your passion for plants into a non-timber forest-products (NTFP) business.

This section includes examples of other northerners who are doing this, general information on how you can start your own business, some of the rules you will need to operate under, and contact information for the government departments that regulate the natural-health product industry.

Finally, in Part V: For Reference, I have included biographies of some of the historical figures referenced in the book; information on the Celtic tree alphabet; an illustrated introduction to basic botanical terms; a glossary of botanical terms, a glossary of medicinal actions, an overview of herbal constituents and phytochemistry; a bibliography of books and other resources; quick reference charts for herbal usage, medicinal actions, vitamins and minerals; and an index.

When you take *The Boreal Herbal* into the bush, your garden, and your kitchen, I hope you'll find the book easy to use and that it will become a foundation for a lifetime of exploring the amazing world of edible and medicinal plants.

BEFORE YOU GO HARVESTING

WILD HARVESTING—also known as wildcrafting, gathering, or foraging for plants—is the practice of respectfully harvesting and gathering plants that grow in the wild. The plants that are included in *The Boreal Herbal* are common herbs, berries, and trees that grow in the northern boreal forest and can be harvested for food and medicine.

Safety

Before going out harvesting, you need to be prepared. The weather can change rapidly in the North: When you leave the house it may be warm and sunny, but the next thing you know, it's cool, chilly, and snowing. Dress in layers so that you will be ready for any kind of weather. Wear appropriate footwear, and bring water, snacks, a small first-aid kit, matches or a lighter, and, depending on where you are going, a map, compass and/or GPS. If you are going deep into the wilderness, or even just on a day hike, it's always a good idea to leave a hiking plan with a friend in case you get lost or injured and have trouble returning home.

The northern boreal forest is home to both black bears and grizzly bears (not to mention other large animals like moose and caribou) and you may come across them while out foraging for plants. Be on the lookout for signs that a bear is in the vicinity, such as fresh scat and tracks. Bears don't like surprises, so be sure you make noise by singing or talking. That way, if a bear is nearby it can take off—it doesn't want to see you either. Never approach or feed bears. Learn good bear awareness. A wonderful resource is *Staying Safe in Bear Country—A Behavioural Approach to Reducing Risk*, a video produced by Kodiak Wildlife Products Inc. My good friend, Tanya Handley, has a starring role in the video. At the time it was made, she was charged by more grizzly bears on film than anyone else in the world. Yikes!

A surprise summer snowfall. CA

Part I: *Getting Started*

When wildcrafting, make lots of noise so bears know you are coming. CA

Wild Harvesting Basics

Before you start to gather plants for food and medicine, it's important to make some keen observations. Begin by assessing the area where you intend to gather plants: is it clean and free of pollutants? Busy roadways and industrial areas often provide easy access to many of the plants in this book, but the soil and plants in these areas are often saturated with toxins and chemicals that can do our bodies more harm than good. When out on herb walks I am often asked "How do you know an animal hasn't peed on the plants?" The answer is, "They don't taste like salt!" It is good to sample the plants you are gathering to make sure they are clean.

When you are ready, observe the plant. Be positive of its identification so you don't confuse it with a toxic plant that may look similar. If in doubt, pick a sample and bring it to an expert in your community such as an elder, botanist, plant biologist, or herbalist.

Finally, when harvesting, it's very important to walk softly on the Earth and to gather plants with respect. Observe the plant community you are planning to pick from: Is it healthy and vibrant? Or are there only one or a few plants growing? If there are several, do they look well-established, or are they struggling? To allow the community to prosper, do not overharvest, and do what you can to encourage future growth.

When I gather rose petals, for example, I always leave petals on each stem to ensure that the bees will pollinate the plant and that a rosehip will develop. If I pick petals from a plant, I don't gather the hips from that same plant until the following year.

We are visitors, guests, and receivers of gifts on wild lands and in forests. Be gracious. Ask the plant for permission, leave an offering, be respectful, and leave enough behind that the plant community can continue to grow and exist with vitality. (For more on this see also Sacred-Spirit Plant Healing, page 28).

Before You Go Harvesting

Plant Parts and Harvest Times

When harvesting plants, it's important to gather them at the correct point in their life cycle, and at the right time of day to ensure that you will have the best quality herb to use medicinally and nutritionally.

"Herb" is a term that refers to the whole plant, including leaves, flowers, stems, seeds, and sometimes, roots. The whole herb can be harvested while the plant is in flower. If the flower is not going to be used, then the herb can be gathered before the flowers emerge, but after the leaves have appeared because this is when the plant will be most potent.

The aerial parts of plants grow above the ground and include leaves, stems, and flowers. Flowers and leaves are generally high in volatile oils that can be captured if the plant is picked at the optimal time. The best time to gather the aerial parts of the plant is in the morning, after the plants have rested and before the heat of the day evaporates their volatile oils. The midday sun temporarily wilts the plant's energy, yet it also draws oils and resins into aerial parts. Because of this, some herbalists suggest evening harvesting. As you become familiar with plants and spend time observing them, you will better understand the best time to gather them.

The optimum time to gather leaves is when they are fresh, young, and tender: full of energy, oils, and juices.

Flowers are gathered just before or as they are fully expanded, in the pubescent stage when their colour, aroma, and volatile oils are at their most potent. If you miss this stage, pick when they are wide open and at their peak.

The optimum time to gather conifer tips is when they emerge as fresh, juicy, new-sprout greens—do this quickly because this stage does not last long.

Tree buds, such as those of the balsam poplar, emerge in the late autumn and can be gathered anytime, but I find that they have the highest resin content in the early spring.

Roots can be gathered in the spring before leaves start to develop and before the plant goes into flower, or in the autumn after flowering is finished. Be sure to leave plenty of root stock so that plants can continue to flourish.

Fruits and berries should be gathered when they are ripe. Cranberries, rosehips, and crowberries are better picked after the first light frost; this makes them sweeter.

Seeds can be gathered when they are fully ripened. Dry seeds in a basket as they require very little drying.

Bark can be gathered in the spring or late autumn. In most cases it's the inner bark that is used. If you are using the outer bark you can harvest this at anytime. Never strip around a tree, as this will kill it. It's best to prune a branch from the tree instead of cutting into the trunk.

Wise Harvesting Tips

Gather in unpolluted areas

Make noise to alert bears and other animals of your presence

Harvest only where there is a large and healthy community of plants

Be absolutely certain of the identity of the plant you are gathering

Harvest herbs when the plant is in its prime

Harvest plants respectfully and consciously

Harvest only from plants that are healthy

Harvest only what you can process

Do not overharvest, leave plenty of roots, leaves, fruits, seeds, and flowers so that the plants can continue to grow and flourish

Harvesting yarrow. CA

Part I: *Getting Started*

Collecting fir sap. CA

Pitch can be harvested anytime. I like to gather spruce pitch in the winter when it's frozen, because it isn't so sticky and it's a bit easier to work with. If your fingers get sticky from pitch or plant resins use a bit of vegetable oil or butter to rub it off. Sap is harvested in the spring when it is flowing freely through the tree.

Wild Harvesting Equipment

Here is a list of basic equipment for gathering wild plants:
- A basket with a handle, a bucket, or large paper bags
- A good sharp knife (like a Swiss Army knife)
- A hand pruner or garden clippers
- Scissors
- A trowel, shovel, or hand spade, for digging roots
- Gardening gloves
- An accurate field guide that identifies northern plants (such as *The Boreal Herbal*)

The Boreal Forest

The northern boreal forest—also called "taiga" or "snow forest"—is unique. As Herb Hammond, a forest ecologist with the Silva Forest Foundation, has observed, the boreal forest is "Earth's largest terrestrial ecosystem and extends unbroken except for oceans around the northern pole of the Earth."

The boreal forest forms a circumpolar band, and is the largest biome in the world. It covers more than 14-million square kilometres through Russia, Scandinavia, parts of Asia, and northern North America, including Alaska and Canada, and is characterized by its cool soil temperatures, a short growing season, and low annual precipitation. The Canadian boreal forest covers a land mass about 6-million square kilometres and stretches across all the provinces and territories. Conifers are the dominant tree species of the boreal forest.

Within this large, circumpolar ecosystem there are many smaller, diverse ecosystems, all of which are influenced by some of the most challenging conditions on the planet. Up North, the long winters are made up of short days with little sunlight and heavy snowfall. These contrast with the short summers that have many long, sometimes never-ending, sunlit days. But even in the middle of summer, if you look under the moss in a stand of coniferous trees, you will feel the cold radiating up from the earth.

If you want to dig deeper to learn more about this amazing biome, I have included some excellent book titles and organizations working to protect the boreal forest in the reference section of this book (see page 424).

HEALING PLANT BASICS

MOTHER EARTH HAS provided herbs as healing tools for humans since long before recorded history. All cultures throughout the ages have used plants as food and medicines for healing and for maintaining physical, spiritual, and mental health. When pain, injury, or disease struck early civilizations, they had no choice but to turn to the plants.

Indigenous cultures of the boreal forest carry a rich tradition of using plants for healing. Passed down from generation to generation, these rich oral traditions have given much to the modern pharmacopoeias of the world.

What Defines a Herb?

Herbs are plants valued for their flavour, fragrance, medicinal, spiritual, and healthful qualities. Herbs are made up of a whole medicinal plant or its parts, such as the leaves, roots, bark, fruits, flowers, bulb, resin, and seeds.

The Doctrine of Signatures

Herbal usage and treatments were developed empirically by trial and error. Many ancient herbalists believed that plants had been stamped with the image of their properties by God and that those who gathered the plants might understand a plant's use by its visual characteristics. This system of belief is called the Doctrine of Signatures.

For example, a leaf shape might correlate to the shape of a human organ. Lungwort is a good example: its leaves are shaped like lungs and on the back have little hairs that resemble the little hair-like villi found in human lungs. Other parts of a plant, such as flowers, roots, stalks, or seeds, were also examined for hints to their medicinal uses.

The colour of the juices within the plant were also important. For example, the roots of yellow dock are yellow, so it was used to help remove excess bile.

Today, there are many traditional herbalists who still ascribe to this doctrine.

Wild boreal roses look like hearts—what would the Doctrine of Signatures say about this? BG

Wild Food and Medicine Plants of the North

Part I: *Getting Started*

Stinging nettle is rich in vitamins and minerals. BG

Plant Names

The binomial (two-name) system of naming plants was developed by Swedish botanist Carl Linnaeus and consists of two words: the genus and the species. Genera (plural of genus) are usually Latin, but may also be Greek or Latinized names of places and people. Species are Latin words that describe the colour, shape, size, habitat, smell, or Latinized names of places and people.

Using Plants Medicinally

The many different actions and uses of plants can cause confusion as to how to use them. Just when you think that you have a herb figured out, you discover that it may have five new and sometimes conflicting actions.

How can one plant be so many different things? If you think about plants in the same way you think about people, you soon come to realize that each plant has as many different personality traits as a person does. Like humans, plants are not static, inorganic objects, but living, breathing, evolving beings. The best way to get to know plants is to work with them, study them, and, most importantly, incorporate them in daily life.

Each plant is composed of hundreds of different organic chemicals that make it unique. The specific medicinal action of a plant is the result of a complex, synergistic interaction of its various chemical constituents. The main usage of a plant may be its dominating chemical, and its secondary uses may be derived from some of its other constituents.

For example, yarrow is well known for its diaphoretic properties that help break up a fever. Secondarily, the bitter taste of the plant is known to aid digestion and is used for mild stomach indigestion, bloating, flatulence, and nausea. Yarrow is also indicated for use to help lower blood pressure, and it also works as an antiseptic that can be used for infections, such as cystitis, and can be used topically to help heal wounds.

Using Plants Nutritionally

Hippocrates said, "Let your food be your medicine and let your medicine be your food." This statement demonstrates that the distinction between food and medicine has not always been as demarcated as it is today. In fact, if you really ponder Hippocrates's quote, all foods—even water—could be considered medicine.

One of the most obvious ways to follow this advice is by using herbs and spices when cooking meals. Herbs and spices can be used for their flavour, colour, scent, or medicinal value. Whenever I make soup I always use herbs like nettle, kelp, cayenne, and garlic, not only for their flavour, but for their medicinal and nutritional value.

Healing Plant Basics

Samuel Thompson, an early 1800s herbalist, described this relationship for himself. He considered his medicinal herbs to be of the same value as his foods. He was quoted as saying, "The fuel which continues the fire or life of man is contained in two things, food and medicines (herbs), which are in harmony with each other."

Herbs can be incorporated into our diets in many ways. One of the easiest ways is by making an infusion. An infusion is basically a strong tea made from the delicate parts of the plant, including leaves, flowers, and seeds.

The nutritional benefits of using infusions are that the vitamins and minerals in the plant are drawn into the preparation. In this form, the body finds the nutrients easy to absorb. Also, the medicinal components of the plant move almost immediately into the bloodstream to start their healing action. (See Herbal Infusions, page 281).

For example, after steeping nettle in boiled water for twenty minutes, the infusion is teeming with nutrients such as calcium, magnesium, chlorophyll, iron, vitamins A, C, and D, zinc and potassium—all of which can be quickly and easily absorbed by the body.

What Are the Benefits of Using Herbs in Our Diets?

The benefits of using botanicals are that they support our mental, emotional, and physical bodies, and protect our vital energy. Herb use helps prevent serious disease from entering our bodies, and has very few, if any, negative side effects.

Herbs are also easy to access. They can be grown, wild harvested, or bought in herb shops, health-food stores, and grocery stores.

Beyond this, I believe that using herbs promotes health—not just within ourselves and our families—but within the broader community and ecosystem, by teaching us reverence and respect for the planet and the diversity of life it supports.

Oxalates

Docks, sorrels, and lamb's quarters are all plants that I cover in this book that contain oxalates, an acid also found in spinach, beet greens, Swiss chard, potatoes, rhubarb stalks and, in very high concentrations, in rhubarb leaves (though this is not why the latter are considered poisonous).

When you buy these common plants in the grocery store they do not come with a warning label, however most wild-plant publications advise foragers to be cautious when eating plants containing oxalic acids. This is because a study showed that oxalic acids can leach calcium from the body and block its absorption, and it was believed that this could lead to the formation of kidney stones or gall stones.

I believe that it is worth noting that there are no documented cases of poisoning from these plants and that you would have to eat an awful lot of these plants over an extended period for this ever to become an issue. Oxalic acids give the wild greens in this book a wonderfully sour, lemon-like flavour, making it unlikely that they would ever be eaten in large quantities on a regular basis.

Not only that, but oxalic acids are known to also bind to heavy metals, such as cadmium, lead, and mercury, and help remove them out of the body—so some people may actually benefit from eating them. These plants also contain very high concentrations of calcium and other minerals, so may in fact give to the body some of the very elements they are said to remove.

This said, everyone's body chemistry is different, and some people are more sensitive to acids than others; what is good for one person may not be good for another. As with anything you put into your body, I encourage you to take the time to observe how you feel when you eat these plants and decide, based on your physical reaction to them, whether or not they are something you should incorporate into your diet.

SACRED-SPIRIT PLANT HEALING

SACRED-SPIRIT PLANT HEALING is a spiritual practice that I call on to connect deeply to plants. This practice helps me to work reverently with the sacred, life-force energy of plants.

Sacred-spirit plant healing involves meeting the plant spirit or energy to influence healing of the human mind, body, spirit, and heart. This energy can be engaged on many levels.

The essence of sacred-spirit plant healing is:
- Plants are energetic beings with a perceptive intelligence and awareness
- All systems of life on Earth and in the cosmos are interrelated
- The human species can heal through connection and identification with nature and the plants that nourish, house, clothe us, and keep our air and water clean

Connecting to the Sacred Spirit of the Plant for Healing

Learning about plants' energetic natures comes from direct experience. I believe that we can connect with plants in the same way we connect with other people, or animals. I also believe that we may have a certain affinity for a plant or a tree that is potentially meant as our "power plant"; sometimes we may have more than one.

Power plants are trees or other plants that come to us on our inner journey for the purpose of helping us in our pursuit of health and awareness. In some cases, the plants have a message to share with us, such as a recipe for a remedy that needs to be prepared and used in daily life. In other cases, there's a call to visit and develop a relationship with a particular tree or plant. We may sense the energy of the tree or plant and may want to work with it to help us bring healing into our lives.

When we create a connection with a plant through listening, observing, and interacting with it, we start to comprehend the vastness of the ancient intelligence and knowledge that plants and trees hold, and the gifts that are shared with us.

Plants can sense our intentions and actions, so staying in continual communication and resonance with them allows for a clean exchange, whether we are connecting to the plant energetically, or intending to harvest all or part of it for medicine.

Sacred-Spirit Plant Healing

Acknowledging and thanking a plant's spirit is crucial. If we treat them kindly with reverence, compassion, and respect, the plants will remain vibrant and full of life-force energy. If we are not grateful, we not only compromise the energy of a plant, but the food or medicine it's offering.

I believe there are two things that are important when connecting spiritually with a plant and its community. The first is to have an energetic exchange with a focus on your intentions; the second is to gift the plant and its community.

Many indigenous cultures make an offering of tobacco, cornmeal, matches, and prayer to the plants they intend to gather.

Gwich'in Elder Ruth Welsh says, "My mom taught me when you wake up the day you are going gathering to say your prayer, ask for guidance to keep you calm, and say prayers for the people who are wanting these medicines. You don't always have to leave tobacco (this came from the south), you can leave tea, but the prayer is the most important."

In my practice, I have chosen to leave a piece of my hair, a song, a stone, or sometimes just a smile and a blessing. We each have our own unique and special way of being grateful and saying thank you.

Consciously bringing ourselves into a meditative and peaceful presence allows us to be energetically porous to the plant's spirit. By residing in our spiritual place we are more attuned energetically, and therefore open to the information that may be imparted by the plant.

How to Meet Power Plants

Plants are living, breathing organisms as well as great healing allies. Each plant has an essence or spirit-like quality. There are many methods of meeting your power plants and connecting with them spiritually. Sometimes it can happen spontaneously while walking in the forest or meandering in your garden. You may feel an energetic pull, or the subtle fragrance of a plant may draw your attention, or you may feel like you are not alone, or have the feeling you are being observed.

Have you ever intuitively been pulled toward a certain flower or plant? This is the plant energy beckoning you. The most traditional way of attuning to a plant is to be present with it. Be observant, look at it, touch it, smell it. Exploring a plant with an intuitive sense is a vital part of the process. Meditating, communing, and dreaming in a plant's presence can bring some interesting insights.

The Spirit of the Plants

The Spirit of the plants has come to me

In the form of a beautiful dancing green

woman

Her eyes filled me with peace

Her dance filled me with peace

The spirit of the plants has come to me

And has blessed me with great peace.

Her eyes filled me with peace

Her dance filled me with peace.

The spirit of the plants has come to me

In the form of a beautiful dancing green

woman.

—Lisa Thiel © 1984, www.sacreddream.com

Part I: Getting Started

How to Tune Into Plants

The first step in an attunement or tuning in is to choose the plant you want to connect with. Find a quiet spot and sit with your plant. Make your own special offering to the plant.

The energy or spirit of the plant is alive and will aid you on your journey or meditation. Each plant has many different ways to help facilitate the healing process. What you will learn in experiencing tuning into the plant's energy is that listening to your intuition is the key ingredient.

It's amazing what plants can teach us if we learn to "listen" to what they are "saying." To do this, sit quietly for about ten to twenty minutes and just study your plant. Look at all of its subtleties, and open your mind and heart to what it's telling you. Look at its structure, touch the leaves, flowers, and stem. Feel the texture and smell the aroma of the plant. Observe your feelings and emotions. What are the thoughts that come into your mind? Write them down, no matter how insignificant they may seem.

After you have finished writing, spend some more time with your plant. You can close your eyes at this point and meditate with the plant. A plant attunement can take thirty minutes or longer. Be sure not to rush through the process. Every person's experiences are different when working with plant energies. There are no wrong or right ways.

After you are done, think about the connections you made. What did you hear? What thoughts came into your mind? What emotions, physical feelings, associations, and thoughts came while in the plant's presence? Did some event occur in your vicinity during this process? Did you see a bird fly by? Did you hear the buzzing of a bee? What did this mean to you? Did this connect to any spiritual or cultural tradition you know of?

The midnight sun. BG

Other Ways to Spiritually Connect with Plant Energy

Sacred Tree Aura

During dusk or dawn select a tree that is offset with the backdrop of the sky. Relaxing your eyes, focus on the trunk or centre of the tree. Let yourself relax, shift your focus to the sky. While keeping your focus on the sky, start to subtly make note of the outline of the tree. Allow your vision to go slightly out of focus if it needs to. Notice the edge of the tree and the outline of the branches. You may see a white glow around the perimeter of the tree that looks like a sun dog around the sun on a cold winter's day. You may also observe energy streaks radiating outward from the branches. Observe this energy and breathe it into your core.

Spirits of the Woods Meditation

As you enter a forest it can feel like going through a sacred portal or doorway to an energetically vibrant and mysterious world. After spending time in the forest, a person can feel a sense of harmony within or an energetic cleansing.

This spirit of the woods meditation is a wonderful way to connect with nature. I always feel like I am among community when I am meditating in the forest, as if the trees and I are recognizing each other as kindred spirits.

Sacred-Spirit Plant Healing

Heart of nature. BG

Find a spot in nature where you feel comfortable, sit down with a straight spine, and start to feel the energies of your surroundings. Try and perceive the surroundings with all of your senses—sight, smell, taste, hearing, touch, energy, and intuition.

Slowly inhale the fresh oxygen into your body and slowly exhale, releasing the air from your lungs. As you bring new air into your lungs, start to notice any subtle shifts in your energy. As you exhale, notice any subtle shifts in the energies of the plants that surround you. Notice the exchange of energies.

Once you settle in, gently close your eyes, continuing to breathe in harmony with nature. Now start to shift your focus from being the observer to being the observed. Simply notice how this feels. Check in with your senses. Does it feel like you are being observed? If so, by whom? Start to tap into this energy. What was your experience?

"The only real valuable thing is intuition."

—Albert Einstein

Give a Tree a Hug

I am a tree hugger! Hugging trees helps me slow down, become grounded, and is quieting to my essence. When I'm hugging a tree, I like to imagine that I'm hugging the tree heart to heart. I imagine that our roots are one and that our crowns are both extending toward the cosmos.

Trees are the ancient guardians of this Earth, the elders of the land; they have so much to offer us. While hugging a tree, remember to not just receive, but also to be gracious and reciprocate, offering your gratitude for its teachings and majestic beauty.

PRESERVING WILD PLANTS

YOU FEEL EXHILARATED when you come home with a basket full of fresh, wild plants that you gathered with wisdom, love, and grace. For us herbalists the gathering of the herbs is part of the medicine, and how we approach harvesting wild plants affects the overall vibration of the remedies we make. Our relationship to the plant brings the medicine alive; we are a conduit for the medicines to take form, thanks to a co-creative, organic, and intuitive relationship with the energy that flows from grace.

Wild, gathered plants can be prepared in a variety of ways using many methods. The type of medicinal preparation depends on the plant part used, the medicinal qualities it possesses, and how you plan to use it. You can transform berries, leaves, and flowers into tasty jams and jellies, aromatic teas, and remedies, and still preserve their nutrients and healing qualities.

Some medicinal values of plants work best preserved and used as tinctures or teas, while others work best as poultices, ointments, capsules, or as baths, inhalations, and in massage. To ensure that the medicine you are using is going to help you with your needs,

Wild mint and red clover make excellent teas. BG

Preserving Wild Plants

it's wise to choose the herbal preparation that optimally releases the plant's healing constituents. For example, if you want a remedy that is high in vitamins and minerals, the natural choice would be to make a tea or nutritive herbal vinegar, instead of an alcohol tincture that during preparation doesn't draw out as many nutrients from the plants.

It usually takes twice as long to process fresh plants for medicine making and food preparation than it does to gather them. Always remember this so you don't over-harvest the plant communities, otherwise they will end up as compost (though even as compost they still contribute to the cycle of life).

Herb Garbling

"Garbling," aside being a fun word to say, can be a very meditative practice. Garbling describes the removing of stems, twigs, leaves, and flowers. This practice also helps remove any impurities, such as old, dried-up plant matter, or insects that may also have plans for the plant. If you do find insects, gently help them get back to their natural habitat.

Mindful garbling leads to great herbal remedies. BG

Garbling can be done before, during, or after the process of drying. Observing the plant you are going to garble is helpful. Generally, plants with tender leaves, such as plantain and chickweed, should not be broken up before drying, otherwise you may bruise the plant's membrane and cause browning or darkening of the leaves during the drying process.

Garbling takes a fair amount of space, so make sure you have a clear kitchen table or counter to work on. For plants like stinging nettle, it's good to have a pair of garden gloves handy.

In *The Herbal Medicine-Maker's Handbook*, James Green writes, "Garbling is an unsung backstage activity akin to the kneading of bread, the curing of firewood, proofreading a manuscript, and the arduous pulling of taffy. Mindful garbling transforms good-quality herb into great-quality herb and great-quality herb into primo herb."

Drying Herbs

Drying herbs immediately after gathering them helps preserve their vitality for use in future herbal preparations such as teas, tinctures, syrups, and spices. The dried herb should look energetic and true to its colour—not all tired and spent, which happens when herbs are dried too hot and too fast, or in direct sunlight.

The optimal conditions for drying fresh herbs are out of direct sunlight, in a well-ventilated area, with low light and low heat. The most common ways to dry herbs include drying bundles, drying racks, drying baskets, drying bags, dehydrator dying, and oven drying.

Part I: Getting Started

"Profuse picking followed by procrastination (or processing fatigue) is a notorious terminator of vast quantities of potentially dried herb. For the sake of the plant communities and to prevent your own disappointment, try not to over-harvest, and be determined to complete your day's work. The reward is great."

—James Green, herbalist and author of *The Herbal Medicine-Maker's Handbook*

Drying Bundles

Herbs can be tied in bundles and hung to dry, out of direct sunlight and away from dust. I use elastic bands or twine to secure the bundles—just make sure not to make them too fat, otherwise the plants in the middle may not dry fast enough, and turn mouldy. As the bundles dry, they shrink a bit, and you may lose some stems to the force of gravity.

Drying Racks

If you are harvesting an abundance of wild plants, it might be worth making drying racks. In the past I have taken old wooden screen-door frames and covered them with light cotton sheets secured with staples to the frames. These create airy yet taut surfaces. If you want to build your own frames, make sure the screening you use is not aluminum, as it can taint and discolour the herbs (even through a sheet). Suspend screens out of direct sunlight in a well-ventilated, dry space that is free of contaminants such as dust, fumes, and cooking odours. I also have drying racks made from bamboo rice baskets—these come in a set of four—I've tied them together and hung them from the ceiling in my drying space—they look beautiful and have good air ventilation.

Drying Baskets

Wicker baskets are great for drying small amounts of botanicals. In the summer months I always have some plant or another drying in baskets around the house. To create airflow through the bottom of the basket, sit the basket on top of a cup or mug and be sure to turn the herbs daily.

Drying Bags

Small amounts of herbs can be dried in paper bags. Make sure to give the bags a shake a couple of times a day to prevent mould from growing.

Dehydrator Drying

I love my Excalibur dehydrator; it's a quick and efficient way to dry herbs and fruits. It also comes in handy for making raw foods, such as fruit leather, crackers, and chips. The benefit of a dehydrator is you can control the temperature to ensure quick drying without cooking the herbs. (For more information visit *www.excaliburdehydrator.com*.)

Various drying techniques can be used to preserve your harvest. BG

Preserving Wild Plants

Oven Drying

Roots and barks can be cut up and dried in an oven at a very low temperature. But for the more delicate plant parts, such as flowers and leaves, I would use the air-drying method as heat releases their volatile oils.

Drying Berries

Dried berries add a fruity flavour to your wild teas and can be ground into a powder to add extra flavour and nutrients to cereals, pancakes, baking, or even salads. As with herbs, berries can be dried in an oven or in a dehydrator, but it's best to first soften the membranes of the fruit skin. To do this, simply dunk berries into boiling water for a minute using a colander or a strainer, then let them drain completely and pat them dry with a cloth or paper towel. Spread berries out on a drying rack in the dehydrator, or on a glass baking dish in the oven at the lowest heat. That said, if you want to preserve the maximum flavour and antioxidant activity of your wild berries, it is best to freeze them and use them as needed.

Top: Using a dehydrator is a convenient way to preserve rosehips. BG

Above: Drying dandelion root for making dandelion coffee. BG

Storing Herbs and Berries

For optimal results, dried herbs and berries should be stored in a dark, cool, dry area. Light, heat, moisture, and exposure to air will deteriorate dried plants and decrease their shelf life.

Dried herbs can be stored in canning jars and kept in a cupboard. If you are reusing jars, make sure there is no aromatic residue left inside, as this could compromise the smell and taste of the herb. I like using clear glass jars for certain herbs such as fireweed flowers, rhodiola root, and nettle leaves because the herbs are so pretty to look at and glass makes it easy to see what's inside the jar—this is handy, especially if I use the herbs on a daily basis.

Dark glass jars are difficult to find but are preferable for longer-term storage of botanicals. A woman told me she painted the outside of clear glass jars to make them light tight. Paper or plastic bags are okay for short-term storage. If you have large quantities of dried herbs you may have no other choice but to store them in a large plastic tub with a lid.

Dried leaves and flowers can be stored for up to one year; roots, seeds, and bark up to three years. Powdered herbs lose their properties quicker than herbs that have had little processing.

Tinctures made with alcohol will last for up to ten years and vinegars for at least one year. Syrups kept in the refrigerator will last from six months to a year.

Labelling

Label your plant storage jars and containers clearly. You may wish to include:
- Common name of the plant
- Latin name of the plant
- Location of harvest
- Date harvested
- Date dried
- It's helpful to include what the herb is used for and recommended dosages; this way, you may feel more inspired to use them.

Be thankful for the gifts that the Earth and all the elements have given you. Give the herbs back to the Earth by composting after use.

THE KITCHEN APOTHECARY

UNDERSTANDING THE ACTIONS of the plants that grow in your region, and knowing how to prepare them, will allow you to make a diverse collection of herbal remedies that can aid you in addressing many health issues.

Creating a kitchen apothecary is easy. You probably already have the tools and some of the ingredients you will need to make most of the remedies in this book.

Essential Tools
- Stainless-steel pots
- Spoons and spatulas
- Measuring cups (glass)
- Glass canning jars

Other Useful Tools
- Cheesecloth or muslin cloths (for straining the plant matter)
- Stainless-steel strainers
- Coffee grinder just for herbs and spices
- Stainless-steel double boiler
- Blender
- Funnels (various sizes)
- Scale (optional)

Make sure all your tools are clean, dry, and sterilized, if possible. This will help protect your preparations from bacteria and from developing mould. Have sterilized tools assembled before you start to make a preparation.

Common Ingredients
- Beeswax
- Olive oil
- Vodka or brandy
- Apple cider vinegar
- Distilled water or spring water

The Kitchen Apothecary

Measuring

The Simpler's Measurement Method

The term "simpler" was in use long before the words herbalist, "wise wo/man," or healer came into use. It referred to people who lived close to the Earth and used plants for health and healing, and who were observant, creative, and used their intuition and inner wisdom for making medicinal preparations.

I use the simpler's method of measurement for many of the recipes in this book. The reason for this is that the simpler's method of measurement is an adaptable and easy system for making herbal remedies. This is because it is based on ratios and the units of measurement are referred to as "parts." For instance, "3 parts stinging nettle, 2 parts raspberry leaf, 1 part red clover" is a very common 3:2:1 blending ratio. This measurement lets you make your formulation in any volume you need as long as you use that ratio consistently throughout the recipe. For example, if you wanted to make just enough of a blend for a large pot of tea you could use 3 tablespoons stinging nettle, 2 tablespoons raspberry leaf, and 1 tablespoon red clover, but if you wanted to make the herbal blend in bulk, you could use either cups or kilograms and the measurements would still be true.

Medicine Chest

Herbal medicine is the people's medicine! Therefore it's only natural that we should want to have healthy pantries and medicine chests filled with dried wild foods, preserves, nutritious and medicinal teas, and healing remedies. These can be easily prepared in your home or camp kitchen.

The first step in creating a household collection of remedies is to identify what your needs are. This will help you define what type of collection you want to make. A comprehensive collection contains remedies that can be used on a daily basis for nutritional purposes, prevention of disease, or for common ailments such as colds, coughs, and flus, bug bites, cuts, stomach aches, toothache, etc.

Write down your needs and a list of remedies you want to have on hand. This will help you decide what herbs to gather and how you want to prepare them.

The information and recipes I have provided are a good foundation for developing your creativity and blending your own formulas. Grouping herbs together often leads to a comprehensive remedy that can assist you in your healing process. That said, using herbs one at a time and not in a formula is very beneficial and sometimes a preferred way of healing with herbs. It is also a good way to get to know a plant!

Top: Ground dandelion root is a good coffee substitute. CA

Above: Aromatic beeswax. CA

Wild Food and Medicine Plants of the North | 37

Part I: *Getting Started*

When To Take Herbs

- **On an empty stomach:** For conditions that have excess mucus and for cleansing and detoxification.
- **Before a meal:** For nervous conditions, to help digest food and help reduce fat for weight loss.
- **After a meal:** For indigestion and gas, and for preventing mucus buildup.
- **Taken with food:** For health-compromised individuals with conditions that have weakened their health.
- **Between meals:** For conditions relating to urinary- and nervous-system conditions.

Determining Dosage

Determining dosage of herbal medicines is based on an individual's physiology, age, sex, temperament, illness, stress, digestion, and whether or not the person is pregnant or lactating. As each person has his or her own unique physiological and biochemical constitution, the person's constitution should be supported by the action of the herbs.

As you experience herbs for the first time, try to keep simple notes so you can keep track of your experience. Listen to your body: if the herb is feeding you and you are experiencing positive results, keep using it; if you experience any unusual reactions, stop using the herb. An exception to this rule is if you are on a herbal detoxification program and you experience flu-like symptoms, in which case you could be having what is called in the natural-health world a "healing crisis." This usually passes after twenty-four to forty-eight hours.

When using a herb for the first time, it may be advisable to use only a small amount in case there is an immediate, adverse reaction. A way to test if you are allergic to a plant remedy is to rub some of the remedy or the herb onto the skin of your inner arm, and wait twenty-four hours, if you experience a skin rash, itchy eyes, itchy throat or swelling, do not use this plant remedy. As with other allergens such as milk products or plant pollens, even the most common herbs can affect different people in different ways. If you are taking herbs for an extended period of time, it may be advisable to periodically take a break from them for a couple of weeks. This will allow you to determine whether or not your body still needs the herb.

Please remember that when using any of the recipes in this book to reread the plant profile in Part II: Plant

My home apothecary. BG

The Kitchen Apothecary

Profiles and note any special cautions, uses, and actions of the plants. You are your best educator! It's important to note that if you have serious symptoms occurring with your health, do not try to self-diagnose—go see a health-care provider to help you gather information and figure out what is going on.

When determining the dosage of herbs to use as medicine, there are some general rules. However, every rule has its exception, so it's wise to apply common sense, listen to your intuition, and make observations.

Caution: When taking prescription medications, or even over-the-counter pharmaceuticals, make sure that the herb you are taking does not contraindicate the drug. This will help you avoid any adverse herb/drug interactions.

Adult Dosages
A General Rule of Thumb*

For chronic long-term problems like insomnia, pain, arthritis, and allergies the standard dosages are as follows:
- **Capsules:** 2 capsules, 3 times a day
- **Syrups and elixirs:** ½–1 teaspoon, 3 times a day
- **Tea:** 3–4 cups a day, for a few weeks
- **Tinctures/juice concentrates:** ½–1 teaspoon, 3 times a day

For acute problems that come on suddenly and need quick attention, such as headache, toothache, cut, bruising, topical bleeding, sore throat, or the first stage of the common cold:
- Capsules: 1 capsule every hour, until symptoms retreat
- Syrups and elixirs: ½ teaspoon throughout the day, until symptoms subside
- Tea: ¼–½ cup throughout the day, up to 3 cups a day
- Tincture/juice concentrates: ¼–½ teaspoon every hour, until symptoms subside

Herbs and Children

I have raised my children on herbs and natural remedies, but there have been times when I was absolutely grateful for the help of my doctor and the medical community. Parents are most familiar with their child's health and are best able to determine when to seek medical help. Allopathic medicine is excellent in a crisis or for illnesses that need emergency intervention.

I remember when my eldest daughter, Ceilidh, was six years old and she had an accident that ripped the flesh of her gums away from her lower teeth. It was very frightening for her

"Do not subscribe to the 'more is better' viewpoint. Higher than recommended doses could produce adverse or paradoxical effects. Because of their nutritive and supporting nature, many herbs may take a few days to a few weeks for the desired effect to be achieved. When tonifying herbs are used for chronic conditions, it is better to take them in small doses over an extended period of time. As a general rule, dosage should be related to body weight. Large people require larger doses than small people."

—David Hoffmann, herbalist, *The Holistic Herbal*

---

*Based on information from the Science and Art Herbalism Home Study Course by Rosemary Gladstar

Part I: *Getting Started*

A bouquet of daisies for Mom. CA

and us, and we rushed her to the hospital forty kilometres away as fast as we could. But before I jumped in the car I quickly grabbed some yarrow leaves from a nearby plant, chewed them briefly to release their healing properties, and gently stuffed them behind her lower lip. She was bleeding profusely and I knew the yarrow, a styptic herb, would help stop the bleeding, and it did. By the time we saw the emergency doctor the bleeding had stopped, which made it easier for her to see what my daughter's injuries were. She didn't need stitches, so I made a herbal antiseptic and tissue-healing mouthwash for her and she healed quickly.

If your child is ill, try to gather as much information as possible, by asking your child questions, making observations, and providing lots of TLC. A trip to see your health-care provider may also help provide answers about what is happening.

When my children were babies, I treated them with herbs in the bath, steams, homeopath remedies, and mild herbal oil massages. The bumps and bruises of toddlerhood were soothed with salves, poultices, and washes.

As my children grew older I introduced them to mild tisanes, used herbs in our foods, made herbal popsicles, iced tea, and offered remedies such as rosehip tea or syrup when a cold was starting.

As young women, my daughters now often reach for tinctures that treat what ails them fast and easily.

It's always wise to give herbs to your children when they are healthy so that you know how they will respond to the herbs when they are ill. Similarly, if you have more than one child, it's important to make sure that the herbs you're choosing are safe for them as individuals: what might be good for one child may not be good for the other. Always remember that children require much smaller dosages of remedies than adults; their bodies are sensitive and respond naturally and quickly to the healing energies and properties of herbs.

Children's Dosages*

There are no absolute guidelines for dispensing herbs for children, but there are certain recommendations and rules that are taught at many herbal schools. There are at least three different methods for determining a child's dosage, and as you will see they are quite different.

Clark's Rule

Divide the child's weight in pounds by 150 (the average weight of an adult) to give the fraction of an adult dosage. For a 50-pound child, divide 50 by 150, which equals 0.33 or ⅓. Therefore, the child's dosage is ⅓ of the adult dosage.

---

*The information in *The Boreal Herbal* is not meant to replace advice from your health-care provider.

The Kitchen Apothecary

Cowling's Rule
Divide the number of the child's age by 24. For example, the dosage for a child who is four years old: 4 ÷ 24 = .16 or ⅙ of the adult dosage.

Young's Rule
Add 12 to the child's age. Divide the child's age by this total. For example, the dosage for a four-year-old: 4 + 12 = 16; 4÷16=0.25 or ¼ of the adult dose.

Suggested Dosages for Children*
Herbal Teas
If the adult dose is 1 cup (250 mL):
- Younger than 2 years: ½ teaspoon (2 mL) to 1 teaspoon (5 mL)
- 2 to 4 years: 2 teaspoons (10 mL)
- 4 to 7 years: 1 tablespoon (15 mL)
- 7 to 11 years: 2 tablespoons (30 mL)

Above: As children, my daughters loved spending time in the raspberry patch. CA

Note: A nursing mother may take an adult dose and the herbs will be transmitted to her baby through her breast milk, filtered and diluted to the appropriate strength.

Tinctures, Syrups, and Herbal Juices
If the adult dose is 1 teaspoon (5 mL) or 60 drops (should be diluted in water):
- 3 to 6 months: 3 drops
- 6 to 9 months: 4 drops
- 9 to 12 months: 5 drops
- 12 to 18 months: 7 drops
- 18 to 24 months: 8 drops
- 2 to 3 years: 10 drops
- 3 to 4 years: 12 drops
- 4 to 6 years: 15 drops
- 6 to 9 years: 24 drops
- 9 to 12 years: 30 drops

Note: Children under the age of twelve months shouldn't be given syrups made with unpasteurized honey. If tinctures are made of alcohol, dilute in boiling water to let the alcohol dissipate, then serve cool.

---

*Based on information from the Science and Art Herbalism Home Study Course by Rosemary Gladstar

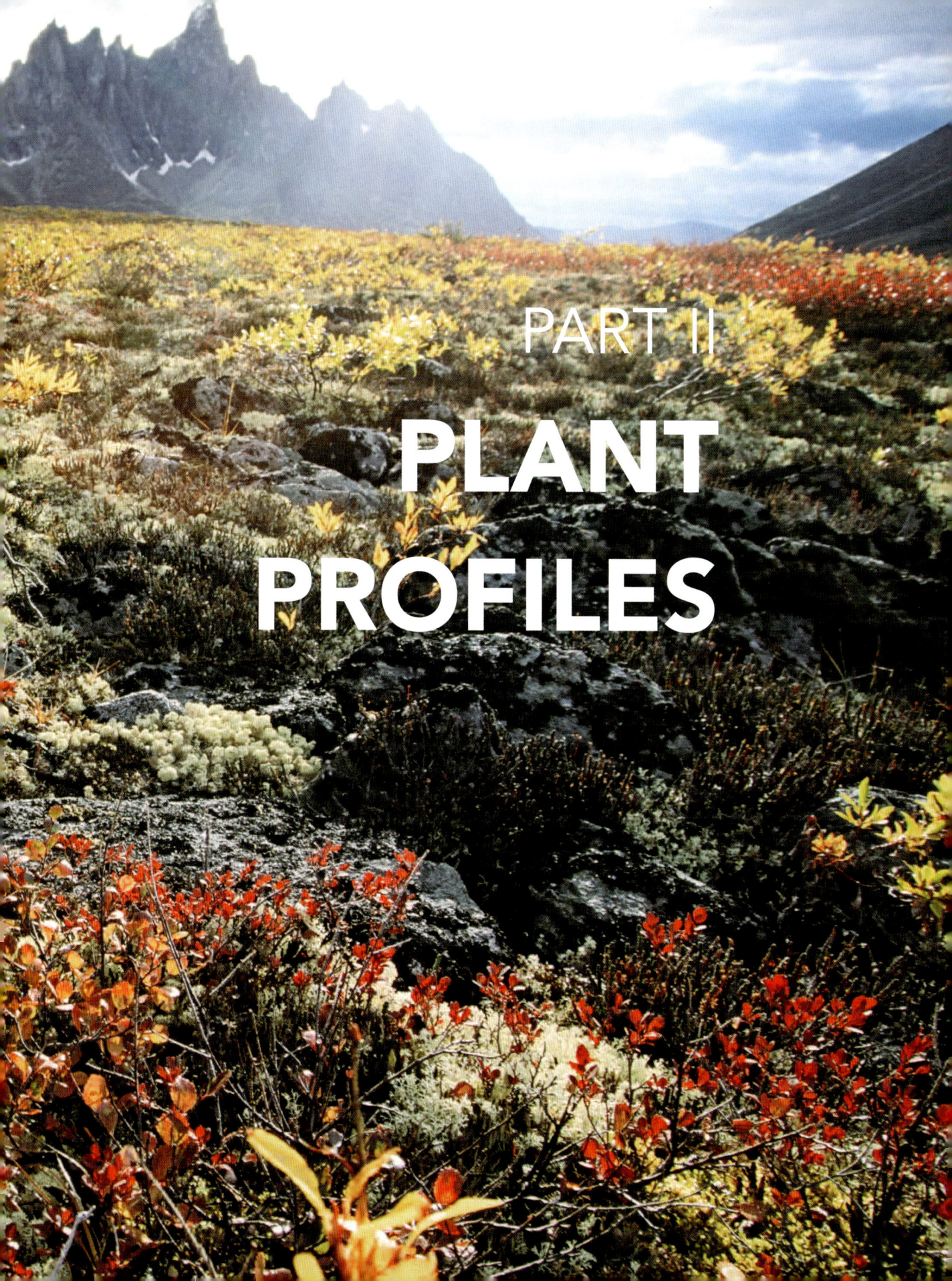

PART II
PLANT PROFILES

HERBS

HERB PROFILES AND RECIPES

Arnica 47
   Arnica Footbath 49
   Arnica Poultice 49
   Arnica Ointment 50
Bearberry 51
   Bearberry Tea 54
Bedstraw 55
   Bedstraw Ear Oil 58
   Bedstraw Deodorant 58
Wild Chamomile 59
   Wild Chamomile Tea 62
Chickweed 63
   Chickweed-Garlic
      Green Dip 66
Red Clover 67
   Clear-Skin Tincture 72
   Susun Weed's Fertility-
      Enhancing Infusion 72
Coltsfoot 73
   Coltsfoot Cough Syrup 76
Dandelion 77
   Dandelion Wine 81
Devil's Club 83
   Ancient Aromatic Panax Oil ... 86
Dock 87
   Wild Dock
      Sorrel-Potato Soup 90

Fireweed 91
   Fireweed Jelly 94
Gentian 95
   Gentian Tincture 97
Goldenrod 98
   Golden Flower Body-and-
      Massage Oil 101
Horsetail 102
   Horsetail Tooth Powder 105
Labrador Tea 106
   Labrador Tea-Flower Oil 110
Lamb's Quarters 111
   Lamb's Quarter Omelette 114
Lungwort 115
   Lungwort Fomentation 117
Mint 118
   Wildly Mint Jelly 122
Stinging Nettle 123
   Nettle Tincture 128
Wild Onion 129
   Wild-Onion Spread 131
Plantain 132
   Plantain Herbal Healing Oil .. 136
Rhodiola 137
   Rhodiola Decoction 140
River Beauty 141
   River Beauty Summer Salad .. 143

Rose 144
   Wild-Rose Petal Healing
      Ointment 148
Wild Sage 149
   The Hiker's Herb 153
Shepherd's Purse 154
   Shepherd's Purse Tincture ... 156
Mountain Sorrel 157
   Mountain Sorrel Iced Tea ... 159
Sheep Sorrel 160
   Borealis Green-Goddess
      Dressing 163
Strawberry Blite 164
   Strawberry Spinach
      Smoothie 166
Sweetgrass 167
   Sacred-Spirit
      Sweetgrass Cream 170
Twinflower 171
   Twinflower Vibrational
      Essence 173
Valerian 174
   Valerian-Root Tincture 177
Yarrow 178
   Yarrow Oily Skin Treatment ... 183

Pages 42–43: Berry picking in the Tombstone Mountains, Yukon. FM

Pages 44–45: Harvesting fireweed. CA

ARNICA

Arnica cordifolia (heart-leaved arnica)
Arnica angustifolia (narrow-leaved arnica)

Other Names and Etymology
Lamb's skin, alpine arnica, wolf's bane, leopard's bane, heart-leaf arnica. *Arnica angustifolia* (narrow-leaved arnica): The genus name *Arnica* is derived from the Greek *arnakis*, meaning "sheep skin" in reference to the feel of its soft, hairy leaves.

Family
Asteraceae (aster, daisy, or sunflower family)

Arnica flowers have a bright and sunny disposition. CA

Part II: *Herbs*

> "...flowers, because they are a proud assertion that a ray of beauty outvalues all the utilities of the world."
>
> — Ralph Waldo Emerson, author, "Gifts" (1844)

Botanical Descriptions

Arnica cordifolia or *A. angustifolia*: Perennial herb growing from a slender rhizome up to 5 cm long. Arnica has an erect stem that stands 10–50 cm high. Its leaves are opposite, simple, entire or toothed. Sunflower-like in appearance, arnica flowers consist of bright yellow disks surrounded by showy ray petals, with a circular cup of bracts at the base. (Note: There are ten arnica species in the Yukon and they all have similar healing properties.)

Habitat and Range

Moist mountainous areas, usually shading themselves among poplars, aspens and conifers. *Arnica cordifolia*: North America, cordilleran; extends northward into southern Yukon, southwestern district of Mackenzie, disjunct eastward to eastern-central Saskatchewan, southern Manitoba and Lake Superior, and south to California, northern Arizona and New Mexico. *Arnica angustifolia* or arctic-alpine arnica: Greenland to Alaska and in the Yukon north to the Arctic coast.

Plant Part Used

Fresh flower heads

Harvest Time

Pinch the flower head off the stem when in full flower; generally from early June to late July. Harvest before the heat of the day. Allow the flowers to wilt for a few hours so the excess moisture evaporates, then infuse in oil immediately because the flowers will turn to a white seed fluff as they begin to dry.

BG

ARNICA'S BRIGHT YELLOW FLOWERS look like shining jewels against the boreal forest floor, reaching up like miniature sunflowers in a farmer's field. Every summer I head to the mountains to gather the flower heads to infuse into a golden healing oil. The oil is transformed into a cream that can help to decrease the pain and inflammation of life's bumps and bruises. Like so many before me, I have called on arnica's medicine more times than I can remember to reduce pain and promote healing. Arnica ointment is a handy first-aid staple to have in your first-aid kit or sports bag! I once slammed my fingers in the car door. Luckily, I had some arnica cream in my bag. I applied it and within half an hour it was as if the incident hadn't happened; my fingers were fine. Once you welcome the healing properties of this plant into your life, it's hard to live without it.

Arnica

Medicinal Actions
Anti-inflammatory, nervine, stimulant

Medicinal Preparations
Cream, homeopathic, liniment, oil, poultice, salve, soak

Medicinal Uses
Arnica's extraordinary bright yellow flowers are used topically for a wide range of conditions. The active constituents of arnica stimulate and dilate the blood vessels near the surface of the skin. This in turn improves circulation to the injured area and promotes the healing of bruises, sprains, strains, muscular inflammation, aches, pains, rheumatic joint pain, inflammation from insect bites, and swelling due to fractures.

Arnica can be taken internally only as a homeopathic preparation or flower essence to help minimize bruising, pain, and trauma.

Arnica flowers turn into a white seed fluff as they dry, so it is best to let them wilt just enough to allow excess moisture to evaporate, then immediately infuse them in oil. BG

Other Uses
In the *Flower Essence Repertory*, authors Patricia Kaminski and Richard Katz recommend using arnica flower essence for "Conscious embodiment, especially during shock or trauma; recovery from deep-seated shock or trauma." They go on to say that it helps with, "Patterns of imbalance and disconnection of the higher self from the body during shock or trauma; disassociation, unconsciousness."

The Solar Plexus Chakra Vibrational Essence I created includes arnica for use with the third chakra, our centre of personal power. Energetically, arnica helps us to heal the deep wounds that can take away personal power.

Cautions
Ingesting large amounts of arnica can cause dizziness, tremors, and heart irregularities. It may also irritate mucous membranes and cause vomiting. Large doses can even be fatal. Never take arnica as a tea or tincture, or use topical preparations on broken or bleeding skin. Do not add excessive heat after applying arnica to skin.

Arnica Footbath
After a long day of hiking, soak your tender feet in a warm arnica footbath. Add a handful of fresh flowers in a basin, pour in boiling water, let cool to toe temperature, and then soak feet for 15–30 minutes.

Arnica Poultice
Mix a handful of fresh arnica flower heads and leaves with enough boiling water to cover. Let arnica steep until preparation has reached body temperature. Wrap warm herbs in a layer of cloth and place over affected area.

Part II: *Herbs*

Arnica flowers make a great remedy for the bumps and bruises of life. BG

Arnica Ointment

1 cup (250 mL) arnica flowers
1½ cups (375 mL) sunflower oil
½ cup (125 mL) olive oil
1 teaspoon (5 mL) vitamin E
1 oz. (30 mL) beeswax

ADDITIONAL RECIPES

Muscle-and-Pain-Relief Oil, page 307

Muscle-Ease Cream, page 317

Place arnica flowers and oil in a double boiler. Warm slowly on a medium heat. Let simmer 20 to 40 minutes. Stir often. Strain flowers out and wipe the pot clean, so that there are no petals left. Add beeswax to pot and let melt. Add strained oil and stir. Once blended, pour into a jar. Cap jar only after the ointment has cooled down and solidified.

See Topical Treatments section for more in-depth details.

Arnica

Medicinal Actions
Anti-inflammatory, nervine, stimulant

Medicinal Preparations
Cream, homeopathic, liniment, oil, poultice, salve, soak

Medicinal Uses
Arnica's extraordinary bright yellow flowers are used topically for a wide range of conditions. The active constituents of arnica stimulate and dilate the blood vessels near the surface of the skin. This in turn improves circulation to the injured area and promotes the healing of bruises, sprains, strains, muscular inflammation, aches, pains, rheumatic joint pain, inflammation from insect bites, and swelling due to fractures.

Arnica can be taken internally only as a homeopathic preparation or flower essence to help minimize bruising, pain, and trauma.

Arnica flowers turn into a white seed fluff as they dry, so it is best to let them wilt just enough to allow excess moisture to evaporate, then immediately infuse them in oil. BG

Other Uses
In the *Flower Essence Repertory*, authors Patricia Kaminski and Richard Katz recommend using arnica flower essence for "Conscious embodiment, especially during shock or trauma; recovery from deep-seated shock or trauma." They go on to say that it helps with, "Patterns of imbalance and disconnection of the higher self from the body during shock or trauma; disassociation, unconsciousness."

The Solar Plexus Chakra Vibrational Essence I created includes arnica for use with the third chakra, our centre of personal power. Energetically, arnica helps us to heal the deep wounds that can take away personal power.

Cautions
Ingesting large amounts of arnica can cause dizziness, tremors, and heart irregularities. It may also irritate mucous membranes and cause vomiting. Large doses can even be fatal. Never take arnica as a tea or tincture, or use topical preparations on broken or bleeding skin. Do not add excessive heat after applying arnica to skin.

Arnica Footbath
After a long day of hiking, soak your tender feet in a warm arnica footbath. Add a handful of fresh flowers in a basin, pour in boiling water, let cool to toe temperature, and then soak feet for 15–30 minutes.

Arnica Poultice
Mix a handful of fresh arnica flower heads and leaves with enough boiling water to cover. Let arnica steep until preparation has reached body temperature. Wrap warm herbs in a layer of cloth and place over affected area.

Part II: *Herbs*

Arnica flowers make a great remedy for the bumps and bruises of life. BG

Arnica Ointment

1 cup (250 mL) arnica flowers
1½ cups (375 mL) sunflower oil
½ cup (125 mL) olive oil
1 teaspoon (5 mL) vitamin E
1 oz. (30 mL) beeswax

ADDITIONAL RECIPES

Muscle-and-Pain-Relief Oil, page 307

Muscle-Ease Cream, page 317

Place arnica flowers and oil in a double boiler. Warm slowly on a medium heat. Let simmer 20 to 40 minutes. Stir often. Strain flowers out and wipe the pot clean, so that there are no petals left. Add beeswax to pot and let melt. Add strained oil and stir. Once blended, pour into a jar. Cap jar only after the ointment has cooled down and solidified.

See Topical Treatments section for more in-depth details.

BEARBERRY

Arctostaphylos uva-ursi (common bearberry)
Arctous alpina (alpine bearberry)
Arctous rubra (red bearberry)

Other Names and Etymology
Kinnikinnick, stoneberry, mealberry, mountain tobacco, upland cranberry, bear grapes. *Arctous* is Greek for "bear" and *Arctostaphylos* translates as "bear's bunch of grapes," while the Latin word *uva-ursi* means "grape of the bear." The Van Tat Gwich'in word for alpine bearberry is *jiindée*. In their community of Old Crow, Yukon, the plant is also commonly called "bird's-eye berry."

Family
Ericaceae (heath family)

Botanical Descriptions
Arctostaphylos uva-ursi: A wiry, prostrate, evergreen perennial shrub that forms large mats that hug the earth. The small, deep green, leather-like leaves alternate, are 1–2 cm in length, shiny, and oblong. They grow on a red, shredded, woody stem that lies just above the surface of the soil. Bearberry flower-clusters emerge in the spring. Each flower is

Left: *A. uva-ursi* flowers. PL

Right: Bearberries can be eaten in a survival situation. BG

Part II: *Herbs*

Bearberry leaves can be gathered from early spring through late autumn. BK

a tiny pinkish-white bell with 10 stamens. Its fruit is dull red, mealy inside, and contains one hard-walled seed. The berries have a pithy texture and very little flavour when eaten raw. *Arctous alpina*'s (alpine bearberry) leaves are obovate and textured turning a deep red in the autumn. The berries are edible but bland.

The berries of *A. rubra* (red bearberry) are red and almost translucent, and the leaves are bright green, thin, and highly textured. They grow up to 15 cm high. In the autumn the leaves are a blazing reddish orange colour.

Habitat and Range

Arctostaphylos uva-ursi: Likes sunny, sandy and dry slopes, rocky areas, sandy soils, gravel ridges, riverbanks, and is ground cover in coniferous forests. Circumpolar, circumboreal, wide-ranging; in the NWT and Yukon found northward to about the treeline. *Arctous alpina*: Likes acidic, rocky and gravelly situations and rocky tundra. *Arctous rubra*: Grows in peat-like soils near creeks, among open spruce forests, and rocky tundra. Circumpolar and wide ranging in North America, all Canadian provinces and territories, and all the western (cordilleran) northern and northeastern states.

Plant Parts Used

Leaves, berries, flowers

Harvest Time

The thick evergreen leaves of common bearberry can be gathered early spring through late autumn. In the winter, if needed, they can also be harvested from under the snow.

> In fall, bears will ingest massive amounts of the bearberries, which has a numbing/paralyzing action on the intestine. Bears will often follow this meal with *Carex*, a rough-edged sedge that travels right through their intestines, dragging with it tapeworms and other parasites paralyzed by the bearberry.

W HEN MY CHILDREN WERE YOUNG they would play on the hillside that our house sits on. The southern facing slope is covered with a bearberry carpet that leads down to a cold-water creek. In the early summer they would be drawn into the blooming mass of delicate, bell-shaped "pinkie fairy flowers," and occasionally pop the sweet, hillside delicacies they call "honeysuckles" into their mouths.

Medicinal Actions

Leaves: Anti-catarrhal, antilithic, antimicrobial, astringent, bitter, demulcent, diuretic, enuresis, styptic, tonic, urinary antiseptic

Medicinal Preparations

Cold tea infusion, oil, ointment, sitz bath, steam, tea/infusion, tincture

Bearberry

Medicinal Uses

The medicinal deep green leaves of the bearberry plant were first documented in *The Physicians of Myddfai*, a thirteenth-century Welsh herbal reference book.

The leaves can be made into a tincture and are predominantly used as a urinary antiseptic for urinary-tract infections, including cystitis, urethritis, and prostatitis. The leaves' antimicrobial actions help to kill bacteria in the urine. A decoction of the leaves makes an excellent mouthwash for mouth infections and can also be taken orally to help with diarrhea. The anticatarrhal properties make the decoction an effective treatment for bronchitis to thin out excessive-sticky mucus. The antilithic properties help prevent the formation of and assist in the removal of stones from the urinary system.

One of bearberry's main constituents is arbutin, which acts as an antibacterial in the genitourinary tract and can change the colour of urine to a harmless green.

Bearberry leaf also acts as a mild vasoconstrictor to the endometrial lining of the uterus therefore helping to alleviate pain associated with menstruation. The leaf decoction can also be used as a douche for vaginal ulceration and infection, and as a sitz bath after birthing. An infusion of leaves can be mixed with cranberry tea or juice for bladder infections. Bearberry leaves are considered a powerful tonic for the sphincter muscle of the bladder, and is claimed to help with bladder-control problems and as a remedy for bedwetting.

The Van Tat Gwich'in use the whole plant and berries as a tea or juice for chest pains, stomach ailments, and infections.

When making a bearberry infusion, the European Scientific Cooperative on Phytotherapy (ESCOP) recommends making a cold-water infusion instead of a hot infusion because it contains fewer tannins. To prepare, crush leaves and steep overnight in cool water. In the morning, drink as is or warm up the drink, but do not boil; strain before drinking. This process increases levels of arbutin while decreasing potential tannin irritation. For urinary tract infections take ⅓ cup to ⅔ cup (80 to 160 mL) of tea at a time at least three times a day.

It's worth noting here that the cold infusion is an effective urinary disinfectant if the urine is alkaline. To tell if your urine is acid or alkaline you can purchase pH urine test strips. If it's not alkaline, drink a pinch of sodium bicarbonate dissolved in water to alkalize the urine. Vitamin C (ascorbic acid) and fruit juice should be avoided due to their acidifying nature, but calcium and potassium citrates will help.

Bearberry's berries can be used to treat constipation, but should only be used for a short period of time.

Witch's Broom

If you look closely you may notice that many bearberry leaves are covered with purple-brown spots. The spots are from a fungus, which in its early stages lives on the bearberry leaves for half a year and on the spruce the other half. Witch's broom—a thick, yellow tangle of branches, twigs, and needles close to the trunk of the spruce tree is created by the fungus migrating back and forth between the two plants. In early summer it releases a mild ammonia- and fungus-like scent into the air. The fungus can't survive without dividing its life cycle between both plants. When harvesting for medicine, it's best to gather leaves that do not have the fungus on them.

BG

Part II: *Herbs*

Food Use
In a survival situation, the pithy red berries can be eaten for their high vitamin C content and carbohydrates which also can provide much-needed energy to tired hikers. Boiling the berries helps to soften them for easier seed removal.

Nutritional Profile
Berries: High vitamin C and carbohydrates. Leaves: Trace minerals such as calcium, iron, magnesium, manganese, and potassium. Dried leaves contain up to 28,000 IU of vitamin A per 100 g and trace amounts of zinc and vitamin C.

Other Uses
Many northerners call bearberry "kinnikinnick," for its historical use in a blended smoking mixture made by First Nations. The mixture included dried bearberry leaves, Labrador tea, and wild sage leaves.

The dried bearberry leaves were traditionally used in aboriginal pipe ceremonies. Other plants that were mixed in included wild sage, red osier dogwood, and tobacco.

According to herbalist Robert Rogers, "Bearberry flower essence is for those who seek to strengthen and increase their psychic abilities."

The tannins in the bearberry leaves create an ash-coloured dye that can be used for tanning hides.

Cautions
Large doses should be avoided during pregnancy. Do not take for longer than two weeks at a time. Raw berries can be toxic if eaten in large quantities.

> "In all things of nature there is something marvelous."
>
> —Aristotle, philosopher (384 BC–322 BC)

Bearberry Tea

Over the years, many women have relayed stories to me of how bearberry leaf tea or tincture helped them when they were struck by a bladder infection. A cold infusion can be made as explained above but if you need quick relief prepare it as follows.

1 teaspoon (5 mL) of fresh or dried bearberry leaves
1 cup (250 mL) of water

Bring water to a boil. Remove from heat, add water to the leaves, cover, and let sit for 15 minutes or until cooled.

ADDITIONAL RECIPES

All-Purpose Botanical Ointment, page 313

Rejuvenating Facial Steam, page 329

Urinary-Tract Tincture, page 294

BEDSTRAW

Galium boreale (northern bedstraw)
Galium triflorum (sweet-scented bedstraw)
Galium trifidum (small bedstraw)

Other Names and Etymology
Our Lady's bedstraw, goose grass, gravel grass. *Galium triflorum*: Cleavers, fragrant bedstraw. *Galium* is from the Greek word *gala*, which means "milk." *Boreale* means "of the North."

Family
Rubiaceae (madder family)

Botanical Descriptions
All Galium species share the signature green whorled leaves that grow on their squarish stems. The fragrant flowers grow in branched cyme clusters; petals are pure white to yellow and hairy, fruits or nutlets come in pairs and are about 2 mm in length. The plants connect to the earth through their creeping rhizome. *Galium boreale* (northern bedstraw): Grows to 30–70 cm high and has stiff square stems (which becomes obvious when rolled between the fingers). The slender leaves are clustered in groups of four at each node. Leaves are 3–4 cm long, each having three distinct nerves. Flowers are small but showy in dense clusters at the tops of plants. There are four pure white petals. The dry fruit look like tiny hairy nutlets and are sticky like Velcro. *Galium triflorum* (sweet-scented or fragrant bedstraw): Plants with weak, mainly

Bedstraw flowers and leaves are anti-inflammatory.

Wild Food and Medicine Plants of the North

Part II: *Herbs*

Bedstraw flowers have a sweet, hay-like scent. BG

simple stems, 5–10 cm long, from a creeping rhizome; leaves in whorls of six, elliptic lanceolate, distinctly cuspidate. Inflorescences on axillary peduncles, mostly three-flowered; flowers small, greenish white, pedicellated; fruit densely covered with hooked bristles. Entire plant sweet-scented in drying. *Galium trifidum* (small bedstraw): Stems slender, retrorsely scabrous, ascending, freely branched, often matted. Leaves in whorls of four, linear to linear-oblanceolate, blunt, retrorsely scabrous-margined. Flowers small, solitary or three together; corolla three-lobed, white; pedicels at least somewhat scabrous; fruit smooth.

Habitat and Range

Galium boreale: Grows along riverbanks and dry, open or gravelly places. Circumpolar; in the Yukon north to about latitude 65° north and disjunct to the Porcupine River valley. *Galium triflorum:* Openings in rich woods and hot-spring meadows. Circumpolar; in North America from Newfoundland to Alaska, south to Florida, California, and Mexico; in the Yukon territory known only as far north as latitude 63° north in the McArthur and Ethel lakes area. *Galium trifidum*: Common, although no doubt often overlooked, among tall sedges in wet woodland bogs. Circumpolar; in North America, wide ranging from Newfoundland to Alaska, south to Maine, Michigan, Illinois, South Dakota, Colorado, and Oregon; in the Yukon territory, north to about latitude 65° north and disjunct to nearly latitude 69° north.

Plant Parts Used
Leaves, flowers and seeds

Harvest Time
Flowering plant: summer. Seeds: late summer. Roots (for dye): autumn

LEGEND HAS IT that bedstraw's name comes from its use as bedding for baby Jesus. Bedstraw's European cousin, cleavers, was used as rennet or as a milk-coagulating substitute. Early settlers used the aromatic bedstraw to stuff mattresses and pillows. I like infusing the fresh sweet, hay-like scented flowers in a light carrier oil like sunflower or almond oil. This makes a lovely body oil that can be used to nourish the whole body. The leaves can be gathered in the early summer by cutting the stems. The leaves and flowers can then be removed by garbling the plant when dry (see page 33) for later use in teas or tinctures. The leaves can also be tinctured when they are fresh.

Bedstraw

Medicinal Actions
Alterative, anti-inflammatory, antioxidant, astringent, diuretic, hepatic, laxative, tonic, vulnerary

Medicinal Preparations
Compress, dream pillow, facial steam, tea/infusion, tincture

Medicinal Uses
Herbalist Robert Rogers says that "all species of Galium contain asperuloside, an anti-inflammatory and mildly laxative agent. Asperuloside produces coumarin that is responsible for the sweet, hay-like scent. It can also be converted to prostaglandins, which are hormone-like compounds that stimulate the uterus and blood vessels."

The polyphenols in *G. boreale* act as antioxidants to eliminate free radicals in the body. The polyphenols help prevent premature aging and are said to help prevent cancer, among other things. According to Rogers some tannins may even have antibiotic benefits. Bedstraws are very effective for the treatment of all urinary and reproductive-organ inflammations including cystitis, as well as hepatitis.

Bedstraw tea has been used as a weight-loss aid. It's said to speed up the metabolism to reduce weight.

Galium acts as a lymphatic tonic, diuretic, and blood cleanser and can help those with compromised health. Sweet-scented bedstraw *(G. triflorum)* contains blood-pressure lowering substances.

Bedstraw leaves and flowers can be used topically to make sweet-smelling hot compresses to stop bleeding, soothe sore muscles, and alleviate irritating skin conditions such as eczema. Some skin conditions are the result of poor lymphatic drainage and respond to both internal and external treatment. Mashed bedstraw can be rubbed into the scalp to help stimulate hair growth and can also be used as an astringent facial cleanser.

Herbalist Matthew Wood, author of *The Book of Herbal Wisdom*, says that bedstraw is beneficial for "gathering of the nerves," and "inflammation of the nerve endings, tickling and itchy skin." He goes on to say that, "It is a specific in the condition, Dupuytren's contracture and Morton's neuroma, when the tendons tighten up under the middle finger or toes."

Food Uses
The tender young green plants can be fried in butter or olive oil and eaten alone or added to salads.

The leaves and roots can be infused in water to make a tea. Bedstraw is in the coffee family and its little seeds can be dried, roasted, and ground for a coffee substitute.

Bedstraw contains a milk-curdling enzyme, and can curdle milk for cheese making.

> "Northern Bedstraw is common throughout the Yukon. It has a dusty odor and some contend it induces a form of hay fever."
>
> —Martha Louise Black, *Yukon Wild Flowers* (1940)

Tender young bedstraw leaves can be fried up in butter or oil. CA

Part II: *Herbs*

Drying bedstraw bundles after the harvest. BG

Cosmetic Uses

Sweet-scented bedstraw can be sun infused or prepared in a low-temperature slow cooker with jojoba or sunflower oil in a ratio of one part flowers and leaves to two parts oil. The end result is sweet-scented vanilla-like oil that can be used as a carrier oil for lymphatic massage blends. The oil can also be used to treat earache (see recipe below).

Other Uses

First Nation peoples have combined the roots with highbush cranberries to produce a red dye.

Cautions

Continual use can irritate the mouth. It's recommended that people with poor circulation and diabetes avoid using this plant medicinally.

Bedstraw Ear Oil

Place fresh bedstraw flowers and leaves in a pot with double the amount of olive oil. Warm on low heat for 30 minutes, let sit to cool, strain and bottle. Warm a dropperful to body temperature and release into the sore ear as needed.

Bedstraw Deodorant

ADDITIONAL RECIPE

Detoxifying Infusion, page 283

Bedstraw flowers and leaves can be used as an underarm deodorant. Simply make a decoction by simmering leaves and flowers for up to fifteen minutes, cool down, strain, and bottle. Put in fridge and apply daily to the armpits with a cotton ball. Or you can add 1 part alcohol (such as vodka) to preserve to 4 parts decoction and put in a spray bottle to mist under armpits or on the bottom of the feet (for more on this, see Decoctions page 286). Adding usnea lichen (see page 264) tincture to this mixture will help eliminate odour because of its anti-bacterial properties.

WILD CHAMOMILE

Matricaria discoidea

Other Names and Etymology
Pineappleweed, false chamomile. Wild chamomile's scientific names are from the Latin *matrix*, meaning "mother" and *caria* meaning "dear" and refers to the plant's many uses for women and children. While many northerners have known the Latin name for this plant to be *Matricaria matricarioides*, Yukon biologist Bruce Bennett says this name was actually misapplied to the northern boreal plant *Matricaria discoidea*.

M. discoidea

Family
Asteraceae (aster, daisy, or sunflower family)

Botanical Description
Many-branched, leafy annual that smells like pineapple when crushed; stems 10–30 cm or higher. Leaves highly dissected, almost fern-like. Flowering heads yellowy green, numerous, conical, in a cup of bracts; ray flowers absent. Unlike its cousin German chamomile (*Matricaria recutita*), it doesn't have white petals. In size and colour, the flower looks like the yellow centre of domesticated chamomile.

Habitat and Range
Wild chamomile lives in disturbed areas with poor compacted soils. It can be seen blossoming along footpaths, garden edges, natural places where people gather, and roadsides, in spring and early summer. A cosmopolitan weed, it grows in North America from Newfoundland to British Columbia, south to Mexico. It appears to have been introduced to the Yukon, Alaska, and NWT.

Fragrant wild chamomile heads are wonderful and soothing when added to a dream pillow or sachet. PL

Part II: *Herbs*

Plant Parts Used
Flowers and leaves

Harvest Time
The optimum time to gather wild chamomile is as soon as the tiny, yellowy-green disc flowers appear but they can also be gathered throughout the summer.

WILD CHAMOMILE, or "pineappleweed" as it's widely known, likes to grow where humans walk and create pathways. Brushing or stepping on this sunny plant releases a scent similar to fresh pineapple fruit or ripe apples, and infuses the air with an essence that gently awakens the olfactory system to create a feeling of happiness and relaxation.

The plant always seems to rebound and look good no matter how much it gets trampled. (I know a few people like that, too!)

I find it difficult to walk by a patch of wild chamomile without pinching a flower head for the pure aromatherapy effect. The oil from the heads has come in handy as a spit poultice when I've gotten a nettle or insect sting. Being from the Asteraceae family, this plant has a sunny disposition with a bright energy emanating from it.

Medicinal Actions
Anti-inflammatory, antispasmodic, aromatic bitter, carminative, diaphoretic, galactagogue, hypnotic, nervine, (mild) sedative, stomachic, vermifuge

Medicinal Preparations
Bath, cream, hydrosol, infused oil, poultice, salve, tea/infusion, wash

Medicinal Uses
A close relative of *Matricaria recutita* (German chamomile), which is frequently used in commercial herbal teas, wild chamomile can be enjoyed in a fresh, mild tea. Because of its carminative properties, wild chamomile is good if you suffer from gas, heartburn, and/or mild gastrointestinal upset. Drinking the tea before bed will help you get a restful night's sleep, as it soothes the nervous system. The dried flowers make a wonderful sachet to put under your pillow at night; the soothing fragrance is calming and will help you to fall asleep.

Matricaria is also used as a remedy for menstrual cramping. A teaspoon of the tea can be diluted in a cup of water and administered in small amounts to a colicky or teething baby (for more on dosage, see page 38).

It's a gentle cleanser and helps rid the body of pinworms, a condition that makes children cranky, irritable, grind their teeth, and scratch their bottoms.

Wild Chamomile

The hot tea makes a nice beverage if you have a cold or the flu.

Topically, wild chamomile can be used in a poultice or as a wash to help alleviate the inflammation of a wound or for achy, sore muscles. It can also be infused in carrier oil for use in a salve or cream, or in jojoba oil for perfume or massage oil.

The fragrant flower heads are wonderful and soothing when added to a dream pillow or sachet. They can also be used in a foot soak, facial steam, or in a full bath.

In his book *Edible and Medicinal Plants of the Rocky Mountains and Neighbouring Territories,* Master Herbalist Terry Willard says, "As a treatment for diarrhea, the whole plant was decocted. It is similar to chamomile in many of its medicinal qualities but much milder. It is used for stomach aches, flatulence, as a mild relaxant and for colds and menstrual problems. Externally it can be used for itching and sores."

The Dena'ina of Alaska gave wild chamomile tea to new mothers and babies to cleanse and heal the system, and start the flow of breast milk. It was also used by menstruating women, as a laxative, and as an eye wash for treating snow blindness.

> "Perfumes are the feelings of flowers."
>
> —Heinrich Heine, *The Harz Journey* (1826)

Food Uses

The flower heads can be eaten raw. While a few are nice, I would hardly sit down and make a meal of them unless I had to! I have added the heads to salads, and a tea of the buds can be added to smoothies and even to cake or muffin recipes. I've heard that in the past the flowers were crushed and spread over meats to prevent spoilage.

Nutritional Profile

There is very little information about the nutrients found in wild chamomile. I suspect that that the breakdown is similar to chamomile and that this herb contains vitamins A, C, and many minerals such as calcium, magnesium, and potassium.

Other Uses

Steve Johnson, founder of the Alaskan Flower Essence Project and author of *The Essence of Healing,* says that pineappleweed is indicated for use "when we feel a lack of harmony with our physical environment; unaware of the support and nurturing that is available from nature; and a weak nurturing bond between mother and child." He adds, "The essence helps us maintain a calm awareness of ourselves and our surroundings so that we can remain free from injury and risk; promotes harmony between mothers and children, and between humans and the earth."

According to herbalist Robert Rogers, "Pineappleweed is the flower essence of mother and child. It may be used frequently whenever there is disharmony between the two, for colic or vomiting in the baby, or general nervous tension for mother and child due to tiredness and fatigue."

Part II: *Herbs*

ADDITIONAL RECIPES

After-Dinner Elixir, page 297

Afternoon Floral Tisane, page 281

All-Purpose Botanical Ointment, page 313

Aromatic Herbal Balm, page 314

Boreal Healing Bath, page 325

Flower-Power-Peace Bath, page 325

Good-for-Gout Infusion, page 284

Good-for-Nausea Infusion, page 284

Good-Sleep Infusion, page 283

Living Fruit Leather, page 377

Sleep-Aid Capsule, page 301

Cautions
It may cause allergies in people with sensitivities to plants in the Asteraceae family.

Wild Chamomile Tea

To make a soothing and relaxing cup of tea, add a teaspoonful or a pinch of dried wild chamomile blossoms into a tea strainer or loose into a cup. Pour boiling water over the herb, cover, and allow it to steep for about 10 minutes.

To dry wild chamomile for tea: Gather the fresh blossoms before the heat of the day, gently wash in cold water and pick out any deadheads. Spread out evenly in a wicker basket or on a drying rack, out of direct sunlight, until they feel dry. Store in a sealed jar.

Drinking wild chamomile tea before bed will help you get a restful sleep. BG

CHICKWEED

Stellaria media

Other Names and Etymology
Star lady, common starwort, clucken-weed, mischievous Jack, hen's inheritance. *Stellaria* is Latin for "star" and refers to the star-shaped flower.

Family
Caryophyllaceae (pink family)

Botanical Description
Chickweed is a trailing or matted ground-cover herb that can reach lengths of 10–60 cm. Lower leaves are broad with long sparsely hairy petioles, whereas upper leaves are usually smaller and lack petioles. The tiny, dainty flowers are sparsely scattered along the plant; chickweed can continue to flower until snowfall in the North. The five white petals are deeply cleft; the petals are shorter than the green hairy sepals. The seeds are tiny reddish brown. Each plant has the capability to produce over 15,000 seeds per year.

S. media

Chickweed is a tasty salad green. BG

Up North, chickweed can flower until the snow falls. BG

Habitat and Range
Introduced from Europe to New England in the 1600s, chickweed is now widespread throughout the circumpolar north including the NWT, Yukon, and Alaska. It's considered a weed and found in disturbed soils such as garden beds and edges, and on lawns.

Plant Parts Used
Leaves, stems, flowers, seeds

Harvest Time
Summer through late autumn

I CAN'T COUNT HOW MANY TIMES I have called on the virtues of chickweed for food and medicine. Called the "magic healer," chickweed has multiple uses and is definitely a herb that should be honoured. It's delicious! I graze on it throughout the day when I'm gardening. I have a really good patch of chickweed that I use in the summer for salads, green drinks, and common skin irritations. I also use it dry and fresh for a multitude of Aroma Borealis healing salves and ointments. It is grown commercially throughout the world as a medicinal herb. Chickweed is strong enough to help draw out infection from a wound or boil, yet gentle enough to use as a wash or poultice for eye irritations and on the inflamed nipples of breastfeeding moms.

Medicinal Actions
Antipyretic, anti-inflammatory, anti-rheumatic, bitter, demulcent, diuretic, emollient, expectorant, laxative, stomachic, vulnerary

Medicinal Preparations
Bath, compress, cream, juice, oil, poultice, salve, tea/infusion, tincture, wash

Chickweed

Medicinal Uses

Chickweed contains saponins that benefit the digestive system as a digestive aid, in regulating intestinal flora, absorbing toxins from the bowel, helping to regulate colon bacteria and yeasts, and lowering the bowel transit time (making it a good herb to use if you are constipated).

The saponins are also known for helping to dissolve fat cells and lower cholesterol, so it's commonly used in tea and tinctures as a weight-loss remedy.

As a mild diuretic it can temporarily increase the elimination of urine. It's a demulcent herb that can soothe irritated tissue both internally and externally. It's often used as a tea or tincture for bladder, kidney, and urinary troubles.

It has mild expectorant properties, making it a good choice for an irritating summer cough. According to Michael Tierra, the author of *Planetary Herbology*, chickweed "moistens and aids the expectoration of phlegm from the lungs, relieves sore throats, lowers fevers, and helps to treat stomach and duodenal ulcers."

Along with its anti-inflammatory properties, chickweed can help to increase circulation and as a result is used by herbalists as an anti-rheumatic and to help reduce the swelling of sprains and strains.

Externally it can be used as a wash, poultice, compress, oil, salve, or cream to help reduce inflammation of irritated and itchy skin. It's commonly used for insect bites, boils, skin infections such as abscesses, and for dissolving skin tumours. As a compress it's an excellent remedy for new mothers to soothe dry, cracked nipples and to draw out mild breast infections.

In her book *Healing Wise*, herbalist Susun Weed says when chickweed tincture is taken it has "secret dissolving powers." She adds, "Ovarian cysts, dermoid cysts, lumps in the breast and elsewhere can't hold their own against her slippery ways when a dropper full (1 mL) is taken 4–5 times a day, persistently, for many months."

I have used chickweed effectively over the years as an eyewash and compress for conjunctivitis (pink eye), weeping eye irritations, eyes inflamed by allergies, and eye wounds.

> "But a weed is simply a plant that wants to grow where people want something else. In blaming nature, people mistake the culprit. Weeds are people's idea, not nature's."
>
> —Author unknown

Food Uses

Tender and nutritious chickweed is a super salad green. It's excellent to add to soups, stews, pastas, sandwiches, and, my favourite, wild-weed spanakopita.

I often use chickweed in place of parsley, and add it to pesto when I don't have enough basil. Chickweed also excels in salad dressings and dips. Combining fresh chickweed with water in a blender creates an excellent green drink. I like to freeze the fresh juice in ice-cube portions to use throughout the winter as a base for soup stocks and smoothies.

As its common name indicates, chickweed also makes an excellent food for fowl such as chickens, ducks, and turkeys.

Part II: *Herbs*

ADDITIONAL RECIPES

Boreal Skin Oil, page 307

Chickweed Shooter, page 339

Cream of Wild-Weed Soup, page 364

Extra-Green Pesto Sauce, page 356

Green Sauce, page 360

Herb and Berry Antioxidant Smoothie, page 340

Kalerific!, page 338

Moisturizing Lip Balm, page 314

Nourishing Vinegar Tonic, page 296

Skin Disinfectant Spray, page 295

Sunshine Daydream Smoothie, page 337

Wise-Weed Smoothie, page 337

Wild-Weed Dressing, page 361

Wild-Weed Salad, page 365

Wild-Weed Spanakopita, page 367

Nutritional Profile

Chickweed is high in vitamins C and A, and minerals such as calcium, iron, magnesium, manganese, niacin, phosphorus, potassium, and zinc. It's also high in useable plant protein.

Cosmetic Uses

Chickweed excels at healing acne, rashes, and rosacea. Its demulcent and moisturizing properties help to alleviate dry, itchy skin. It can be added to facial steams, used as a compress, as a wash, and can be made into a healing oil, salve or cream to help soothe, draw out, and reduce redness due to irritations.

Other Uses

Chickweed is well known in herb circles for its healing charms. According to Gurudas, the author of *The Spiritual Properties of Herbs*, "chickweed's signature relates to the fact that the plant is found all over the planet and is used in joining together harmonious thought forms that are shared worldwide, like world peace. It shows that worldwide we are all one." He goes on to say that, "this is a good plant for the third, fourth and fifth chakras. And rebalances the higher and lower levels and creates more energy in the heart chakra."

Cautions

Excess consumption can lead to diarrhea. During pregnancy only use topically and in small amounts as food (but not as a tea, tincture, or juice).

Chickweed-Garlic Green Dip

High in antioxidants this dip has many uses. It's great as a condiment on almost anything! I like it as a dip for crackers and carrots, and as a favourite potluck offering.

1 bunch chopped chickweed
¼ cup (60 mL) chopped garlic
2 tablespoons (30 mL) apple cider vinegar
½ teaspoon (2 mL) sea salt
Olive oil to cover

Mix chopped chickweed, garlic, vinegar and salt together in a blender; blend until smooth, and pour into a bowl. Cover with olive oil to prevent browning.

Stir before serving, eat, and enjoy. Store in airtight container in the fridge.

RED CLOVER

Trifolium pratense

Other Names and Etymology
Cow clover, meadow clover, wild clover, wild shamrock. The plant was named *Trifolium pratense* by Carl Linnaeus in 1753.

Trifolium is from Greek and means "three leafed." *Pratense* is Latin for "found in meadows."

Family
Fabaceae (pea family)

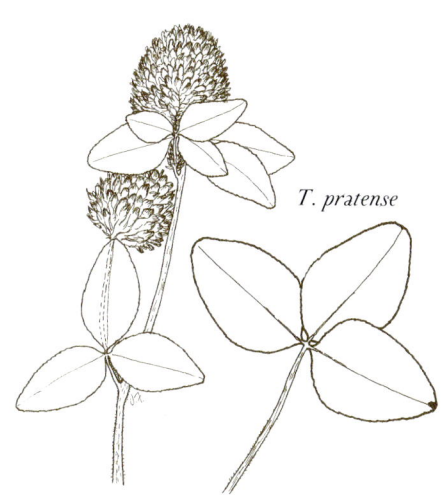

T. pratense

Botanical Description
Short-lived perennial; stems up to 80 cm high, erect or spreading. Three leaflets per leaf (but the lucky ones have four!) oval-elliptic, 2–5 cm long. Lower leaves, long petioled; upper leaves, short petioled or stalkless. Clusters of flowers globose; flowers 1.2–2 mm long, red or pink, fragrant.

Left: You can suck honey-sweet nectar from red clover petals. BG

Below: As a child, I spent hours in the clover patch searching for four-leaf clovers. BG

Part II: *Herbs*

Habitat and Range
Grows wild all over the world, including the Yukon. It's easily cultivated in the harsh boreal climate and is a welcome volunteer in my garden. It's an excellent fertilizer for poor soils as it helps to fix nitrogen, which increases soil fertility.

Plant Parts Used
The flowering tops with the top-two leaves

Harvest Time
Flowers and leaves: when the flowers have just bloomed

I WAS A CLOVER KID! Like many kids, I could spend hours in a patch searching for the prized four-leaf clover, and the eternal optimist in me always seemed to find one. Each leaf of the coveted four-leaf clover has a meaning like fame, wealth, faithful lover, and health. Like the honeybee, I was very attracted to the pink globe-shaped flower head and plucked it apart, sucking the nectar from each petal. As an adult, whenever I find myself in a clover patch, gathering flowers for medicine, I still keep an eye out for the good-luck floral emblem of Ireland.

Medicinally, clover has a long history dating back to the ancient Druids who placed a high value on its virtues. The three-leaf clover can also represent the triple goddess of mother, maiden, crone; or the Christian trinity.

Herbalist Robert Rogers notes that looking closely at the florets with a hand lens reveals the anatomical structure of the external female organ system—a doctrine of signatures whose meaning cannot be mistaken.

Red clover comes in many shades of pink. CA

Medicinal Actions
Alterative, antioxidant, antispasmodic, astringent, bitter, cholagogue, diuretic, expectorant, galactagogue, nervine, sedative, tonic

Medicinal Preparations
Flower essence, infusions, oil, plaster, poultice, salve, tea, tincture

Medicinal Uses
Red clover is considered the "queen of blood purifiers." It's used internally or topically to help clear up inflammatory skin conditions like acne, rashes, psoriasis, and

Red Clover

eczema. It's helpful for dry, irritable or inflamed skin because it's considered soothing, cooling, and moistening. It's especially useful for children with skin problems like eczema.

As an organic source of sodium, it helps to alkalize the system and restore the acid-alkaline balance to the body.

The respiratory system also benefits from red clover as it cleanses and soothes the lungs and the air passages and is a mild expectorant. It's therefore good in a tea, tincture, or cough syrup for treating coughs, asthma, bronchitis, sore throats, whooping cough, and/or wheezing. It's also helpful for sore throats, colds, and fevers.

Red clover also helps to increase lymph flow, detoxifying and reducing inflammation in the body caused by lymphatic congestion. This helps explain its use in treating sore joints, arthritis, rheumatism, and gout.

Red clover is also stimulating to the digestive organs and can help with inflammatory-bowel disorders, improve digestion, and help to relieve gas.

Due to its blood-thinning properties, herbalists also recommend red-clover tincture or tea to help regulate high cholesterol levels.

In our high-speed world, many of us find ourselves tired all the time. Red clover might be helpful as a bedtime infusion as it's a useful tonic to soothe the nerves and may help with exhaustion and headaches. Men can benefit from using red clover as it helps to promote prostate and heart health.

Red clover is also an excellent woman's herb. It can be used to promote menstrual flow and can be used for women with dysmenorrhea (painful menstrual cramping). Red clover combined with herbs such as raspberry leaf, nettle, crampbark (highbush cranberry bark), and yarrow can really help alleviate menstrual pain. Red clover can also help to regulate the menstrual cycle by providing the body with much-needed minerals and helping with hormonal balancing. However, if you flood when you menstruate, you should avoid red clover or use it in moderation because of its blood-thinning properties.

Red clover can be used topically as a poultice for mastitis (breast inflammation) that may be experienced during nursing, and as a mild infusion to promote healthy breast milk. It also works well in a salve or a sitz bath for hemorrhoids caused by pregnancy and birthing.

For fertility issues, Susun Weed, herbalist and the author of *Wise Woman Herbal for the Childbearing Year,* writes, "One of the most cherished of the fertility-increasing plants is red clover." She continues, "It is especially helpful if there is scarring of the fallopian tubes, irregular menses, abnormal cells in the reproductive tract, or 'unexplained' infertility…"

Four different water-soluble antioxidant isoflavones have been identified in red clover. One of them, genistein, is one of the most talked-about isoflavone plant chemicals

> The Fairy: O, what a great big bee
>
> Has come to visit me!
>
> He's come to find my honey.
>
> O, what a great big bee!
>
> The Bee: O, what a great big Clover!
>
> I'll search it well, all over,
>
> And gather all its honey.
>
> O, what a great big clover!
>
> —Cicely Mary Barker, author,
>
> *Flower Fairies of the Wayside* (1948)

Part II: *Herbs*

Red clover is high in minerals and makes a refreshing iced tea. PL

because it simulates the effects of estrogen in the body. Isoflavones are helpful for women experiencing symptoms of menopause, such as hot flashes, depression, irritability, mood swings, osteoporosis, and heart health. Studies on red clover have validated that the isoflavones slow down bone loss and boost bone mineralization in menopausal women.

There is an ongoing debate about whether red clover is a risk for women who are predisposed to estrogenic cancers. While there has been very little proof supporting or denying this theory, women who are at high risk of developing estrogenic cancers such as breast, uterine, and ovarian cancers, should be cautious about using red clover. Red clover's detractors argue that red clover mimics the effects of estrogens, and high levels of estrogens sometimes contribute to these cancers. Because of this, they recommend that women who experience uterine fibroids, who are ingesting drugs that increase estrogen levels in the body (such as birth-control pills or cancer-therapy drugs like tamoxifen) avoid using red clover.

However, it's worth noting that red clover's advocates believe that the herb's isoflavones can be helpful in preventing and treating estrogen-dependent cancers by competing for estrogen receptor sites with xenoestrogens (estrogen-mimicking compounds that are released as toxins from plastics and heavy metals).

I recommend following developing research studies on red clover, but also to listen to your intuition, your body, know your family cancer history, know your risk factors, and ask questions of your doctor, naturopath, or herbalist.

The U.S. government's National Cancer Institute has validated red clover's many anti-cancer properties. The plant has been studied as a natural cancer treatment for well over one hundred years and is used to treat cancer by thirty-three cultures around the world. In the old medicinal literature, red clover is touted for its anti-tumour properties and was used internally and topically as a poultice for external cancers, like skin cancer.

Dr. Samuel Thomson (1769–1843) successfully used red clover in a cancer plaster and drawing salves.

Red clover blossoms were used by Dr. John Christopher (1909–1983) in his anti-cancer remedy, the Basic Trifolium Compound, that was similar to the controversial Hoxsey Method anti-cancer formula (which the U.S. Food and Drug Administration banned the promotion of in 1960).

Dr. Christopher was once overheard saying, "I just can't help myself… red clover is the antidote for cancer."

Rene Caisse (1888–1978) was a Canadian nurse who also used red clover blossoms (along with other herbs such as rhubarb root, sheep sorrel, kelp, watercress, burdock, slippery elm, and blessed thistle) in her famous cancer-fighting Essiac Formula that is still in use today. Red clover is also a key ingredient in Jason Winters Tea, which is sold in health food and herb shops.

Red Clover

Food Uses

Sprouted red-clover seeds make a great addition to salads and sandwiches. The sprouts are crispy, tasty, and nutritious, and make a great midday snack.

Gathering seeds from the wild can be a bit labour intensive, so you might want to buy seeds from a herb seed company like Richters (www.richters.com).

Red clover has great nutritional value, although it would be challenging to make it into a main dish. Enrich the nutrient value of your meal by adding cooked, dried, or fresh clover to grain dishes, salads, stir-fries, soups, sauces, breads, and muffins.

A red-clover iced tea with a splash of lemon is a refreshing and cleansing drink throughout the year, but especially in summer. Red clover is fun to use in herbal popsicles for the kids. They love it! Simply place a blossom at the bottom of popsicle moulds before filling with iced tea. When you take them out—voila!—a flowered popsicle.

Nutritional Profile

Vitamins A, C, E, K, and B-12, and minerals such as calcium, magnesium, chromium, iron, manganese, niacin, phosphorus, potassium, selenium, silicon, thiamine, trace amounts of zinc, and small amounts of protein.

Other Uses

Dreaming of clover brings fortuitous health and prosperity. The expression, "living in the clover," conveys this. Carrying a three-leaf clover gives you protection from unwanted negative energy. Worn over the right breast it will bring you success in all undertakings.

According to Patricia Kaminski and Richard Katz, authors of *Flower Essence Repertory*, "Red Clover flower essence's positive qualities are: self-aware behavior, calm and steady presence, especially in emergency situations." They go on to write that the "Patterns of imbalance are susceptible to mass hysteria and anxiety, easily influenced by panic or other forms of group thought."

They later add that "Red Clover flower essence is a powerful cleanser and balancer; it is especially related to the psychic properties of the blood, where the spiritual ego of each individual resides. Red Clover infuses strong forces of self-awareness so that the individual can think in a calm and steady way, and act from his/her own center of truth."

Cautions

Red clover is an amazing herb but use it mindfully. While no serious adverse effects have been reported, use it in moderation for short periods of time. Those who have breast cancer or other hormone-sensitive cancers should avoid taking it internally. Also those on birth-control pills, and on blood-thinning medication, or tamoxifen should avoid it.

Sprouting red clover seeds is easy. BG

Part II: *Herbs*

Trifolium hybridum, also known as "alsike clover," is naturalized from Europe and found throughout the Yukon and Alaska. It makes a nice light tea. BG

Clear-Skin Tincture

1 part yellow dock root
2 parts nettle leaf
2 parts red clover flower
Vodka

Add fresh or dried herbs to a clean jar, cover with vodka, and shake every day for 4 to 6 weeks. Strain, bottle, and label. (See Tincture Method, page 293.)

Susun Weed's Fertility-Enhancing Infusion

Take one ounce (28 g) dried, red clover blossoms (fresh won't work for this application) and put them in a one quart (1 L) canning jar. Fill the jar with boiling water, screw the lid on tight, and let it steep at room temperature overnight (or for at least four hours). Drink throughout the day.

ADDITIONAL RECIPES

Boreal Bitters, page 297

Good-Sleep Infusion, page 283

Herb-and-Berry Antioxidant Smoothie, page 340

Lung-Tonic Infusion, page 283

Moon-Time-Infusion, page 284

Wild-Greens Stir-Fry, page 370

Wild-Weed Salad, page 365

COLTSFOOT

Petasites frigidus (sweet coltsfoot)
Petasites frigidus var. *sagittatus* (arrow-leaved coltsfoot)

Other Names and Etymology
Arctic sweet coltsfoot, sweet coltsfoot, Arctic butterbur, palmate-leaved coltsfoot, western coltsfoot, coughwort, arrow-leaved coltsfoot. *Petasites* comes from the Greek word *petasos* meaning "large-brimmed hat." *Frigidus* means "cold." *Sagittatus* means "arrow." An old name for the common coltsfoot was "the son before the father" because the flowers appear and wither before the broad, sea-green leaves are produced.

Family
Asteraceae (aster, daisy, or sunflower family)

Left: Coltsfoot has been used as a cough remedy for at least 2,500 years. BG

Below: An essence made from coltsfoot flower helps us breathe deeply. CA

Wild Food and Medicine Plants of the North | 73

Part II: *Herbs*

Botanical Descriptions
Coltsfoot is a stout perennial herb with milky sap. Coltsfoot flowers early in spring, usually before leaves emerge. Clusters of small white flowers, with dandelion-like heads, on a stalk; seeds have feathery bristles for wind dispersal. Stems 10–50 cm high, leaves 2–18 cm high arising from base, variable in shape, toothed margins, usually hairless and dark green on upper surface, dense white wooly hairs on lower surface. There are several subspecies of coltsfoot in the North that are differentiated by leaf shape. *Petasites frigidus* var. *sagittatus* has a heart or arrow-shaped leaf. (Note: *Petasites* is very closely related to the introduced European genus *Tussilago* that is also known as coltsfoot.)

French apothecary shops would advertise their presence by painting the distinctive shape of coltsfoot leaves on their signs. BG

Habitat and Range
Coltsfoot grows in moist habitats, from lakeshores to woods, to alpine slopes. It's found across North America, including in the Yukon, NWT, and Alaska.

Plant Parts Used
Roots, stalks, leaves

Harvest Time
Roots: early spring. Stems: late spring. Leaves: early summer

THE FIRST TIME I saw wild coltsfoot growing was on an early summer morning along the Lappe River near Canol Road in the southern Yukon. Though I'd only seen a photograph of it before, I instantly recognized the unique-looking plant. I gathered some leaves and made a cough syrup with it that evening over the campfire (see recipe page 76). I kept the syrup throughout the winter months, stored in the refrigerator, and used it as needed.

Coltsfoot has been used as a cough remedy for at least 2,500 years in both western herbalism and traditional Chinese medicine. In eighteenth-century France apothecary shops advertised their presence by painting a picture of the herb on their signs.

Medicinal Actions
Anti-catarrhal, antispasmodic, demulcent, diaphoretic, diuretic, emollient, expectorant, pectoral, sedative, tonic

Medicinal Preparations
Oil, poultice, smoking mixture, syrup, tea/infusion, tincture

Coltsfoot

Medicinal Uses

Dried coltsfoot leaves are commonly prepared as a tea or a cough syrup for coughs and upper-respiratory problems. The antispasmodic and sedative effects of coltsfoot are soothing and calming to the respiratory system because they help block the nerve impulses that initiate a cough.

Back in the seventeenth century, physician and astrologer Nicholas Culpeper observed, "The plant is under Venus, the fresh leaves or juice, or syrup thereof is good for a hot dry cough, or wheezing, and shortness of breath."

Today we know that coltsfoot contains a substance called mucilage that coats and soothes the throat. In chest colds this helps to relieve pain in the intercostal muscles (between the ribs) due to coughing, and to treat symptoms of asthma, bronchitis, whooping cough, dry hacking cough, laryngitis and hoarseness, lung cancer symptoms, mouth and throat irritations, and wheezing. Coltsfoot also stimulates the lungs to expel phlegm.

Coltsfoot leaves have also been employed in smoking mixtures to relieve cough. Externally, a poultice or salve of flowers and leaves can be applied to the skin to treat eczema, stings, bites, and other skin inflammations. The fresh or dried leaves can also be soaked in warm water and applied directly to heal open sores or ulcers.

Herbalist Robert Rogers writes, "In northern Alberta, the Cree know it as *puskwa* or commonly refer to it as 'wolverine's foot' or 'owl's blanket' due to its insulating value. The furry inside of the leaves are gathered by birds to line their nests. The Slave call the plant *ya yenoshetia*, meaning 'bear eats it.' In Alaska, the Dena'ina call Arctic and palmate-leaved coltsfoot, owl's blanket, or *k'ijeghi ch'da*. The roots are soaked in hot water and tea taken internally for tuberculosis, sore throat, stomach ulcers and such. It appears to stop the internal bleeding associated with these conditions."

> "The smoking of the leaves for a cough has the recommendation of Dioscorides, Galen, Pliny, Boyle, and other great authorities, both ancient and modern, Linnaeus stating that the Swedes of his time smoked it for that purpose. Pliny recommended the use of both roots and leaves. The leaves are the basis of the British Herb Tobacco, in which Coltsfoot predominates."
>
> —Mrs. M. Grieve, herbalist, *A Modern Herbal* (1931)

Food Uses

The young tender stalks of *P. frigidus* should be harvested before the flower buds appear. Boil them until tender and season with some salt, butter, olive oil, or hemp oil. The salty young leaves can be steamed and eaten. Coltsfoot is an acquired taste, and the felt-like texture of the leaves feels unusual in the mouth.

Traditionally the ash of the plant was used as a salt substitute. The stems and green leaves were rolled up into balls, dried, and then placed on top of a very small fire rock and burned. The black powdery ashes were then gathered and stored for use in cooking.

Part II: *Herbs*

Symbols of love on the forest floor! BG

Other Uses

Coltsfoot leaves were traditionally smoked to help generate visions. The Bailey Flower Essences indicate the use of coltsfoot flower essence for "those who have a deep fear of their own power, a symptom created out of a lack of self-love and a lack of trust in life."

According to Herbalist Robert Rogers, "Coltsfoot… flower essence may help us breathe deeply, ground our bodies and subsequently gain perspective. Deep breathing helps us merge back into our own skins, release grief, and open to the present moment. Coltsfoot is an essence of orientation."

Coltsfoot is said to be influenced by Venus. As such it can be used magically to attract love and bring peace.

Cautions

Petasites species contain naturally occurring compounds called pyrrolizidine alkaloids and can be harmful if eaten in large quantities for long periods of time. The concentrations are often highest in the rhizomes and stalks, and lowest in the leaves, though this may vary depending on where the plants grow. Pregnant and nursing women and children under the age of six should avoid coltsfoot. The German Federal Ministry of Health's Commission E approved the use of fresh or dried coltsfoot leaf in products to treat dry cough, hoarseness, mild throat and/or mouth inflammations. German authorities, however, recommend that preparations containing coltsfoot leaf not be taken for more than four to six weeks each year.

Coltsfoot Cough Syrup*

2 cups (500 mL) of fresh coltsfoot leaves
4 cups (1 L) water
1 cup (250 mL) honey (local if available)
2 tablespoons (30 mL) of alcohol such as brandy or vodka
 or apple cider vinegar

ADDITIONAL RECIPES

Cough-and-Cold Medicinal Syrup, page 300

Lung-Tonic Infusion, page 283

Add coltsfoot leaves to boiling water; simmer down to half the liquid amount. Let sit and cool. Strain. Add honey for sweetness and alcohol to help preserve the remedy. Store in a dark bottle in a cool place, preferably in the refrigerator. Use as needed. A teaspoon can also be added to herbal teas.

---

*Based on a recipe from *Discovering Wild Plants* by Janice Schofield.

DANDELION

Taraxacum officinale
Taraxacum alaskanum
Taraxacum ceratophorum

Other Names and Etymology
Lion's tooth, dent-de-lion, dandy-lioness, Irish daisy, fairy clock, puffball, wild endive. In Latin *taraxos* means "disorder" and *achos* means "remedy." *Officinale* signifies that it is an official medicinal plant.

Family
Asteraceae (aster, daisy, or sunflower family)

Botanical Descriptions
An easy plant to identify, dandelion is a herbaceous perennial with a rosette of jagged, irregular-lobed green leaves produced from a long, thick, fleshy taproot that can descend more than one metre into the soil. The stem is hollow with a white, latex-like substance lining the interior. The bright yellow flower heads are composed of ray florets and surrounded by two rows of floral bracts. The outer row is often rolled back towards itself. The flowers open to sunshine and close if it is dark or cloudy. The dandelion is a prolific seed producer with up to 200 elliptical seeds (attached to fine white barbed hairs) produced per head. A single plant can produce more than 5,000 seeds per year.

Dandelion blossoms will leave a fresh-smelling, dark yellow resin on your fingers. BG

Part II: *Herbs*

> "If dandelions were hard to grow, they would be most welcome on any lawn."
>
> —Andrew V. Mason

There are four native dandelion species in the Yukon, *T. ceratophorum* and *T. lyratum* are the two most commonly found. The common dandelion (*Taraxacum officinale*) and the native species have different-looking bracts. The bracts on the introduced dandelion are bent downwards toward the stem, whereas on the native species the bracts are erect. The native dandelions aren't as robust as the weedy dandelions.

Habitat and Range
Disturbed soils, garden edges, meadows, roadsides.

Taraxacum officinale is one of the most widespread vascular plant species in the world: circumpolar including NWT and Yukon, Canada, United States, New Zealand, Australia, India, South America. *Taraxacum alaskanum* grows in undisturbed natural areas, woodlands to the Arctic tundra, and is an alpine plant.

Plant Parts Used
Roots, stems, leaves, flowers

Harvest Time
Roots: in spring before the plant flowers; in autumn after the first frost. Young leaves: in early spring and throughout the summer (as they age the leaves become bitter tasting). Flowers: when in full blossom. Gathering dandelion blossoms is fun and easy and leaves a fresh-smelling, dark yellow resin on your fingers. Gathering 44 cups (11 L) of dandelion flower heads leaves you with about 24 cups (6 L) of flower petals once you have pulled off the green sepals. I find it is important to process the flowers immediately after gathering because the flower heads start to close up, making it harder to remove the petals. The 24 cups (6 L) of petals gives you enough petals to make all the dandelion flower recipes in this book and enough left over for tea to enjoy in the winter months!

Top: Dandelion blossoms are a good source of the "sunshine vitamin," vitamin D. BG

Above: Dandelion leaves are delicious in soups, salads, and entrées. BG

78 | The Boreal Herbal

Dandelion

HERBALISTS and natural-medicine practitioners have used dandelion as medicine since time immemorial.

The bitter flavonoid compounds in dandelion roots and leaves give the plant its diuretic properties, which in turn help to purify the blood and liver, relieve muscle spasms, and reduce inflammation. The leaves have more diuretic action than the roots.

Although misunderstood by many a gardener, dandelions actually restore minerals and other nutrient-rich ingredients to the soil, help create drainage channels in compacted soils, and are an important early-blooming bee plant.

Call me naïve, or maybe I was bushed, but one summer a few years back I went to a local store and saw a colourful bag with a dandelion on it and the words, "Weed Food." "Wow," I said to the guy working there, "how progressive: weed food!" He looked at me and said, "Ya, it kills them!" Yikes, why call it food?

Dandelion roots and leaves contain rich amounts of usable minerals. BG

Medicinal Actions
Antibilious, anti-rheumatic, anti-inflammatory, antispasmodic, astringent, bacteriostatic, bitter, cholagogue, cholerectic, digestive, diaphoretic, diuretic, fungistatic, galactagogue, hepatic, laxative, stomachic, tonic

Medicinal Preparations
Decoction, extract, flower essence, oil, salve, tea/infusion, tincture, vinegar

Medicinal Uses
Dandelion root is used by herbalists to treat a variety of liver and digestive disorders, such as hepatitis and jaundice. *The Botanical Pharmacy* authors Heather Boon and Michael Smith note that, "It is considered to be both a cholerectic, promoting the production of bile, and a cholagogue, causing contraction of the bile duct initiating the flow of stored bile." They go on to say that, "studies have shown that administration of dandelion increased bile flow in dogs and rats and aided in the management of a variety of gall bladder conditions."

Dandelion root's digestive and bitter properties are helpful when used for indigestion, spleen disorders, relieving heartburn and constipation, and stimulating the appetite. If taken before a meal, dandelion will help increase the production of hydrochloric acid in the stomach, increasing bio-availability of nutrients, especially calcium. Some studies report that the inulin found in dandelion root may assist helpful bacteria growth in the digestive tract.

In addition, dandelion root's anti-inflammatory properties are used to treat rheumatism, gout, and eczema. The root may also help with lowering cholesterol and high blood pressure. The constituent inulin found in autumn dandelion is being looked at more

Part II: *Herbs*

Symmetry. BG

closely for its potential in managing diabetes. Dandelion is also good for treating the symptoms associated with PMS.

In keeping with its French name, *pissenlit*, which means "to piss the bed," dandelion leaf supports the kidneys as a natural diuretic that helps increase urine output, allowing the body to reduce water retention in cases of edema. This contributes to supporting improved digestion and liver function and may help with high blood pressure and weight loss. Most diuretics result in a loss of potassium through the kidneys. Since dandelion is high in potassium it helps counteract this effect. Dandelion has also played a historical role in the management of kidney stones.

When infused, dandelion flower makes a beautiful golden oil. According to Sat Dharam Kaur, N.D., in her book *The Complete Natural Medicine Guide to Breast Cancer*, "Both the flower and the root of the dandelion can be used in an oil to reduce breast cysts, clear long-held emotions, and improve liver function."

Dandelion stem is filled with a rich latex-like substance that can be used topically to help get rid of warts.

Food Uses

High in the electrolytes sodium and potassium, young dandelion roots can be eaten as a nutritious vegetable. Like carrots and other root vegetables the roots can be boiled, baked, or diced up and added to soups and stews.

Nourishing vinegars can be made with the roots, leaves, and flowers to use in the kitchen, for salad dressings, sauces, and marinades.

The dried and powdered roots make a pleasant caffeine-free coffee substitute.

Dandelion leaf is a nutritious vegetable high in calcium and vitamin C. The leaves can be eaten fresh in salads, steamed like spinach, added to stir-fries, soups, and stews, and can also be used fresh or dried for use as a tea.

Dandelion blossoms are rich in the "sunshine vitamin," vitamin D. The flowers are wonderful on their own, fried up in a bit of butter and garlic, or dipped into a savoury batter and baked or fried. Yum! They can also be added to salads for a splash of colour.

Nutritional Profile

Dandelion roots and leaves are abundant in rich amounts of usable minerals such as calcium, iron, copper, magnesium, manganese, phosphorus, potassium, selenium, silicon, and zinc. They are also an excellent source of vitamin A, the B complex vitamins, C, and D. They also supply beneficial carotenoids, fatty acids, flavonoids, and phytosterols. Dandelion leaves and roots contain high amounts of potassium. The leaves contain as much as 290 mg of potassium per 100 g, and the roots as much as 1,200 mg per 100 g. The roots are also exceptionally high in vitamin A, with up to 14,000 IU per 100 g of dried root.

Dandelion

Cosmetic Uses

The skin is said to act as the "third kidney," helping to cleanse and detoxify the body. Dandelion-root decoction or tincture, or an infusion of the leaves, can help to keep skin clear of blemishes. Add dandelion flowers and leaves to a facial steam to help relieve congested pores. They can also be added to baths and foot soaks.

Other Uses

The dandelion's sunny disposition is used energetically to balance the third chakra, also known as the solar plexus chakra. The element of fire represents this chakra. The dandelion symbolizes the will to be rooted while being bright and determined enough to step into your centre of personal power and light. It can be used as a flower essence, in sun tea (flowers and leaves), or through plant-spirit meditation (see page 30 for instructions).

Steve Johnson, founder of the Alaskan Flower Essence Project and author of *The Essence of Healing*, writes, "Dandelions can help us remain connected to our inner qualities of gentle strength and endurance during times of turmoil and stress." Ted Andrews' book *Nature-Speak* says, "Dandelion reminds us to find beauty where we haven't looked; look beyond the surface."

The magenta pigment of the dandelion root was traditionally used to dye tartans.

Cautions

Be aware that roadside dandelions may be full of contaminants from vehicles and lawn dandelions may be full of harmful pesticides; both of which can do more harm than good.

People with allergies and sensitivities to plants in the Asteraceae family may want to be cautious when first using dandelion flowers as food.

Dandelion Wine

The following recipe was shared by Yukon artisan vintner Vanora Millar who explained that: "According to English tradition, dandelions must be picked on St. George's Day, April 23rd, at the height of a sunny afternoon. In the north, it's more like June. In either case, the flowers must be fully open and winemaking begins immediately."

4 cups (1 L) dandelion petals, all green parts removed
7 cups (1.75 L) sugar
3 ½ cups (875 mL) sultana raisins, rinsed in hot water and finely chopped
6 cups (1.5 L) boiling water
1 teaspoon (5 ml) yeast nutrient
1 heaped teaspoon (5 ml) tartaric or citric acid
1 pint (2 cups) boiled water
1 Campden tablet
1 package wine yeast

ADDITIONAL RECIPES

After-Dinner Elixir, page 297

Detoxifying Infusion, page 283

Boreal Bitters, page 297

Dandelion-and-Birch Cornbread, page 347

Healthful Weed Pie, page 368

Dandelion-Petal Cake, page 371

Dandelion Jelly, page 350

Dandelion-Petal Mustard, page 359

Dandelion-Petal Pancakes, page 345

Dandelion-Petal Syrup, page 355

Roasted Dandelion Root Ice Cream, page 374

Dandy-Lioness Oil, page 308

Dandelion Dressing, page 362

(continued next page)

Part II: *Herbs*

ADDITIONAL RECIPES

Dandelion-Root Coffee, page 285

Flower-Power-Peace Bath, page 325

Flower-Delight Tempura, page 375

Good-for-Gout Infusion, page 284

Green Sauce, page 360

Pine Forest Smoothie, page 338

Spring-Cleanse Dandelion Decoction, page 286

Spring-in-the-Boreal Smoothie, page 336

Sunshine Daydream Smoothie, page 337

Toon-Town Smoothie, page 337

Wild-Greens Stir-Fry, page 370

Wild-Weed Dressing, page 361

Wild-Weed Salad, page 365

Wild-Weed Spanakopita, page 367

Put all ingredients, except for wine yeast, into a sterilized primary fermentor.

Add more hot water if necessary to bring the volume up to one gallon (approximately 3.75 L) and stir to dissolve sugar. Cover with a sterile plastic sheet and allow the dandelion mixture to cool to 70–75°F (21–24°C).

Add yeast according to directions on package.

Stir daily with a sterile, long-handled spoon, covering loosely with plastic sheet after each stir.

Ferment for 5 or 6 days at 70–75°F (21–24°C) until specific gravity is 1.040 (you'll need a hydrometer to test for this).

Strain out the pulp and press.

Siphon into carboy and attach fermentation lock.

Rack in 3 weeks and again in 3 months.

Top up with rhubarb wine if you have some kicking around, or with a light, white wine, if necessary.

When wine is clear and stable, bottle.

Age 6-12 months. Your wine will always be better after a year than after 6 months. Rhubarb needs a year, and is better after two.

Pure liquid sunshine! BG

DEVIL'S CLUB

Oplopanax horridus

Other Names and Etymology
Alaskan ginseng, Tlingit aspirin, devil's walking stick. *Oplo* is derived from the Greek word *hoplon*, meaning "weapon"; *panax* is from the Greek *pan*, which means "everything." *Horridus* means "bristly" or "wild." The Dena'ina word *heshkeghka'a* means "big, big prickle."

Family
Araliaceae (ginseng family)

Botanical Description
Coarse, strongly scented shrub, 1–1.5 m high; stems densely covered with spines and prickles. Leaves alternate, long-petioled; blades 10–30 cm wide, palmately five- to seven-lobed,

O. horridus

Left: Devil's club leaves are huge. BG

Below: Though beautiful, devil's club berries are inedible. CA

Part II: *Herbs*

leaf veins prickly beneath. Flowers in a large terminal cluster, 10–20 cm long; petals greenish; terminal clusters of inedible red fruit 4–5 mm long, two-seeded.

Habitat and Range
Subalpine thickets. Devil's club grows in the understorey of moist, shady, and dense conifer forests. It grows throughout British Columbia, southern Alaska, western Alberta, south to Oregon, Idaho, and Montana. Grows sparsely in the extreme southeast and southwest Yukon in the Labiche, Beaver, and Alsek river areas.

Plant Parts Used
Early shoots, roots, stems, dried inner bark

Harvest Time
Shoots: early spring, just after they appear. Roots and inner bark of recumbent stem: spring or early summer. (You will need a pointed shovel to dig up the roots, and strong clippers or a machete for the stem. Wearing gloves and a long-sleeved shirt is recommended.)

Devil's club does not reproduce very quickly and is sensitive to human impact. The plants are slow growing and take many years to reach seed-bearing maturity. Remember to only gather what you need, with as little impact as possible.

There is something sacred-feeling about a thicket of devil's club. BG

AS YOU WALK toward a community of devil's club, the plants' sweet, earthy aroma will grab all of your senses: it looks and feels like you have travelled back to prehistoric times. When in the presence of devil's club there is a real sense of sacredness. The first time I went to gather this plant I couldn't do it; I personally wasn't ready. At that moment I inherently understood how magical and potent this medicine was. It took me a few visits, prayers, and offerings to tune into the subtle energies of this much-revered plant. When the time came and I was ready to gather, it was with ease and grace. Not one scratch. The respect I have for this northern ginseng is the same I have for any wise, engaging, and inspired teacher.

Devil's Club

Medicinal Actions
Adaptogen, analgesic, antibacterial, antifungal, antiviral, cathartic, diaphoretic, emetic, expectorant, hypoglycemic, laxative, stomachic

Medicinal Preparations
Cream, decoction, oil, poultice, salve, tea/infusion, tincture

> "He who wants the rose must respect the thorn."
>
> —Persian proverb

Medicinal Uses
Dried devil's club root and the inner bark of the stem are used as a tea and tincture to strengthen the body and to help it adapt to stresses and illnesses.

The inner bark is used for the prevention of tuberculosis. Devil's club works well for colds, flu, and bronchitis because it supports the body in removing excess amounts of mucus from the respiratory system, and it promotes perspiration, which helps with the elimination of toxins through the skin.

A poultice of the inner bark can be used topically on a nursing mother's breasts to stop excessive milk flow. (Note: Wash poultice residue off the breast before feeding a baby.) A decoction has also been used as a gentle eyewash for the treatment of cataracts.

The Tlingit people use the inner bark of devil's club in a decoction or infusion to treat pain associated with arthritis, rheumatism, and general joint pain. The inner bark can also be used topically as a poultice, in a salve, cream, or liniment for pain or for open wounds, infections and ulcers. The inner-bark tea, when drunk in a decoction, helps lower blood-sugar levels, which helps control diabetes. (Note: Consult your doctor or naturopath before use if you have diabetes.)

Herbalist Robert Rogers points out that a recent study at the Tang Center for Herbal Medicine Research at the University of Chicago found that at low dosage, the herb induces a self-programmed death in chemotherapy-resistant strains of hormone-sensitive cancers, and at higher doses kills cancer cells directly.[*]

Devil's club is also used for stomach ulcers, is a mild laxative, and can help soothe a nervous stomach.

The berries can be mashed into a pulp and then rubbed onto the scalp to help get rid of head lice and to make hair shiny.

Note: To prepare the stems for making medicine, use gloves to remove the outer bark and then whittle off the entire inner-bark layer. The roots should be washed and then whittled or chipped away. Dry the shredded roots and stems in a basket, dehydrator, or on drying racks.

Food Uses
The early young shoots can be eaten, but only within a few days after they have sprouted from the soil. They can be eaten raw or peeled and then cooked. The spring roots can be chewed after peeling the outer layer of bark off; they have an almost nutty flavour.

---

[*]Sun, Shi. 2010. Improving anticancer activities of *Oplopanax horridus* root bark extract by removing water-soluble components. *Fitoterapia*.

Part II: *Herbs*

Other Uses
According to Steve Johnson, founder of the Alaskan Flower Essence Project and author of *The Essence of Healing*, "Devil's club flower essence clears ambivalence about being present on the earth; helps one to express one's truth firmly and clearly from the heart."

Devil's club is a significant plant to coastal First Nations and the Tlingit of the southern Yukon.

Cautions
When harvesting devil's club, be aware that the prickly spines can break your skin and quickly cause an infection. The inner bark is emetic and in large doses can cause vomiting.

Ancient Aromatic Panax Oil
Good as massage, bath, or body oil.

ADDITIONAL RECIPES

Muscle-and-Pain Relief Oil, page 307

Muscle-Ease Cream, page 317

Pain-Aid Tincture, page 294

Respiratory Steam, page 329

1 part prepared devil's club root or inner bark of the stem
2 parts sunflower oil or other light carrier oil

Place prepared devil's club with oil in a double boiler.

Slowly warm oil and botanicals and simmer for 30–60 minutes, stirring frequently. Strain through cheesecloth and let the mixture cool. Bottle. (For more on oil infusions, see page 305.)

Although covered with sharp prickles, the inner bark of devil's club can be used topically as a poultice, salve, or cream to treat pain. BG

DOCK

Rumex arcticus (Arctic dock)
Rumex crispus (yellow dock or curled dock)

Other Names and Etymology
Rumex arcticus: sour dock. *Rumex crispus*: sour dock, narrow dock, yellow root. The Van Tat Gwich'in name for Arctic dock is *tri'itthoh*. The Latin word *rumex* means "lance" in reference to the shape of the leaves. *Crispus*, also from Latin, means "leaves having crisped at the edges," while *arcticus* means "of the Arctic."

Family
Polygonaceae (buckwheat family)

Botanical Descriptions
Rumex arcticus: Is a hairless perennial herb. Stems erect, 20–100 cm or more high, from a stout fleshy base. Leaves arise from the base; blade dark green, somewhat fleshy, oblong-oval or more slender, 7–30 cm long, 2–5 cm wide. Flowers reddish and small, 2–3 mm long. Dry fruits angular, three-sided and winged with seeds 3-4 mm long. *Rumex crispus*: Is a perennial herb; stems erect, 30–100 cm high, thick yellow taproot. Basal leaves oblong with long petioles, 8–30 cm long, wavy-crinkled margins; leaves becoming smaller up the stem. Flowers numerous, in clusters with large leafy bracts; 1.5–2 mm long. Dry fruits angular, three-sided and winged with seeds 2–3 mm long, brown, lustrous.

Sour young dock leaves make a good spinach replacement. BG

Wild Food and Medicine Plants of the North | 87

Habitat and Range
Rumex arcticus: Common in damp turfy places. An arctic alpine plant found across North America. Rare in British Columbia, found throughout the tundra regions of the Yukon. *Rumex crispus*: Introduced from Europe; in North America widely distributed across Canada and the United States; rare in the Yukon, known around Dawson and the Carcross Valley; likely in other disturbed habitats.

Plant Parts Used
Leaves, stems, seeds, and roots

Harvest Time
Leaves and stems: spring. Seeds: early autumn. *Rumex crispus* roots: autumn.

ARCTIC DOCK GROWS all over the Yukon and is heavily concentrated in the Old Crow and Dawson regions. The Van Tat Gwich'in consider it an important medicine plant and use it for colds, chest pain, infections, and sore throats. They harvest the roots in the autumn; dry, clean, boil, and strain to make medicinal juice. They will often store the juice for use throughout the long winter.

Medicinal Actions
Rumex crispus: Alterative, antimicrobial, astringent, diuretic, hepatonic, laxative

Medicinal Preparations
Oil, poultice, salve, syrup, tea, tincture

Medicinal Uses
Rumex crispus (yellow dock): The carrot-shaped root is highly prized in herbalist circles and has a multitude of medicinal uses. I find tincturing the fresh or dried root the best way to preserve this plant for internal and external use.

Yellow dock syrup, tea, or tincture helps relieve symptoms of upper-respiratory complaints including asthma, bronchitis, sore throats, and even emphysema. It's helpful for excretions of yellow phlegm. Dock root also acts as a liver stimulant and blood cleanser and is helpful in treating liver problems, swollen lymph nodes, jaundice, and flare-ups associated with hepatitis. The roots are also helpful for balancing hormones and helping to reduce menstrual cramping and flooding. It has a mild laxative effect, so is good for constipation.

Herbalist Robert Rogers says that yellow dock root is great for women because it's good for uterine fibroids as well as early and heavy menses with dark and painful clotting. He

Dock

also says it works well for menopausal constipation and for digestive weakness when taken daily in small doses in water before meals.

Herbalist Matthew Wood suggests yellow dock root for fractious, irritable sleep especially in pubescent girls.

Topically, the powdered root and leaves are good on skin eruptions because it acts as a rubefacient, stimulating a nourishing flow of blood to the skin. I like to dab a bit of root powder on pimples. It's also used for psoriasis, eczema, boils, skin rashes and sores, warts, and as an antifungal wash for ringworm. The leaves and roots can be prepared for use in salves, oils, and poultices. Because yellow dock is an alterative (blood cleanser) it works well internally for the skin, especially in cases of hormonal acne.

As a mouthwash, yellow dock root decoctions are astringent and are therefore good for healing mouth, gum, and throat inflammations.

Some herbalists believe the more yellow the root, the stronger the medicinal value.

Food Uses

Young dock leaves make a thirst-quenching snack while out hiking. The sour young leaves of all the docks can be gathered for use as a cooking vegetable and make a good spinach replacement. One of the recommended ways to prepare the leaves is by boiling them once to remove any bitter taste, strain, add more water, and bring to a boil a second time, cooking until the greens are tender. I prefer to stir fry the leaves in a light vegetable oil or butter with garlic and hemp seeds. The greens and roots are both full of vitamin C.

Yellow dock root *(R. crispus)* is an excellent nutritive tonic and considered by some herbalists to be the finest organic source of iron. This makes it useful in treating anemia and other conditions concerning weakness and fatigue. It works, in part, by releasing stored ferritin from the liver. Ferritin is a protein that stores iron for later use by the body. It also binds to heavy metals like lead and arsenic, so helps remove these toxins from the body.

Iron-rich dock seeds are nutritious and tasty. Gather in late autumn, hang upside down in a paper bag, and the seeds will naturally fall off. The seeds can then be roasted, or used dry, and added to muffin, cracker and bread recipes, as well as to soups, stews, and sauces.

Dock plants turn a brilliant red in the autumn. PL

Nutritional Profile

Dock leaves are high in protein, calcium, iron, potassium, and various vitamins. The roots are protein rich and high in vitamins A, C, niacin (vitamin B-3), thiamine (vitamin B-1), and minerals such as calcium, iron, magnesium, phosphorus, potassium, selenium,

Part II: *Herbs*

> "You must weed your mind as you would weed your garden."
>
> —Astrid Alauda

trace amounts of zinc, and trace elements such as manganese. *Rumex crispus* (yellow dock) is particularly high in iron. Dock seeds are a good source of protein.

Other Uses

In *Flowers of the Yukon*, author Mickey Lammers says the roots of the yellow dock are used for wool dying.

Cautions

Dock leaves contain oxalates, an acid that is also found in spinach, Swiss chard, potatoes, and rhubarb. Cooking and freezing the plants will help break down the acids. (For more information on oxalic acid, see page 27).

Dock leaves can irritate the digestive tract and cause irritation in the mouth and throat if consumed in large amounts, so eat them in moderation, using only young leaves.

Dock should be avoided by people with a tendency to stone formation, renal dysfunction, Type 1 or 2 diabetes, or people with severe electrolyte abnormalities.

ADDITIONAL RECIPES

Boreal Bitters, page 297

High-Iron Syrup, page 299

Nourishing Vinegar Tonic, page 296

Wild Dock Sorrel-Potato Soup

2 tablespoons (30 mL) butter
4 cups (1 L) water
4 cups (1 L) dock or sheep sorrel leaves, chopped (or any mixed wild edible green)
2 large potatoes, peeled and cut
1 cup (250 mL) minced wild chives (or garlic or onion chives)
1 teaspoon (5 mL) salt
Dash of sea salt and pepper
1 cup (250 mL) sour cream or plain yogourt
1 tablespoon (15 mL) minced wild chives, for garnish

Melt the butter in a soup pot with ½ cup of water. Add the wild greens and salt. Simmer, covered, for 5 minutes over low heat. Add spuds and sauté for 10 minutes, stirring occasionally. Pour the rest of the water in and bring it to a boil. Lower the heat and simmer until the potatoes are tender, about 25 minutes. Stir in sour cream or yogourt just before serving and garnish with chives.

FIREWEED

Chamerion angustifolium

Other Names and Etymology
Willowherb, rosebay-willowherb. Formerly known as *Epilobium angustifolium*.

Family
Onagraceae (evening primrose family)

Botanical Description
Perennial herb with erect fibrous stems up to 1.5 m high, arising from far-reaching horizontal roots. Leaves alternate, lance-shaped, 5–20 cm long, hairless, paler, and veiny beneath. Flowering stalk tall, flowers opening at the base in early summer; the last flowers bloom at the terminus of the stalk, heralding the end of summer. Petals magenta rose-pink, 10–12 mm long. Long pink seed capsules 4–10 cm, developing at the base of the flowering stalk while upper flowers are still opening.

Habitat and Range
Disturbed soils, recently burned areas. It's considered a pioneer species. It's also found

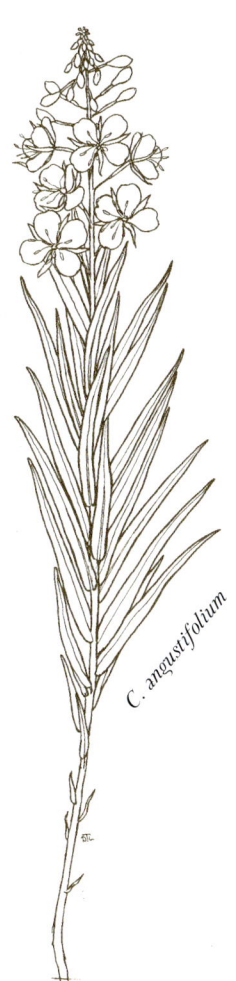

C. angustifolium

The official flower of the Yukon Territory, fireweed is edible at all stages of growth, from when it first sprouts to flowering maturity. L to R: CA, BG, BG

Wild Food and Medicine Plants of the North

Part II: *Herbs*

Fireweed is an important source of nectar for northern bees. PL

in open woods, on hillsides and on stream banks. Circumpolar distribution. It ranges from southern Greenland to Alaska, south to California. It's widespread in Canada and in NWT and Yukon.

Plant Parts Used
Roots, leaves, flowers

Harvest Time
Spring: leaves and roots. Summer: leaves and flowers. Autumn: roots

THE FLAMBOYANT FIREWEED is the follower of boreal forest fires and is the official flower of the Yukon. It grows in blazing-magenta abundance all over the territory and throughout the rest of Canada and Alaska as well. Fireweed is a tremendously hardy plant: one year after the volcanic eruption of Mount St. Helens in Washington state, 81 percent of all the seedlings present on the mountainside were fireweed. Actually, the most prominent plant visible after any boreal forest fire is almost always fireweed. The fruit of the fireweed can contain as many as 500 seeds and a single plant can produce as many as 80,000 seeds per year. Despite its high seed production, fireweed propagates most successfully through its rhizomes, or spreading root system. It also helps stabilize the soil and reduce erosion while the burn area regenerates.

Fireweed is also a very beneficial plant to humans. The flowers, leaves, and shoots are edible, and many parts of the plant have medicinal uses. As a pollination and honey plant, it also plays an important part in the ecological hierarchy of the northern boreal forest, providing large amounts of nectar for bees.

We also shouldn't overlook one of its most pleasing benefits: it creates a colourful and aromatic vista throughout the boreal forest that stimulates and renews energy on all levels.

Medicinal Actions
Anti-inflammatory, anti-irritant, antimicrobial, antiseptic

Medicinal Preparations
Cream, oil, poultice, salve, tea/infusion, tincture, vinegar

Medicinal Uses
Medicinally, fireweed tea has a mild laxative effect and is used by First Nation peoples to dispel intestinal worms, for digestive upsets, and topically for burns and other skin ailments. More recently, in experimental studies, tannins have been identified in fireweed

Fireweed

that can reduce benign prostatic hyperplasia, which is a non-cancerous enlargement of the prostate gland that can interfere with urination. Fireweed also has potent anti-inflammatory qualities. Topically, fireweed extract can be used in creams, salves, or poultices for dry, irritated skin, such as eczema or psoriasis.

Food Uses

Fireweed is sometimes called "wild asparagus" because the young spring shoots are often prepared and eaten like asparagus. These shoots are high in vitamins A and C, and are a healthy addition to spring salads. For centuries First Nation peoples throughout the circumpolar north have eaten the spring shoots raw, as a steamed vegetable, or steeped in a relaxing tea.

I use fireweed leaves throughout the summer; the larger leaves are excellent added to a stir-fry with other wild greens or just fried up with a bit of butter, garlic and other garden herbs.

I love to make a beautiful fireweed jelly (see following page): its taste is delicate and lovely and the colour always reminds me of summer.

Nutritional Profile

Vitamins C and A

Cosmetic Uses

Because of its anti-inflammatory properties, fireweed is excellent for skin care to help reduce redness and acne.

Other Uses

Steve Johnson, founder of the Alaskan Flower Essence Project and author of *The Essence of Healing*, says "Fireweed flower essence is used for shock or trauma; energy stagnation on any level; feeling burned out; weak connection to the earth." He goes on to say, "Its healing qualities are that it strengthens the grounding connection to the earth; helps break up and move out old energy patterns that are being held in the etheric body so that new cycles of revitalization and renewal can be initiated."

In Russia, fermented and dried fireweed leaves are brewed for *kapporie*, or "kapor tea." Kapor tea is high in both vitamin A and C and is used to relieve stomach aches. BG

White Fireweed

It's said that white fireweed is an indication of radiation from a nearby uranium deposit.

The Alaskan Flower Essence Project has created a white fireweed essence that is indicated for "deep emotional shock and trauma, profound alienation from the body after an experience of sexual or emotional abuse."

Its healing properties are that it "calms the emotional body after a traumatic or shocking experience; helps us release the imprint of painful emotional experiences from the cellular memory so that rejuvenation can begin."

PL

Wild Food and Medicine Plants of the North | 93

Part II: *Herbs*

Rose-Bay Willow-Herb

On the breeze my fluff is blown;
So my airy seeds are sown.

Where the earth is burnt and sad,
I will come to make it glad.

All forlorn and ruined places,
All neglected empty spaces,

I can cover—only think!—
With a mass of rosy pink.

Burst then, seed-pods; breezes, blow!
Far and wide my seeds shall go!

—Cicely Mary Barker, author,
Flower Fairies of the Wayside (1948)

Fireweed Jelly

2½ cups (625 mL) fireweed petals, fresh or dried
2 cups (500 mL) water
2 cups (500 mL) cane sugar
½ cup (125 mL) white grape juice
½ cup (125 mL) red grape juice
1 package (57 grams) powdered pectin
2 tablespoons (30 mL) rosewater (optional)

Place fireweed petals, water, and ⅔ cup (160 mL) sugar in a saucepan and bring to a boil. Reduce heat and simmer for 5 minutes. Remove from heat and leave to stand overnight so the fireweed petals can release their fragrance into the syrup-like sugar water infusion.

Strain the petals; pour the petal infusion into a pot. Add the grape juices and pectin. Boil hard for 1 minute. Add the rest of the sugar and stir. Boil the mixture hard for 1 minute more or until you can't stir it down. Remove from heat.

Test for setting by placing a teaspoon of hot jelly on a plate, let cool—the surface should wrinkle when pushed with your finger. If it's still runny, put back on heat and continue boiling and testing until jelly sets. It should make a delicate soft jelly.

Add rosewater, stir. Pour jelly into measuring cup and then pour into sterilized jars, sealing lids properly. (See page 349 for tips for successful canning and heat processing).

Fireweed

that can reduce benign prostatic hyperplasia, which is a non-cancerous enlargement of the prostate gland that can interfere with urination. Fireweed also has potent anti-inflammatory qualities. Topically, fireweed extract can be used in creams, salves, or poultices for dry, irritated skin, such as eczema or psoriasis.

Food Uses

Fireweed is sometimes called "wild asparagus" because the young spring shoots are often prepared and eaten like asparagus. These shoots are high in vitamins A and C, and are a healthy addition to spring salads. For centuries First Nation peoples throughout the circumpolar north have eaten the spring shoots raw, as a steamed vegetable, or steeped in a relaxing tea.

I use fireweed leaves throughout the summer; the larger leaves are excellent added to a stir-fry with other wild greens or just fried up with a bit of butter, garlic and other garden herbs.

I love to make a beautiful fireweed jelly (see following page): its taste is delicate and lovely and the colour always reminds me of summer.

Nutritional Profile
Vitamins C and A

Cosmetic Uses
Because of its anti-inflammatory properties, fireweed is excellent for skin care to help reduce redness and acne.

Other Uses
Steve Johnson, founder of the Alaskan Flower Essence Project and author of *The Essence of Healing*, says "Fireweed flower essence is used for shock or trauma; energy stagnation on any level; feeling burned out; weak connection to the earth." He goes on to say, "Its healing qualities are that it strengthens the grounding connection to the earth; helps break up and move out old energy patterns that are being held in the etheric body so that new cycles of revitalization and renewal can be initiated."

In Russia, fermented and dried fireweed leaves are brewed for *kapporie*, or "kapor tea." Kapor tea is high in both vitamin A and C and is used to relieve stomach aches. BG

White Fireweed

It's said that white fireweed is an indication of radiation from a nearby uranium deposit.

The Alaskan Flower Essence Project has created a white fireweed essence that is indicated for "deep emotional shock and trauma, profound alienation from the body after an experience of sexual or emotional abuse."

Its healing properties are that it "calms the emotional body after a traumatic or shocking experience; helps us release the imprint of painful emotional experiences from the cellular memory so that rejuvenation can begin."

Part II: *Herbs*

Rose-Bay Willow-Herb

On the breeze my fluff is blown;
So my airy seeds are sown.

Where the earth is burnt and sad,
I will come to make it glad.

All forlorn and ruined places,
All neglected empty spaces,

I can cover—only think!—
With a mass of rosy pink.

Burst then, seed-pods; breezes, blow!
Far and wide my seeds shall go!

—Cicely Mary Barker, author,
Flower Fairies of the Wayside (1948)

Fireweed Jelly

2½ cups (625 mL) fireweed petals, fresh or dried
2 cups (500 mL) water
2 cups (500 mL) cane sugar
½ cup (125 mL) white grape juice
½ cup (125 mL) red grape juice
1 package (57 grams) powdered pectin
2 tablespoons (30 mL) rosewater (optional)

Place fireweed petals, water, and $2/3$ cup (160 mL) sugar in a saucepan and bring to a boil. Reduce heat and simmer for 5 minutes. Remove from heat and leave to stand overnight so the fireweed petals can release their fragrance into the syrup-like sugar water infusion.

Strain the petals; pour the petal infusion into a pot. Add the grape juices and pectin. Boil hard for 1 minute. Add the rest of the sugar and stir. Boil the mixture hard for 1 minute more or until you can't stir it down. Remove from heat.

Test for setting by placing a teaspoon of hot jelly on a plate, let cool—the surface should wrinkle when pushed with your finger. If it's still runny, put back on heat and continue boiling and testing until jelly sets. It should make a delicate soft jelly.

Add rosewater, stir. Pour jelly into measuring cup and then pour into sterilized jars, sealing lids properly. (See page 349 for tips for successful canning and heat processing).

GENTIAN

Gentianella amarella

Other Names and Etymology
Northern gentian, felwort, bitter gentian.

According to Roman naturalist and philosopher Pliny, who discovered the medicinal virtues of gentian, the plant's name is derived from Gentius, King of Illyria, in 168 B.C. *Amarella* stems from the Latin *amarus*, which means "bitter."

Family
Gentianaceae (gentian family)

Botanical Description
Annual herb with stems 5–40 cm high, with strongly ascending branches. Lower leaves elliptic, 1–3 cm long; cauline leaves more rounded, with base clasping the stem, 1–6 cm long. Flowers numerous, in branching clusters, or hugging the stem at leaf bases. Petals mauve to purple, fused into a tube, 10–15 mm long; hairs in the throat of the tube, mauve to purple; seed capsule cylindrical, as long or longer than petals.

Habitat and Range
Moist stream banks, meadows, and clearings. Boreal North America; from Labrador to Alaska, and south to California. In the Yukon, gentian grows as far north as Dawson. *Gentianella propinqua*, another kind of gentian, is also very widespread in the Yukon.

In late summer, keep an eye out for amethyst-coloured flowers sprouting up from the forest floor. Gentian petals can be mauve to purple in colour. PL

Wild Food and Medicine Plants of the North

Part II: *Herbs*

Gentian is one of the best-known boreal digestive remedies. BG

Plant Parts Used
Flowers and roots

Harvest Time
Roots: early autumn after the flowers have died off

THE ROOTS of this small, unassuming, delicate, purple-flowered plant pack quite a punch. Just a little chew on a gentian root can leave a bitter taste in your mouth for hours.

In traditional Chinese medicine, the use of gentian roots dates back 5,000 years. It was called *lung tan*, meaning "dragon's gall" because of its exceedingly bitter taste.

Gentian root is one of the best-known boreal digestive remedies. The roots are long and thin and easy to gather after the flowers have died off. I just use a hand spade to dig them out.

The northern gentian's close cousins in Europe have a long medicinal history. The German father of natural healing, Sebastian Kneipp, praised this plant, describing the effects of gentian as, "Strengthening and supportive to stomach secretion and as strengthening to the nerves," and recommended that "nerve-weak" people take five drops every day. He said gentian root is one of the best aids to increase appetite, and, "If the food is felt to lie heavy in the stomach and is troublesome a little cordial made with a teaspoonful of the extract in a half a glass of water, will soon stop the disorder." He also said that gentian is very good for stomach cramps.

Medicinal Actions
Analgesic, anthelmintic, anti-inflammatory, antiseptic, bitter tonic, cholagogue, digestive, emmenagogue, expectorant, febrifuge, refrigerant, stomachic

Medicinal Preparations
Syrup, tea/infusion, tincture

Medicinal Uses
Gentian is renowned for increasing the production of digestive fluids and bile. Taking the tincture helps increase the appetite of people who have no desire to eat due to stress or exhaustion. Its astringent properties, along with its bitter constituents, have been used for the chronic diarrhea caused by Crohn's disease. If you suffer from mild heartburn, bloating or flatulence, gentian tincture before a meal can help remedy this. Be careful though if you have hyperacidity or gastritis, the stimulating effect may be irritating.

Herbalist Robert Rogers mentions that gentian is also useful for headaches associated with low blood pressure.

Gentian

Gentian's bitter principals are also said to help rebalance the thyroid gland. Margaret Grieve, in *A Modern Herbal* (1931), says, "It is one of the best strengtheners of the human system."

Food Uses
While not considered a food, gentian is used in alcoholic bitters that are generally taken before a meal to induce the appetite. It is also used to flavour vermouth. The roots have been used as a substitute for hops in beer making.

Nutritional Profile
Roots: Vitamins C, A, and zinc

Other Uses
Gentian is one of the Bach flower remedies. It's indicated for use in overcoming feelings of discouragement and doubt that may be caused by even small obstacles. It promotes the confidence to overcome problems. Used if you are feeling doubt, depression, and discouragement.

Cautions
Excessive use of gentian root can cause nausea and vomiting. Pregnant and nursing mothers, and those with ulcers and high blood pressure, should avoid using gentian.

> "Gentius King of Illyria claimed to have been cured of malaria by the bitter tonic made from the juice of the plants, hence the name."
>
> —Martha Louise Black, *Yukon Wild Flowers* (1940)

Gentian Tincture

Gentian roots
Vodka

After digging up the thin, tiny roots, wash off the soil and chop them up.

Fresh root tincture can be made using a 1:2 ratio (1 part root: 2 parts alcohol) or from the dried roots in a 1:4 ratio. Let sit for up to six weeks, shaking daily.

The recommended dosage is 10–20 drops before a heavy meal.

If drying, it's recommended the roots be dried quickly to keep the medicinal properties intact. I use the herb-drying setting on my dehydrator. Once dry, the roots can be used to make tincture or bitters.

For more on tinctures, see page 291.

ADDITIONAL RECIPE

Boreal Bitters, page 297

GOLDENROD

Solidago species

Other Names and Etymology
Woundwort, blue mountain tea, Aaron's rod.

Solidago is from the Latin *soldare* meaning, "healthy and strong," and *ago* means "to make whole or solid," referring to the fact that the plant is used medicinally to repair torn flesh.

Family
Asteraceae (aster, daisy, or sunflower family)

Botanical Descriptions
Solidago multiradiata: A perennial herb with stems 5–40 cm or more high. Basal leaves 1.5–20 cm long, elongated, tapering to a petiole with flattened "winged" edges; more or less toothed toward the tip, hairless on both sides. Leaves along the stem similar but becoming stalkless and smaller upwards. Flower heads dense with a flat or rounded top; yellow ray flower petals 4–6.5 mm long. Seeds are hairy. *Solidago canadensis*: Lance-shaped oblong leaves, mustard-yellow ray flowers in clusters on the branch. Stems usually more than 35 cm high, from well-developed rhizomes; involucres usually less than 4 mm high. *Solidago lepida*: To the naked eye, *S. lepida* looks extremely similar to *S. canadensis*.

Habitat and Range
River meadows and open woods to moist alpine slopes. There are various *Solidago* species found throughout arctic alpine North America, from Newfoundland to Alaska, south through the mountains of British Columbia to California and New Mexico. *Solidago canadensis* is more of an eastern boreal plant, extending as far west

Eye-catching Canada goldenrod contains quercetin, a powerful antioxidant that is also a natural antihistamine and anti-inflammatory. CA

Goldenrod

as Manitoba, while *S. lepida* is more of a western plant, being found from the Aleutian Islands into Alaska and the Yukon, south to northern California, in the Rocky Mountains to Arizona and New Mexico, and spottily across the Canadian prairies. *S. multiradiata* favours tundra and tundra-like habitats, alpine slopes and meadows. Another kind of *Solidago*, *S. simplex*, is also common in northern boreal regions.

Plant Parts Used
Flowers, leaves, roots

Harvest Time
Aerial parts: gather in summer when plant is just flowering. Roots: early spring or autumn

MY FAVOURITE Canada goldenrod patch is at the water outtake of the Takhini Hot Springs in Yukon. The hot springs' warm water allows this plant to grow to its full potential, as it rises up out of the muddy ground, its golden flowers attracting bees and curious bathers.

Used alone or in combination with other northern wild plants goldenrod is a wonderful addition to oils, ointments, and salves. In the summer, I am always adding goldenrod flowers to salads and to decorate cakes—the blossoms always look wonderful when contrasted with rose petals, fireweed flowers, and juniper berries.

In southern parts of Canada, goldenrod has a bad reputation because it's often confused with allergy-inducing ragweed. In actual fact, the pollen of goldenrod is too heavy to be carried on the wind. And, ironically, *S. canadensis* contains quercetin, a natural antihistamine that can help to relieve allergic symptoms such as itchy watery eyes. Herbalist Matthew Wood states in his book, *The Earthwise Herbal: A Complete Guide to Old World Medicinal Plants,* that goldenrod can be specifically used against cat allergies. This said, those with allergies to plants in the Asteraceae family (such as daisies and sunflowers) may experience a reaction if they are handling the plant or eating it.

Medicinal Actions
Anti-catarrhal, anti-fungal, anti-inflammatory, antiseptic, carminative, diaphoretic, diuretic

Medicinal Preparations
Bath, cream, essential oil, hydrosol, liniment, oil, ointment, poultice, powder, salve, tea, tincture, wash

Medicinal Uses
Medicinally the *Solidago* species are used for upper-respiratory catarrh, to reduce the

Part II: *Herbs*

> "Plants are capable of intent: they can stretch toward, or seek out what they want in ways as mysterious as the most fantastic creations of romance."
>
> —Peter Tompkins and Christopher Bird,
>
> *The Secret Life of Plants* (1973)

production of mucus in the bronchial tubes, and to promote tissue repair.

It also helps with problems of the urinary system. As a diuretic it increases the production of urine, while its anti-inflammatory actions help treat conditions such as cystitis and urethritis. It's also carminative, helping to reduce gas pain and weakness of bowels and bladder. It acts as a kidney tonic when taken as a tea, and, in Europe in particular, it is often included in medicinal tea blends to help treat and prevent kidney stones and stop inflammation of the urinary tract.

Solidago canadensis contains quercetin, a powerful antioxidant that is also a natural antihistamine and anti-inflammatory. Research shows that quercetin may help as a preventative for prostate cancer. Its anti-inflammatory properties may also help to reduce pain from disorders such as arthritis. Quercetin may also help reduce symptoms from fatigue, depression, and anxiety.

Herbalist Robert Rogers notes that work done by two different sets of researchers found that Canada goldenrod extracts are more powerful in antioxidant activity than green tea or ascorbic acid.

I like to call goldenrod a singer's herb simply because it works great as a gargle for mouth inflammation, laryngitis, and sore throats. The plant contains saponins that are antifungal and act specifically against the candida fungus that causes vaginal and oral thrush.

The leaves can also be used as a spit poultice to relieve burns, insect bites, and stings. The leaves and flowers can be dried, powdered, and added to your first-aid kit for use as a styptic to help stop the bleeding of wounds—hence its common name "woundwort."

A wash or a poultice can be made for pain relief along the nerve endings and for conditions such as rheumatism and sciatic pain. The whole plant can be infused in oil to make a healing oil or salve. I like to combine it with other herbs like yarrow, highbush cranberry bark, and usnea in a topical skin-healing remedy.

If you have a painful toothache, try chewing a bit of the root and then pack it on and around the irritated tooth and gum—you should notice the pain decreasing within a few minutes. The herb's antiseptic qualities may even help clear up the cause of the ache.

The hydrosol of goldenrod is also a strong anti-inflammatory and moderate antispasmodic for sore muscles, stiff neck, tendonitis, and repetitive-strain injuries.

Suzanne Catty, author of *Hydrosols: The Next Aromatherapy*, recommends goldenrod hydrosol for its "strong diuretic properties" and that, "taken internally it may aid the treatment and prevention of kidney stones. Topically, it can be used as a compress for fluid retention and uric acid in the joints and tissues." She adds that energetically, "goldenrod carries the intense vibrations of heat and the sun, opens the solar plexus and diaphragm, bringing a state of calm."

The essential oil is anti-inflammatory and antiseptic. When blended with a carrier oil, such as olive oil or sunflower oil, it's good for topical healing and skin infections, and muscular aches and pains.

Goldenrod

Food Uses
The young leaves can be cooked like spinach and added to soups, stews, and stir-fries. The flowers look wonderful in salads and are fun to add to muffins, cake batters, egg dishes, and as a soup garnish. The young flowers or seeds can be gathered in the autumn, dried, and used as a thickener (for gravies, soups, etc.) throughout the winter.

Other Uses
According to Patricia Kaminski and Richard Katz, authors of *Flower Essence Repertory*, the essence of goldenrod is "good for people who are easily influenced by group or family ties, unable to be true to one's self, and are easily subject to peer pressure or social expectations." So therefore it's good for individuals who want to "develop individuality and an inner sense of self balance within a group or social consciousness."

The flowers produce a mustard-coloured dye, and an orange-and-brown dye can be obtained from the whole plant.

The druids used long goldenrod stems as a divining tool to find water.

Cosmetic Uses
Goldenrod flowers can be brewed into a strong tea and used as a hair rinse to bring out blond highlights. For the best results, rinse hair with the tea and go out into the sun to dry your hair. The flowers can also be added to facial steams and creams to help with acne. In the summer it's fun to add the flowers to the bath along with other wildflowers.

Cautions
Use caution if you have allergies to plants in the Asteraceae family.

Golden Flower Body-and-Massage Oil

1 part goldenrod flowers and leaves
2 parts sunflower oil

Let the plant matter wilt for a few hours to let some of the moisture evaporate away.

Place goldenrod in jar and cover with oil. Seal the jar, label it, and let the plant infuse the oil, out of direct sunlight, for 4 to 6 weeks.

Strain, bottle and label.

Optional: Add essential oil of lavender.

Goldenrod is a wonderful addition to oils, ointments, and salves. Top to Bottom: BG, CA

ADDITIONAL RECIPES

Cough-and-Cold Medicinal Syrup, page 300

Flower-Delight Tempura, page 375

Styptic Powder, page 320

Urinary-Tract Tincture, page 294

HORSETAIL

Equisetum arvense

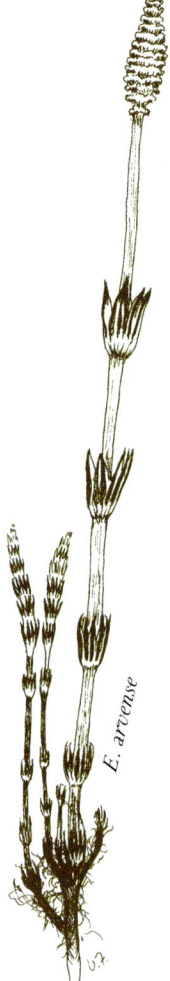

Other Names and Etymology
Field horsetail, puzzle grass, scour weed, shavegrass, scouring rush. The Van Tat Gwich'in name for horsetail is *khehdi'*. *Equisetum* is derived from the Latin roots *equus*, meaning "horse," and *seta*, meaning "bristle."

Family
Equisetaceae (horsetail family)

Botanical Description
A relative of the fern, this primitive plant has jointed, hollow stems that lack broad leaves and flowers. Spores are produced in a cone-like head at the terminus of a brown-coloured fertile stem (the stems lack chlorophyll, hence their colour) that withers after the spores are shed. In contrast, horsetail's sterile stems are green and much more conspicuous and plentiful than its fertile ones. These green stems are generally upright and grow to 50 cm. The three to four angled branches are solid, in whorls at the nodes, spreading

A relative of prehistoric plants, horsetail always reminds me of how old the earth is, conjuring up images of a time before time! L to R: PL, CA

Horsetail

upwards. There are sheaths at the nodes that have four to fourteen short, narrow dark teeth. The first internode on each branch is longer than the adjoining stem sheath. With autumn frosts, the stems wither and die back. These perennials grow from dark brown, hairy, tuber-bearing rhizomes. Horsetail is characterized by its jointed stems that can easily be pulled apart and put back together, leading to its common name "puzzle grass."

Habitat and Range
Horsetail grows in damp open woods, on low open ground, and meadows, and embankments, often in dry sandy soil. In the Arctic it grows on shattered limestone or in pockets of soil with permafrost close to the surface. Circumpolar, it is found throughout the Yukon, NWT, Alberta, and grows in most of North America.

Plant Parts Used
Aerial parts

Harvest Time
Early summer. It's important to look closely at the plant to see if there are crystals on it. Significant deposits of calcium oxalate accumulate on the inner walls of the stems, making them good for scouring dirty camp pots but not good for the kidneys. When its branches are pointing up they are prime for picking for medicinal and food purposes, but be warned: downward-pointing branches are an indication that the plant has developed oxalate crystals. This said, harvesting the plant anytime throughout the summer to make a herbal shampoo, hair rinse, or foot soak is not only safe, but a wonderful treat.

Horsetail shoots can be used as a steamed vegetable or made into a mineral-rich tea. BG

HORSETAIL ALWAYS REMINDS ME of how old the earth is, and conjures up images of a time before time! Its rush-like appearance is a reminder of how plants looked 400 million years ago in the Paleozoic era, when horsetail's ancestors—huge, jointed tree-like plants grew to over 15 m high.

Horsetail was used as a topical and oral herbal remedy by the ancient Greeks and Romans to stop bleeding, heal ulcers and wounds, and treat tuberculosis and kidney problems.

The Van Tat Gwich'in use the leaves to make a cleansing tea, and the dried stems are steeped in hot water for use as a foot wash and as an external wash for hair, skin, and nails. The stems and leaves are also used to scrub pots.

Today horsetail is esteemed in natural health circles for its naturally high silica content and is used in natural health products to promote healthy hair, skin, and nails. Generally, what you find on herb-shop shelves are capsules and liquids.

Horsetail is eaten by caribou, moose, sheep, and bears.

Part II: *Herbs*

> "Everything in nature has a voice, a consciousness. Try to quiet the chatter in your mind enough to notice it. Can you hear the trees, the flowers, the rock, or the ant? What are they saying to you?"
>
> —Theresa Rose, author, *Opening the Kimono* (2008)

Medicinal Actions
Astringent, diuretic, styptic, vulnerary

Medicinal Preparations
Bath, capsule, plant essence, poultice, powder, salve, steam, tea/infusion, tincture, wash

Medicinal Uses
Horsetail stems are rich in silica, which helps our bodies to form collagen—an important protein found in connective tissue, skin, bone, cartilage, and ligaments. Because of this, horsetail is helpful in mending broken bones, treating and preventing osteoporosis, and as a general tonic to keep the framework of the body healthy.

Herbalists also recognize the diuretic properties of horsetail. When prepared in a tea, ingested as a juice, or taken in a capsule, it promotes urination, and is a remedy for bladder and urinary-tract infections, and it helps heal stomach ulcers and remove kidney stones. Horsetail also strengthens the bladder wall.

A warm horsetail poultice can be applied to a fresh cut to help stop bleeding. A poultice or tea benefits fractured bones, sprains, or rheumatic conditions as the silica helps increase bone density, and this makes it a supportive therapy for osteoporosis.

A cool wash can be made to help heal burns, bug bites, and skin eruptions.

Terry Willard, PhD, founder of Wild Rose College of Natural Healing in Alberta, writes in his book, *Edible and Medicinal Plants of the Rocky Mountains and Neighbouring Territories*, "The juice of the plant, especially the sterile stems, is good for anaemia, which may have resulted from internal bleeding from such illnesses as stomach ulcers. It acts by promoting coagulation of the blood." He adds that horsetail tea is also good for excessive menstrual flow.

Food Uses
The spring budding shoots of the antioxidant-rich horsetail can be used as a steamed vegetable or tea. They are high in many minerals including calcium, magnesium, and sulphur.

Nutritional Profile
Acids, ascorbic acid, calcium, carotene, iron, magnesium, niacin, phosphorus, resin, riboflavin, silica, starch, sterols, tannins, thiamine

Cosmetic Uses
Horsetail is a prime ingredient in many nature-based shampoos because it's rich in nutrients and silica, which help strengthen the hair shaft, nourish the scalp, and give the hair an overall healthy sheen. Make a horsetail hair tea to use as a rinse after shampooing.

Horsetail

Other Uses

Steve Johnson, founder of the Alaskan Flower Essence Project and author of *The Essence of Healing*, says that horsetail is good for "distorted communication with other levels of one's consciousness; difficulty communicating with the higher selves of others, including animals."

Aroma Borealis's Root-Chakra Vibrational Essence contains horsetail, because this plant goes back to our roots in humanity. Horsetail essence helps us get in touch with our own ancestral roots and facilitates a feeling of returning home to ourselves, back to the true essence of who we really are. Our root chakra is connected to our home, the nourishment of our physical body, and a heightened awareness and deeper communion with the spirit of the Earth Mother.

Cautions

Long-term use of horsetail is not recommended. Ingesting large amounts may deplete the body of vitamin B1 (thiamin). Herbalist Robert Rogers says that the constituent thiaminase that causes this deficiency is destroyed by tincturing or decocting the herb.

If horsetail is taken regularly, it's recommended you take quality multiple vitamins or at least a B-complex supplement daily. Do not use horsetail if you have edema, gout, heart problems, or kidney inflammation.

Another species of horsetail that grows in the Yukon, *Equisetum palustre*, or marsh horsetail, is—somewhat ironically given the plant's name—poisonous to horses. Although it's not known for certain that *E. palustre* would have the same effect on people, it shouldn't be ingested.

E. palustre

Horsetail Tooth Powder
Makes enough for 15–20 brushings.

- 1 teaspoon (5 mL) dried and ground spring horsetail
- 1 teaspoon (5 mL) dried and ground wild mint
- 2 tablespoons (30 mL) fine sea salt
- 2 tablespoons (30 mL) baking soda
- 4 drops tea tree oil (optional)

Grind horsetail and mint leaves in a clean coffee grinder. Sift through a mesh strainer to get a fine powder. Add salt, baking soda and tea tree essential oil, blend well, and store in an airtight glass jar. To use, sprinkle a pinch of the tooth powder on a moistened toothbrush.

ADDITIONAL RECIPES

Beautiful Hair, Skin, and Nails Infusion, page 285

Cough-and-Cold Medicinal Syrup, page 300

Flower-Power-Peace Bath, page 325

High-Calcium Infusion, page 282

High-Iron Syrup, page 299

Nourishing Vinegar Tonic, page 296

Rejuvenating Facial Steam, page 329

Spirit-of-the-Boreal-Forest Tea, page 281

Styptic Powder, page 320

Urinary-Tract Tincture, page 294

LABRADOR TEA

Rhododendron groenlandicum
Rhododendron tomentosum

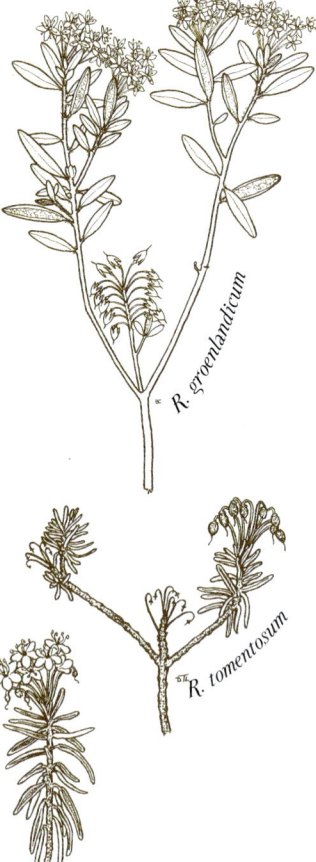

Other Names and Etymology

Trapper's tea, Hudson Bay tea, muskeg tea, marsh tea, Greenland tea, Greenland moss, storytelling tea. It was previously known by the botanical name *Ledum groenlandicum*. The Van Tat Gwich'in name for Labrador tea is *lidii masgit*. *Rhododendron* is from the Greek word *rhodos*, "rose," and *dendron* meaning "tree." *Groenlandicum* refers to Greenland. Its previous Latin name *Ledum* was derived from the Greek *ledos*, meaning, "woolly robe," which aptly describes the underside of the leaf.

Family

Ericaceae (heath family)

Botanical Descriptions

Rhododendron groenlandicum: An aromatic erect shrub, 30–80 cm high. Leathery leaves 2–5 cm long, blunt-tipped with edges curling under. Upper surface of leaves dark green, lower surface covered with densely woolly white hairs when new, turning rust-coloured with age. Flowers 10–12 mm wide, with white petals. (Note: While other members of

A northern staple, Labrador tea leaves can be harvested year-round. Its distinctive white flowers should be gathered as they bloom. L to R: PL, CA

Horsetail

Other Uses

Steve Johnson, founder of the Alaskan Flower Essence Project and author of *The Essence of Healing*, says that horsetail is good for "distorted communication with other levels of one's consciousness; difficulty communicating with the higher selves of others, including animals."

Aroma Borealis's Root-Chakra Vibrational Essence contains horsetail, because this plant goes back to our roots in humanity. Horsetail essence helps us get in touch with our own ancestral roots and facilitates a feeling of returning home to ourselves, back to the true essence of who we really are. Our root chakra is connected to our home, the nourishment of our physical body, and a heightened awareness and deeper communion with the spirit of the Earth Mother.

Cautions

Long-term use of horsetail is not recommended. Ingesting large amounts may deplete the body of vitamin B1 (thiamin). Herbalist Robert Rogers says that the constituent thiaminase that causes this deficiency is destroyed by tincturing or decocting the herb.

If horsetail is taken regularly, it's recommended you take quality multiple vitamins or at least a B-complex supplement daily. Do not use horsetail if you have edema, gout, heart problems, or kidney inflammation.

Another species of horsetail that grows in the Yukon, *Equisetum palustre*, or marsh horsetail, is—somewhat ironically given the plant's name—poisonous to horses. Although it's not known for certain that *E. palustre* would have the same effect on people, it shouldn't be ingested.

E. palustre

Horsetail Tooth Powder
Makes enough for 15–20 brushings.

1 teaspoon (5 mL) dried and ground spring horsetail
1 teaspoon (5 mL) dried and ground wild mint
2 tablespoons (30 mL) fine sea salt
2 tablespoons (30 mL) baking soda
4 drops tea tree oil (optional)

Grind horsetail and mint leaves in a clean coffee grinder. Sift through a mesh strainer to get a fine powder. Add salt, baking soda and tea tree essential oil, blend well, and store in an airtight glass jar. To use, sprinkle a pinch of the tooth powder on a moistened toothbrush.

ADDITIONAL RECIPES

Beautiful Hair, Skin, and Nails Infusion, page 285

Cough-and-Cold Medicinal Syrup, page 300

Flower-Power-Peace Bath, page 325

High-Calcium Infusion, page 282

High-Iron Syrup, page 299

Nourishing Vinegar Tonic, page 296

Rejuvenating Facial Steam, page 329

Spirit-of-the-Boreal-Forest Tea, page 281

Styptic Powder, page 320

Urinary-Tract Tincture, page 294

LABRADOR TEA

Rhododendron groenlandicum
Rhododendron tomentosum

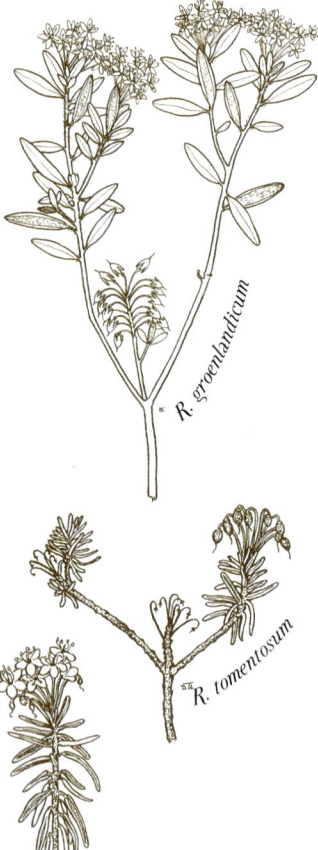

Other Names and Etymology

Trapper's tea, Hudson Bay tea, muskeg tea, marsh tea, Greenland tea, Greenland moss, storytelling tea. It was previously known by the botanical name *Ledum groenlandicum*. The Van Tat Gwich'in name for Labrador tea is *lidii masgit*. *Rhododendron* is from the Greek word *rhodos*, "rose," and *dendron* meaning "tree." *Groenlandicum* refers to Greenland. Its previous Latin name *Ledum* was derived from the Greek *ledos*, meaning, "woolly robe," which aptly describes the underside of the leaf.

Family

Ericaceae (heath family)

Botanical Descriptions

Rhododendron groenlandicum: An aromatic erect shrub, 30–80 cm high. Leathery leaves 2–5 cm long, blunt-tipped with edges curling under. Upper surface of leaves dark green, lower surface covered with densely woolly white hairs when new, turning rust-coloured with age. Flowers 10–12 mm wide, with white petals. (Note: While other members of

A northern staple, Labrador tea leaves can be harvested year-round. Its distinctive white flowers should be gathered as they bloom. L to R: PL, CA

Labrador Tea

the heath family in the boreal forest have petals fused into bell-shaped flowers, this is not the case with Labrador tea.) Dry seed capsules 4.5–6.5 mm long, glandular, rusty brown.

Rhododendron tomentosum: Has smaller leaves than *R. groenlandicum* and many people prefer its taste for tea.

Habitat and Range
Peaty soils, bogs, muskegs, moist conifer forests, meadows. Throughout North America; from Labrador to Alaska, south to New England and Oregon; occurring throughout the NWT, Nunavut, Yukon, Greenland, Russia, and throughout Scandinavia.

Plant Parts Used
Leaves and flowers

Harvest Time
Leaves: early spring throughout the summer, but can be harvested year-round. Flowers: as they bloom

A NORTHERN CAMPFIRE is not complete without a piping-hot cup of aromatic Labrador tea. It not only warms you up, but provides you with a burst of vitamin C.

Labrador tea was and is still used as a substitute for black tea. It's caffeine-free and has a mild narcotic effect. It has an interesting forest-like flavour: A little bitter, a little astringent, a little spicy, a little camphor-like.

Yukon First Nations have a long and intimate relationship with this hardy plant. The leathery green leaves with their woolly, rust-coloured underside are commonly gathered and made as a tea for coughs and colds, relaxation, and sleeping problems. Inhaling the steam that rises from the tea is recommended to clear sinuses.

If you haven't dried Labrador tea in the winter months, not to worry: You can gather the evergreen leaves even when there is snow on the ground. A fresh or dried-leaf infusion is the best method to prepare Labrador tea. Break up the leaves into small pieces (this releases the aromatic oils) and add a pinch per cup. Pour the boiling water over the leaves, cover so as not to release the volatile oils, and infuse for five to ten minutes. The pale greenish-yellow brew has an interesting flavour on its own, or you can sweeten it up with a wee dollop of honey, birch syrup, a few mint leaves, and/or wild chamomile flowers. You can get a couple of cups of tea out of the same leaves—now that's good recycling!

Part II: *Herbs*

Medicinal Actions
Alterative, analgesic, diaphoretic, diuretic, pectoral, tonic

Medicinal Preparations
Essential oil, hydrosol, infused oil, ointment, poultice, sauna herb, spice, tea, tincture

Medicinal Uses
Labrador tea has analgesic properties that help reduce pain when applied either externally as a poultice, infused in oil or ointment, or taken internally as a tea. It's said to be mildly cleansing to the blood and is considered a tonic herb that helps strengthen and tone the whole body. It acts as an intestinal regulator, it's mildly laxative yet soothes diarrhea. Its diuretic effects increase the elimination of urine and its diaphoretic effects aid the skin in the elimination of toxins and promote perspiration. This is helpful during a fever. Also known as a respiratory tonic, it's slightly expectorant and drying to the bronchi and the sinus passages. Only small amounts are needed for coughs and irritations of the lungs.

> "If your knees aren't green by the end of the day, you ought to seriously re-examine your life."
>
> —Bill Watterson, cartoonist and author

In a study published in the *Journal of Ethnopharmacology* in 1992, Dr. Allison McCutcheon and colleagues found the branches of Labrador tea act as an antibiotic against E. coli and *Bacillus subtilis*. Previous studies demonstrated the flowering heads in an extract were effective against both bacteria as well as the yeast *Candida albicans*. Other researchers also found extracts from the leaves active against *Staphylococcus aureus bacterium*.

Research from Quebec that was published by *EthnoPharm* journal in 2007 found extracts from the leaf of Labrador tea has anti-inflammatory and anti-oxidant properties. The twigs were also active against colon carcinoma and lung carcinoma cells.

Other research found Labrador tea extracts appear to possess activity that confirms the traditional aboriginal use of the leaf tea for Type 2 diabetes.

Sitting down in the evening with a steaming cup of Labrador tea helps quiet the nerves and nourish the adrenal glands. Scandinavians infuse the crushed leaves in alcohol for use as a before-bed digestif.

The plant can be used topically as a decoction, poultice, wash, oil, cream or ointment to treat a range of skin problems such as burns, ulcers, itchy chapped skin, stings, scabies, and dandruff.

A decoction, tincture, or an infusion in apple cider vinegar can be made from the leaves to kill (or deter) head lice. Make the decoction, let sit until cool, gently pour and massage into the scalp and leave it on until hair dries. Repeat, a few times throughout the day. A spray can be made with the decoction or tincture to easily administer the remedy to the scalp. Make sure to get behind the ears, the nape of the neck, and the eyebrows. An infused hair oil or pomade can also be made and left on the scalp overnight.

Labrador Tea

The essential oil of Labrador tea can be used topically with a carrier oil to help heal inflammatory conditions of the skin and muscles.

Suzanne Catty, an aromatherapist and the author of *Hydrosols: The Next Aromatherapy*, says Labrador tea hydrosol is "A liver regenerator and cleanser, it detoxifies the organ and seems to improve liver functions generally." She adds that it's also "excellent for recovering from surgery or after a serious illness or infection, as it cleans foreign substances from the system."

Catty suggests that the hydrosol can be used internally in homeopathic-type dosages because it's the most therapeutic of hydrosols, so strong that its recommended dosage is one tablespoon (15 mL) per 6 cups (1.5 L) of water.

The hydrosol, like the tea, supports the immune system and stimulates lymphatic circulation. It can be used as part of a program of recovery from addictions to alcohol, tobacco, and even food.

Food Use
As a spice, crushed or ground Labrador tea leaves add an interesting flavour to meat dishes and salad dressings. The leaves can be used as a substitute for bay leaf in stews, sauces, and soups.

Nutritional Profile
Vitamin C

Cosmetic Uses
Topically the tea or hydrosol can be used to keep skin fresh, clean, and free of blemishes.

Other Uses
Steve Johnson, founder of the Alaskan Flower Essence Project and author of *The Essence of Healing,* says that Labrador tea essence is helpful for "addictions; attempting to balance one extreme with another; extreme imbalance in any area of life; difficulty coming back to center after a traumatic or unsettling experience." He adds, "Its healing qualities help to center energy in the body; relieves stress associated with the experience of extremes; helps us continually learn a new perspective of balance."

The homeopathic remedy Ledum is used for puncture wounds and for arthritic pain that starts in the feet and moves upward and is relieved by cold.

Crushed Labrador tea leaves act as an insect repellent and can help deter rodents from your foodstuffs: add to grain and rice bins either loose or in a sachet. Put in a sachet to deter moths and mice from places where you store wool hats, toques, scarves, and mittens.

Cautions
Use in moderation. Be particularly careful not to use in excess if you are pregnant or have high blood pressure. In larger doses it can be considered cathartic and cause diarrhea.

Labrador tea helps reduce pain when applied externally or taken internally as a tea. BG

Labrador Tea-Flower Oil

1 cup (250 mL) Labrador tea flowers
1½ cups (375 mL) jojoba oil or any oil that you have on hand

Gather flowers before the heat of the day. Place flowers in a jar and cover with oil. Infuse for up to 4 weeks, shaking daily.

ADDITIONAL RECIPES

Aromatic Herbal Balm, page 314

Get-Up-and-Go Infusion, page 283

Muscle-and-Pain Relief Oil, page 307

Muscle-Ease Cream, page 317

LAMB'S QUARTERS

Chenopodium album

Other Names and Etymology
Pigweed, goosefoot, wild spinach, northern spinach, fat hen. *Chenopodium* is from the Greek *chen*, meaning "goose," *podos*, meaning "foot," and *album* meaning "white." Its common name, pigweed, alludes to the fact that pigs like eating it. The leaves are shaped like a goose's foot, and it fattens hens, who in turn make great eggs!

Family
Chenopodiaceae (goosefoot family)

Lamb's quarters leaves are high in vitamins A and C. When eaten with their seeds they form a complete amino-acid complex. BG

Wild Food and Medicine Plants of the North | 111

Part II: *Herbs*

Long Shelf Life

A cache of lamb's quarters seeds were found at an archeological dig near the Oldman River, in southern Alberta, that was occupied by the Blackfoot from 1500-1600 A.D. Archeologists have been able to germinate seeds found at other archeological sites in western North America that were buried for over 1,700 years.

Botanical Description

An annual herb with stems rarely over 60 cm, erect or ascending, simple to many branched. Leaves 1–12 cm long, some shaped like goose feet (it's a member of the goosefoot family, along with spinach and beets). Most parts of this plant have a mealy white coating, at least in part (local environmental conditions dictate the level of "mealiness"). Flowers clustered along the stem, inconspicuous, small and green. Sepal and petal lobes with keeled midribs; seeds 1.1–1.6 mm long, round, shiny black. The early seeds are grey coloured, then turn pinkish-red. (Note: Lamb's quarters is closely related to strawberry blite [*Chenopodium capitatum*]. For more information on this plant, see page 164.)

Habitat and Range

One of the earliest plants introduced to the Yukon, it was first collected in the territory in 1883. Lamb's quarters grows all over the world. It likes to live in disturbed soils, garden edges, compost piles, and along roadsides.

Plant Parts Used

Leaves and seeds

Harvest Time

Leaves: throughout the summer. Seeds: late autumn when they turn reddish

LIVING CLOSE TO GARDENS and disturbed soils, lamb's quarters are a lovely addition to our summer-greens repertoire. Along with nettle it's one of the key greens in my wild-weed spanakopita. Many people disregard lamb's quarters and consider it a nuisance weed, but its tender, young greens have much to offer in terms of food value for humans and livestock. In fact, it gets its common names, "fat hen" and "pigweed" because it's such a good food for poultry and swine—not to mention sheep, goats, and other animals.

Medicinal Actions

Mildly astringent

Medicinal Preparations

Infusion, mouth rinse, spit poultice, tea

Lamb's Quarters

Medicinal Uses

Lamb's quarters tea was traditionally used for stomach ache, gas, bouts of diarrhea, and was used to treat and prevent scurvy.

The cooked leaves and seed heads of epazote (*Chenopodium ambrosioides*), a relative of lamb's quarters, is used in Mexico as a spice for beans, and is believed to keep the digestive system clean and healthy.

Lamb's quarters is also used as an external wash for skin ulcers or as a gargle for ulcers in the throat and mouth.

The lamb's quarters leaves can be made into a spit or traditional poultice to help take the heat out of mosquito bites, wasp stings, bee stings, and nettle rash. In a pinch, a poultice can be used over wounds.

Food Uses

The interesting texture and white coating on the greyish-green, tender little lamb's quarters leaves may not leave you interested in trying them, but they pack a nutritional punch. The leaves are high in vitamins A and C, and when they are eaten with the seeds form a complete amino-acid complex.

Some people call lamb's quarters "northern spinach" because the leaves can be used as a spinach replacement throughout the summer. The fresh leaves can be washed and stored in the fridge for a few days or they can be blanched and frozen to use over the winter months. Drying the leaves and seeds is also an option and they are delicious in sauces for pasta and as a parsley replacement in soups.

Try using the leaves in a stir-fry or as a filling for cannelloni or lasagna. An excellent filling for filo pastry is leaves fried with garlic and mushrooms, then blended with cottage cheese. You can also add them to omelettes or savoury potato pancakes.

Lamb's quarters is also a close cousin to quinoa, a popular ancient grain that can be purchased in health-food stores. Each plant produces up to 75,000 seeds. The seeds are best harvested when they turn a pinkish-red colour. These can then be eaten raw or dried, or ground up and used as an added boost of protein and amino acids in breads, muffins, soups, stews, morning smoothies, and anything else that your imagination can think up. The tricky part of using the seeds is that it's best to separate them from their seed covering (the chaff). To do this, dry the seeds, roll them in your hands with a tray underneath until the little black seeds separate from the hulls. This will give you considerably less volume of seed, but it's worth it. Another way is to put the seeds on a cookie sheet and push a rolling pin over them until the hulls fall from the seeds. And if you don't have patience for this, dry the seeds and use them as is; the chaff is probably good fibre!

The dried seeds can be ground into flour, cooked as a whole-grain cereal, or roasted and used as a coffee substitute. You could also cook them as you would quinoa.

> "A weed is a plant that has mastered every survival skill except for learning how to grow in rows."
>
> —Doug Larson, author and cartoonist

Ground lamb's quarters seeds can add a boost of protein to your morning smoothie. CA

Wild Food and Medicine Plants of the North | 113

Part II: *Herbs*

Nutritional Profile
Calcium, chlorophyll, copper, essential amino acids, flavonoids, fibre, iron, magnesium, phosphorus, potassium, protein, selenium, sodium, zinc, vitamins A, C, E, and K, niacin, folate, and omega 3 and 6 essential fatty acids.

Other Uses
Steve Johnson, founder of the Alaskan Flower Essence Project and author of *The Essence of Healing*, says that lamb's quarters is used for those who have "Perspective limited to what we can understand with the mind; lacking balance and harmony between the mind and heart, the rational and the intuitive." He adds that the healing qualities help to "heal separation between the heart and mind; balances the power of the mind with the joy of the heart."

As medicine for the Earth, lamb's quarters is a "purifier herb," and in its effort to cleanse the soil it absorbs pollutants and concentrates them in its leaves.

Lamb's quarters leaves contain saponins and can be used as a crude (and fun!) soap replacement.

Cautions
Lamb's quarters contain oxalates, an acid that is also found in spinach, Swiss chard, potatoes, and rhubarb. Cooking and freezing the plants will help break down the acids. (For more information on oxalic acid, see page 27.)

Because lamb's quarters absorbs pollutants, foragers should be careful not to harvest lamb's quarters in areas where there may be environmental contamination, such as near roads, mine sites, or industrial areas.

ADDITIONAL RECIPES

Green Sauce, page 360

Healthful Weed Pie, page 368

Pink Drink (High-C Smoothie), page 340

Weed Wise Smoothie, page 337

Wild-Greens Stir-Fry, page 370

Wild-Weed Dressing, page 361

Wild-Weed Salad, page 365

Cream of Wild-Weed Soup, page 364

Wild-Weed Spanakopita, page 367

Lamb's Quarters Omelette
Omelettes are not just for breakfast anymore! For dinner serve with a wild weed salad and for breakfast serve with a green smoothie!

2 eggs, beaten
1 tablespoon (15 mL) milk or unsweetened soy, rice, or almond milk
1 clove of garlic, chopped
Salt and pepper to taste
A pinch of dill weed to taste
½ cup (125 mL) of finely chopped lamb's quarters leaves
¼ cup (60 mL) of crumbled goat feta
2 tsp (10 mL) olive oil

Whisk eggs, milk, salt, pepper, and garlic together. Heat the oil in a small frying pan over a medium heat. Pour egg mixture into the pan, cover for a couple of minutes to allow mixture to cook. Before it is fully cooked, add the lamb's quarters and feta to the top. Cover and cook until done. Use a metal spatula to fold the omelette over and serve.

LUNGWORT

Mertensia paniculata

Other Names and Etymology
Chiming bells, blue bell. *Mertensia* takes its name from German botanist Franz Karl Mertens. *Paniculata* describes the panicle-arrangement of the flowers on the stem.

Family
Boraginaceae (borage family)

Botanical Description
A perennial herb with branching erect stems, 20–70 cm high. Dark green leaves with rough hairs, broad at the base tapering to a long point; upper leaves are sessile (lack petioles). Flowers in drooping clusters; petals fused into funnel-shaped tubes, pink in bud, turning a rich blue in bloom. Seeds are wrinkled nutlets.

Habitat and Range
A common lowland plant of riverbanks, open woods, and clearings, occasionally above timberline. There are two varieties in the Yukon, var. *paniculata* and var. *alaskana*; they can be identified by the different pattern of hairs on the sepals, *alaskana* is also a lankier plant. Lungwort is found in much of boreal North America, from James Bay to Alaska. It's common in much of the Yukon, almost to the Arctic coast.

Plant Part Used
Leaves

Harvest Time
Spring throughout the summer

Lungwort is also known as "chiming bells." Lungwort-leaf tea is good for treating coughs.

Part II: *Herbs*

THE FIRST TIME I noticed lungwort, it was the flowers that drew me in, the pink fading into blue seemed so perfectly aligned with the dark green, hairy leaves. Knowing the plant was in the borage family, I nibbled on a leaf. The flavour was reminiscent of the common *Borago officinalis* (borage) that I grow in my greenhouse and left my taste buds wanting more.

Medicinal Actions
Antidiarrheal, anti-inflammatory, astringent, antihistamine, cell-proliferant, demulcent, emollient, expectorant, galactagogue, pectoral, styptic, tonic

Medicinal Preparations
Fomentation, poultice, salve, syrup, tea/infusion

Medicinal Uses
Herbalist Robert Rogers says that lungwort's "constituents consist of mucins, silicic acid, tannin, saponin, allantoin, quercetin and kaempeferol." He explains that these are "stimulating to the respiratory system, making it beneficial for coughs. Much like its close cousin borage, the astringent qualities of the herb make it effective at relieving diarrhea and hemorrhoids. Allantoin found in lungwort is also found in the herb comfrey, commonly called 'knitbone' because of its ability to help heal tissue and mend bones. A poultice made with the lungwort leaves is used to heal cuts and wounds; and can be used as a tea or fomentation over the lung region for respiratory problems."

> "Lungwort is about the air element. The funnel shape of the flower allows one to almost see the plant breathing. All who work with meditation and the use of Prana would benefit from the use of this plant."
>
> —Gurudas, author of *The Spiritual Properties of Herbs* (1988)

Food Uses
The bristly leaves can be added raw to a salad, shredded for use in cold- or warm-grain salads, added to soups and stews, or eaten as a steamed green.

Other Uses
Steve Johnson, founder of the Alaskan Flower Essence Project and author of *The Essence of Healing*, says that chiming bells *(Mertensia paniculata)* can be used by people who are

Lungwort

"depressed; despondent; disheartened; no joy in one's day-to-day existence; feeling a lack of support and stability at a basic level." He adds that the healing qualities encourage "the experience of joy, peace, and stability at the physical level of our beings; helps us open our hearts to the loving energy of the Divine Mother."

A poultice made of lungwort leaves can be used to heal cuts and wounds. BG

Cautions
None reported.

Lungwort Fomentation

To help break up chest congestion, place the lungwort fomentation over the lung area. I would also use this remedy for someone who is energetically blocked in the heart chakra. This could be helpful and supportive to those who are trying to move with grief.

1 cup (250 mL) fresh lungwort leaves
4 cups (1 L) water

Decoct 1 cup (250 mL) of fresh leaves in 4 cups (1 L) of water (for more information, see Decoctions, page 286.)

When the decoction is done, let it cool slightly (enough that you will not burn your hands when you dip your cloth in). Dip a soft cotton cloth in the preparation, ring it out, and let it cool down to the appropriate temperature (so that it doesn't burn the skin).

Place over affected area and cover with a couple of towels to keep the heat in. Once the fomentation has cooled, remove the cloth, re-soak, and re-apply.

ADDITIONAL RECIPE

Lung-Tonic Infusion, page 283

MINT

Mentha arvensis

Other Names and Etymology
Field mint, wild mint, brook mint, Canadian mint. It was formerly known by the botanical name *Mentha canadensis*.

Mentha is a Latin name used by Pliny and *arvensis* means "of the field."

Family
Lamiaceae (mint family)

Botanical Description
Strongly aromatic perennial herb, stems singular or branching, to 50 cm high. Leaves 1–8 cm long, variously oval-shaped, tapered at both ends, covered with glands (containing the aromatic oils), margins toothed. Flowers small, tightly clustered in axils of middle and upper leaves; petals purplish to pinkish or white, fused into a partial tube 4–6 mm long.

Mint is high in volatile oils, dries well, and, when stored properly, keeps its strong scent and flavour. L to R: CA., BG

Mint

Habitat and Range
Occasional in grassy swales, meadows, moist ditches, riverbanks, and lakeshores. Circumpolar. In North America, from Newfoundland to British Columbia, NWT, and Alaska, south to California. In the Yukon it's found to about latitude 66° north.

Plant Parts Used
Aerial parts

Harvest Time
Throughout the summer. Gather fresh mint leaves before the heat of the day. You can pluck each leaf by hand or use a pair of scissors and give the patch a little haircut. Don't worry it will grow back quickly!

My friend Lois Moorcroft harvesting some of the wild mint she transplanted into her garden in the southern Yukon. BG

THE PLAYFUL SCENT of wild mint is multi-dimensional, Like conventional mint, wild mint's aroma is refreshing but it also has a subtle, earth-like smell and flavour that makes it unique.

Transplanting wild mint into your garden is easy. It will flourish with little help from you. No northern household herbal pantry is complete without wild mint: dried for tea and spice, tinctured as a digestive remedy, and—my favourite—infused in witch hazel to make a deodorant. I recently started using the wild-mint-hydrosol spray in my healing practice, Aura Borealis. It seems to help clients create a fresh outlook, and reduce thought congestion through the healing process. Wild mint has refrigerant properties that are cooling and therefore help promote healing through its anti-inflammatory effects. The hydrosol makes a refreshing spray for women experiencing hot flashes!

Mint is a great "strewing herb": strategically scatter the leaves in places where mice and ants might enter your house.

High in volatile essential oils, it dries well, and when stored properly keeps its strong scent and flavour.

Medicinal Actions
Analgesic, anesthetic (local), antiemetic, anti-inflammatory, antimicrobial, anti-rheumatic, antiseptic, antispasmodic, aromatic, bitter, blood purifier, carminative, diaphoretic, digestive, emmenagogue, febrifuge, galactagogue, nervine, pungent, refrigerant, stimulant, stomachic

Medicinal Preparations
Compress, cream, essential oil, hydrosol, oil infusion, poultice, salve, syrup, tea/infusion, tincture

Part II: *Herbs*

> "It is the destiny of mint to be crushed."
>
> —Waverly Lewis Root, journalist, (1903–1982)

Medicinal Uses

Mint leaves aid digestion and help ease indigestion, gas, heartburn, ulcers, and colic. A mild antispasmodic, mint is also useful for relieving menstrual cramps and nausea. It can also help to bring on delayed menstruation.

An effective vasodilator, mint stimulates circulation and can help reduce heart palpitations. It also alleviates the symptoms of colds and flu by reducing excess mucus and helping to lower fevers. When combined with elder and yarrow, it's particularly good at assisting with fevers by acting as a diaphoretic that helps eliminate toxins by promoting perspiration.

A cold tea can also be used topically as a compress to help reduce fevers, for itchy skin rashes, and to help reduce the pain of arthritis.

Drinking mint tea in the morning is stimulating and helps to clear a tired mind; it can also help deflect oncoming headaches and can alleviate nausea due to motion sickness.

Food Uses

Mint leaves are tasty and refreshing, so don't hesitate to explore all the gastronomic possibilities this plant offers. It's fun to have wild mint ice cubes ready to use in iced teas, or put the leaves in the bottom of homemade popsicles. Fresh or dried, wild mint is also tasty when added to fruit salads, savory dishes such as hummus, and Indian chutneys.

Mint jelly (see page 122) is good to have on hand to add to garden peas, spread on wild game, fowl, or to top ice cream or yogourt.

Wild-mint liqueur is a flavourful addition to any celebration or après-ski with friends.

Wild mint ice cubes are a fun addition to any drink. BG

Mint

Wild mint jelly (recipe on following page) goes well with lamb or on crackers with goat cheese. BG

Nutritional Profile
Plant proteins, vitamin A, C and K, minerals including iron, calcium, magnesium, and trace amounts of zinc

Other Uses
Wild mint and juniper are the key plants used in Aroma Borealis's Crown Chakra Vibrational Essence. The essence is used to help the mind move into sacred thought and clear excess mind clutter that can cloud our judgment. Today we live in a world with a lot of electromagnetic pollution that affects our minds; mint essence helps us to subtly rebalance and be clear and focused. The essence is refreshing, restorative, and aids us with our memory and concentration.

Cosmetic Uses
Fresh or dried wild mint is a wonderful addition to facial steams and foot soaks. A compress can be used for acne eruptions and a facial tonic is easily made with the leaves. Mint helps to stimulate sluggish, oily skin, and is great when added to facial masks, facial creams, shampoos (to treat dandruff), salves, soaps, toothpastes, powders, and deodorants.

Mint acts as a natural insect repellent, it is also good on its own, or combined with other herbs such as yarrow and wild sage.

Cautions
Women who are susceptible to miscarriage should avoid ingesting high doses of wild mint tea or tincture during pregnancy; however, using the dried herb for common culinary use and making it into a light herbal tisane is okay.

Part II: *Herbs*

ADDITIONAL RECIPES

After-Dinner Elixir, page 297

Athlete's-Aid Tincture, page 294

Athletes' Blend Smoothie, page 340

Beautiful Hair, Skin, and Nails Infusion, page 285

Cool Cranberry Juice, page 335

Detoxifying Infusion, page 283

Boreal Bitters, page 297

Cough-and-Cold Syrup, page 300

Cranberry-Mint Muffins, page 343

Get-Up-and-Go Infusion, page 283

Good-for-Fever Infusion, page 282

High-Calcium Infusion, page 282

Moisturizing Lip Balm, page 314

Moon-Cycle Nourishing Vinegar, page 203

Moon-Time Infusion, page 284

Respiratory Steam, page 329

Spirit-of-the-Boreal-Forest Tea, page 281

Wild Rose and Mint Lassi, page 341

Wildly Mint Jelly

Makes a flavourful jelly that is ready to use for desserts, meats, and of course, peas. Unlike commercial mint jelly that is dyed fluorescent green, it has a dark green to gold colour. I have made this recipe with fresh mint and dried mint. The dried mint is more flavourful and darker in colour.

2 cups (500 mL) of packed mint leaves (dried)
4 cups (1 L) of water
¼ cup (60 mL) lemon juice
1½ cups (375 mL) of organic cane sugar
1 package (57 g) of pectin

Place mint leaves and water in a pot, simmer for up to five minutes with the lid on, stirring occasionally. Remove from heat and set aside until completely cool, allowing the flavour to mature.

BG

Strain through a cheesecloth, wringing out every last drop. Pour 3 cups (750 mL) of the strained mint infusion into a clean pot. Add the lemon juice and pectin, stirring to make sure the pectin dissolves. On a high heat bring the mixture to a rolling boil for one minute.

Add the cane sugar, stir to dissolve the sugar, and bring the liquid back up to a rolling boil that can't be stirred down. Boil hard for at least one minute. The jelly is ready when it coats the back of spoon and has a syrup-like consistency. Test for setting by placing a teaspoon of hot jelly on a plate, allow to cool—the surface should wrinkle when pushed with your finger. If it's still runny, put back on heat and continue boiling and testing until jelly sets.

Remove from the heat source. Skim the jelly foam off the top. Pour into hot sterilized jars and seal immediately.

Place in a hot-water bath for 10 minutes to seal. (See page 349 for tips for successful canning and heat processing.)

Makes 2 cups (500 mL) of jelly.

STINGING NETTLE

Urtica dioica

Other Names and Etymology
Burning nettle, common nettle. The word "nettle" may be derived from the Anglo-Saxon *noedl*, or "needle," which speaks to the plant's needle-like sting. As early as 725 A.D. it appeared written as *netlan*, and by about 1200 A.D. it had assumed the name "nettle." *Urtica* is from the Latin *urere*, "to burn," and *dioica* means "two houses," in reference to the male and female flowers on one plant.

Family
Urticaceae (nettle family)

Botanical Description
Perennial herb with tough, simple, or branched stems, to 1 m or higher. Stinging hairs on the stems and leaves are the key feature of this plant. Leaves coarsely toothed, up to 15 cm long. Flowers small, green; in elongated and drooping clusters from upper nodes; seeds small, 1–1.5 mm long, tan to brown.

Transplanting wild nettles in your home garden is easy. The plants will provide you with an abundant source of food and medicine throughout the summer.

Part II: *Herbs*

> "Tender-handed stroke a nettle,
>
> And it stings you for your pains;
>
> Grasp it like a man of mettle,
>
> And it soft as silk remains.
>
> —Aaron Hill, Scottish poet, 1685–1750

Habitat and Range

Grows in thickets near stream banks, disturbed soils, and rich, damp soils. Found from Newfoundland to Alaska, south into the United States. *Urtica dioica* was introduced from Europe; it has only patchy distribution in the Yukon, and is most frequently found around settlements, old historical sites, and homesteads. Found in Old Crow and as far north as Rampart House on the Porcupine River. Another nettle species, *U. gracilis*, is native to the Yukon.

Plant Parts Used

Roots, seeds, leaves

Harvest Time

Leaves: spring and early summer. Roots: early spring and late autumn. Seeds: as they mature. When harvesting nettle it's important to wear gloves because the plants are covered with tiny hairs that contain formic acid. When the hairs come into contact with your skin they sting (hence the name "stinging nettle"!). If you don't have gloves, and want to avoid being stung, grab the stem with pressure to crush the hairs flat and prevent them from penetrating the skin.

If you do get stung, not to worry: the antidote is usually growing nearby. To treat the itching and hives-like response, make a spit poultice of horsetail, dock, or plantain leaf. If these aren't in the vicinity, apply moistened baking soda when you get home.

In the United Kingdom, an annual Stinging-Nettle Eating Championship draws thousands of people to Dorset, where competitors attempt to eat as many raw nettle leaves as possible. Competitors are given 60-cm stalks, from which they strip the leaves and eat them. Whoever eats the most leaves within a fixed time is declared the winner. The competition dates back to 1986, when two neighbouring farmers attempted to settle a dispute over who was responsible for controlling the weed.

Stinging Nettle

FUNNILY ENOUGH, the word "nettle" made its way into the English language as part of the folk expression, "What the nettle!" probably because that is what you say after being stung by one!

Getting stung by nettle is a not so gentle reminder that this plant is vital, healthy, and that it has its defence mechanisms in place.

I always have nettle at hand in a few different forms. I like making the tincture every summer as it's an excellent remedy for allergies. Growing nettles in your home garden is easy and it provides a lot of food and medicine throughout the summer.

In late spring or early summer, prune the leaves off the top half of the plant and gather fresh leaves as they appear. This way your favourite patch will provide you with fresh green leaves throughout the summer. Later in the season, after its drooping flower clusters appear, gather only the top leaves.

Autumn seeds can be gathered from wild plants to seed your garden but as nettle seeds have a slow germination rate, you may have more success by digging up a new spring seedling for transplanting.

There are lots of things you can do with nettles! A few years ago, the guys at the Yukon Brewing Company and I formulated The Aroma Borealis Herbal Cream Ale that sold in Yukon liquor stores for more than four years and included an infusion of nettle, fireweed, rosehip, and mint.

Medicinal Actions
Alterative, antispasmodic, astringent, diuretic, galactagogue, hemostatic, pectoral, rubefacient, tonic

Medicinal Preparations
Capsules, cooking herb, oil infusion, syrup, tea/infusion, tincture, vinegar

Medicinal Uses
Chlorophyll-rich nettle strengthens and supports the whole body, specifically the digestive, respiratory, urinary, and glandular systems. Its antihistamine properties make nettle good to ingest as a tea, capsule, or tincture for the classical symptoms of allergies such as sneezing and itchy watery eyes. People who have eczema can also benefit from including nettle in their herbal repertoire. Nettle leaf helps with the inflammation of arthritis, gout, kidney irritations, and can be made into nourishing syrup or a tea infusion for anaemia.

Nettle-leaf tea is also an effective spring tonic and cleansing herb because it acts as a blood purifier. It helps the efficiency of the kidney and liver, and is a mild laxative and diuretic.

Nettle root is a great preventative remedy for men. In recent clinical trials the extract

Opposite: To avoid being stung by nettle, grab the stem with enough pressure to crush the hairs. BG

Part II: *Herbs*

Beyond food and medicine, nettle can be used to make textiles and even paper. PL

from the root of nettle has been proven as a treatment for symptoms of benign prostatic hyperplasia (BPH), a non-cancerous growth of the prostate gland.

For women, the leaf juice acts as a diuretic and is an excellent remedy for premenstrual water retention. It's also useful for excessive bleeding during menstruation.

When we are under stress we tend to be overly acidic; nettle juice or tea can help balance out the acid-alkaline ratio in the body.

Due to its high vitamin K content, fresh nettle can be used to stop all types of bleeding, though the vitamin K is diminished when the plant is dried. Topically, cooled nettle tea acts as an astringent and can be used as an anti-inflammatory for red and irritated skin. It can be ingested to treat excessive mucus caused by allergy-induced lung irritations.

Herbalist Robert Rogers mentions that the ripe seed tincture clears excess creatinine from urine in seven to ten days and that it's a powerful herb for use in chronic nephritis (inflammation of the kidney).

Nettle is also made into a homeopathic remedy called *Urtica urens* and is used generally for hot, stinging, painful and itchy rashes, itchy blotches, chicken pox, burns and scalds, prickly heat, and rashes due to allergic reactions.

The seventeenth-century physician and astrologer Nicholas Culpeper recommended the use of nettles to "Consume the phlegmatic superfluities in the body of man that the coldness and moisture of winter has left behind." He also prescribed the juice of the leaves as a treatment for gangrene and scabies.

Food Uses

Nettle is a tasty, nutritious herb that adds a nice boost of vitamins, minerals, and protein to any meal. Some people use it as a replacement for spinach, though the texture is very different (but interesting). The leaves can be used in many different ways including fresh as a juice or added to soups and stir-fries, steamed (save the water for tea), and as a nice side dish. I lightly fry the first harvest of the season with garlic, butter, dandelion flowers, and fireweed shoots—a yummy burst of wild flavour!

A juice can be made by adding fresh leaves to a blender and covering with water. Once liquefied, strain out the plant matter, and take the juice in tablespoon portions: Add one to two tablespoons to a cup of water, drinking up to three cups of the diluted juice a day. The leftover juice can be frozen in ice cube trays and then placed in a bag in the freezer until needed for medicinal or nutritive purposes. The ice cubes can either be added to hot water to make a tea or just placed in room-temperature water and drunk when melted. I like to add the cubes to soups and my morning smoothies. Nettles are a great addition to any food recipe. I like to keep ground dried nettles on hand because they are so high in chlorophyll, minerals (including iron), and plant protein. After grinding the dried nettle leaves, keep stored in an airtight, dark container. Storing in the fridge or freezer keeps them really fresh. Ideally, it's best to keep the dried leaf whole and grind it as needed, but sometimes it's more convenient to have it pre-ground.

Interestingly enough, if you get a little kitchen cut, you can apply a dab of the dried nettle powder to help stop the bleeding.

Stinging Nettle

If you're worried about getting stung while eating nettles, don't be! Cooking, drying, chopping, crushing, and juicing neutralize the acid and disable the stinging hairs. Also, soaking nettles in water will remove the formic acid from the plant, allowing them to be handled and eaten without incident.

Nutritional Profile
Nettle is high in calcium, magnesium, chlorophyll, iron, vitamins A, C and D, zinc, potassium, chromium, cobalt, niacin, phosphorus, manganese, and silica. It also contains 2,900 mg of calcium per 100 g of dried nettles, according to chemist Mark Pedersen in his book *Nutritional Herbology*.

Cosmetic Uses
Nettle is used by many producers of natural hair-care products to control dandruff and keep hair healthy and shiny. At home you can make a healthy hair rinse with a strong infusion of nettle tea.

Other Uses
Steve Johnson, founder of the Alaskan Flower Essence Project and author of *The Essence of Healing*, says that stinging nettle helps people who are highly sensitive "stay connected to the Earth and to their feelings; promotes reconnection and grounding after being overwhelmed by too much input; enables us to absorb and process energetic information and input in alignment with our capacity to integrate it; for those who have been hurt deeply in the past and have a tendency to sting and repel those they really want to be close to; helps to heal the alienation that comes from the fear of being hurt again."

Adding nettle to your compost pile will help activate decomposition because nettles contain nitrogen. You can also add it to your compost tea for watering your vegetables.

Growing nettles in your garden can help increase the oil content of herbs such as valerian root, sage, marjoram, mint, and angelica.

Turkey and other poultry, cows and pigs thrive on nettles; adding ground dried nettle to your chicken feed will increase egg production.

Cautions
After stinging nettle enters its flowering and seed-setting stages, the old leaves develop gritty particles called "cystoliths" that can irritate the urinary tract if eaten or ingested as a tea.

Everything Old Is New Again

In Hans Christian Andersen's fairy tale, "The Wild Swans," the princess has to weave coats of nettle to break the spell cast on her brothers.

Today, nettle fibre is being used in the global textile industry. Nettle stems contain a fibre that can be spun or woven into cloth using a process similar to one traditionally used to make linen. Plus, it has naturally antibacterial and mould-resistant qualities, and, unlike cotton, nettles grow easily without pesticides.

While this is really exciting, it also shows that "everything that is old is new again." Nettle already enjoys its rightful place in textile history as a plant that was used in the early part of the twentieth century for making rope and a linen-like fabric.

In Scotland, the mature nettle stems were made into "nettle cloth," which was used for many centuries to make tablecloths and sheets.

In Russia, as well as being used as a fibre, the plant is used to make dyes: the juice from nettle stems makes green; the boiled roots make yellow; and the leaves a yellow-green.

Part II: *Herbs*

ADDITIONAL RECIPES

Allergy-Aid Capsule, page 301

Athlete's-Aid Tincture, page 294

Athletes' Blend Smoothie, page 340

Beautiful Hair, Skin, and Nails Infusion, page 285

Detoxifying Infusion, page 283

Clear-Skin Tincture, page 72

Get-Up-and-Go Infusion, page 283

Good-for-Gout Infusion, page 284

Green Bouillon, page 363

Herb-and-Berry Sweet Summer Days Syrup, page 355

High-Calcium Infusion, page 282

High-Iron Syrup, page 299

Moon-Time Infusion, page 284

Nettle Pasta Sauce, page 358

Nourishing Vinegar Tonic, page 296

Styptic Powder, page 320

Tea-for-Two Pregnancy Infusion, page 284

Living Fruit Leather, page 377

Wild-Weed Spanakopita, page 367

Nettle Tincture

A remedy that helps prevent allergies and is an overall tonic for the body.

1 part nettle leaves
2 parts vodka

Put nettles leaves in a jar. Cover with vodka and let steep for 30 days. Shake every day. (See Tinctures, page 291.)

Stinging nettle growing wild at Rampart House, northern Yukon. BG

WILD ONION

Allium schoenoprasum

Other Names and Etymology
Chives, wild chives. The Van Tat Gwich'in name for onion is *tl'oodrik*. The word "onion" comes from the Latin *unus* meaning the number "one," "unity," or "oneness." *Allium* is from Greek, and means "garlic." *Schoenoprasum* is from the Greek *schoinos*, meaning "rush" or "reed," and *prason*, for "leek."

Family
Liliaceae (lily family)

Botanical Description
A perennial with flower clusters arising from a leafless stem; the underground (food storage) bulb is elongated, egg shaped and papery coated. Flowering stems are 15–45 cm high. Leaves shorter, hollow, and smelling and tasting strongly of onion. Flowers umbellate, petals dark pink to purple, with darker veins.

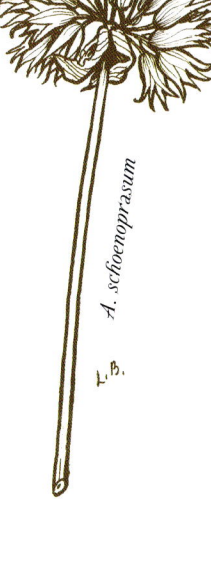

A. schoenoprasum

The small and tasty bulbs can be harvested with the leaves still attached and used as spring onions. BG

Part II: *Herbs*

Habitat and Range
Grows in moist to turfy places along rivers and lakeshores at lower altitudes. Circumpolar, wide-ranging; in North America, from Newfoundland and Nova Scotia to Alaska, south to New York, Minnesota, Colorado, and Oregon; found at lower altitudes throughout much of the NWT and Yukon, north to the Firth River on the north slope of the British Mountains.

Plant Parts Used
Bulbs, flowers, stems, roots

Harvest Time
Spring through autumn

WILD ONIONS ARE a summer-time favourite. The spicy, onion-flavoured greens go hand in hand with summer salads, dressings, and sauces. The pretty, globe-like, purple flowers sit gracefully on top of the dark green, hollow stems, and make salads beautiful.

Medicinal Actions
Anthelmintic, antiseptic, digestive, hypotensive, warming

Medicinal Preparations
Fomentation, food, spice, tea, tincture

Medicinal Uses
Like other plants in this family, such as garlic, wild onion benefits the digestive system and blood circulation by improving the appetite and warming the body. Touted as a mild antiseptic, wild onions were traditionally used to expel worms and parasites. They are great to include in your diet if you feel a cold coming on. Slicing a piece of the bulb and pressing it onto the bottoms of the feet can help break up chest congestion.

Food Uses
The flowers are lovely in salads, and the pungent-yet-tasty leaves are great chopped in salads, soups, with vegetables, omelettes, and in savoury muffins and breads. The hollow leaves can be chopped up and dried or frozen for winter use, when they act as a warming spice for the whole body. The small and tasty bulbs can be harvested with the leaves still attached and then used as spring onions.

> According to herbalist and astrologer Nicolas Culpeper, wild onion is ruled by the planet Mars and is classified as a fire element. It corresponds to the astrological sign Scorpio.

Wild Onion

To make an aromatic and warming oil for salads and sautés, infuse chopped wild onion flowers and chives in olive oil. CA

Nutritional Profile
Vitamins A, C and E, trace minerals such as iron, calcium, copper, magnesium, phosphorus, potassium, selenium and zinc. Also contains sulfur.

Cautions
Large amounts of chives can cause stomach upset.

"He who bears chives on his breath is safe from being kissed to death."

—Marcus V. Martialis, poet (100 AD)

Wild-Onion Spread

One 19-oz. (540 mL) can black beans or cooked dried beans
2 tablespoons (30 mL) tahini (sesame butter)
½ cup (125 mL) hemp or olive oil
1 cup (250 mL) chopped wild chives

In a blender, combine all ingredients and blend until smooth. Serve with crackers or use as a dip for vegetables.

ADDITIONAL RECIPES

Green Sauce, page 360

Wild-Onion Fomentation, page 319

Wild-Weed Dressing, page 361

PLANTAIN

Plantago major

Other Names and Etymology
White man's foot, greater plantain, broad-leaved plantain, ripple grass, snakeweed, Englishman's foot, weybroed (Anglo-Saxon).

The genus name *Plantago* is derived from the Latin *planta*, meaning "sole of the foot," because the leaf is somewhat shaped like the sole of a foot, and because its seeds can be spread by peoples' shoes. It could also be derived from the Greek word *plantus*, which means "flat" or "broad," describing the leaves. There are about 200 species of *Plantago* worldwide.

Family
Plantaginaceae (plantain family)

Botanical Description
Perennial herb, with fibrous roots and leaves growing low to the ground; flowering

Plantain works well as a spit poultice for relieving bee stings and mosquito bites. L to R: PL, BG

Plantain

stems 10–30 cm or higher. Green leaves elliptic to egg shaped with prominent parallel veins; margins entire or undulating, 3–15 cm long. Flower spikes dense, up to 25 cm long, hairless; flowers small (up to 2 mm) and greenish. Seed capsules egg shaped, 4–5 mm long, splitting open near the middle or "equator" of the capsule. Seeds have wavy, thread-like ridges.

Habitat and Range
Grows in disturbed soils, townsites, on riverbanks and garden edges. A cosmopolitan plant, it grows throughout the world; in North America from Labrador to British Columbia, NWT, and Alaska. In the Yukon, plantain is found as far north as the Porcupine River valley.

Plant Parts Used
Leaves, flower-seed spikes

Harvest Time
Leaves: summer. Seeds: late summer or early autumn. When plucking plantain leaves off the plant you will notice a little tug from white, strand-like hairs. Pinching the stems at the base of the leaf makes it easier to collect.

IT'S INTERESTING THAT PLANTAIN always seems to grow closely to nettle. Its soothing and anti-inflammatory properties make it the perfect antidote to nettle stings.

Plantain works well as a spit poultice for bee or wasp stings and mosquito bites. Plantain is such a versatile plant it's well-worth gathering for your herbal pantry and for your first-aid kit. I prefer to make preparations such as tincture from the fresh leaves to use for lung irritations due to colds or seasonal allergies. Dried plantain leaves can be used for teas or topical use. Remember when storing to keep out of direct light because the leaves will start to loose potency and their colour will fade.

Medicinal Actions
Leaves: alterative, antihistamine, anti-inflammatory, anti-microbial, antiseptic, astringent, blood purifier, demulcent, diuretic, emollient, expectorant, immune stimulant, pectoral, styptic, vulnerary. Seeds: mucilaginous.

Medicinal Preparations
Cream, eyewash, oil infusion, poultice, salve, spit poultice, soak, steam, syrup, tea/infusion, tincture

Part II: *Herbs*

Plantain through the Ages

Plantain is one of the herbs in the Old English "Nine Herbs Charm," recorded in the tenth-century *Lacnunga* ("remedies"), a collection of Anglo-Saxon medical texts and prayers. It's said that the charm repelled poison and infection. We could use that on so many levels today!

Nicholas Culpeper (1616–1654), an English botanist, herbalist, physician, and astrologer wrote in his collection of work, *The Complete Herbal*, that "Plantain is in the command of Venus and cures the head by antipathy to Mars, neither is there hardly a martial disease but it cures." He added, "the water is used for all manner of spreading scabs, tetters, ringworm, shingles, etc."

Shakespeare, both in *Love's Labour's Lost* and in *Romeo and Juliet*, speaks of the "plain Plantain" and "Plantain leaf" as excellent for a broken shin, and again in *Two Noble Kinsmen* (written with John Fletcher): "These poore slight sores neede not a Plantain."

Some old European lore states that plantain is effective for the bites of mad dogs, epilepsy, and leprosy. In the United States the plant is called "snake weed," from a belief that it is effective at treating bites from venomous creatures. Plantain was also hung in the house to guard against evil spirits.

Plantain seeds can be used medicinally. PL

Medicinal Uses

Herbalists consider plantain to be a "sacred herb" and it's called the "Mother of Plants."

Plantain's use as a medicinal plant is found in herbal texts dating back to the 1390s. Because plantain grows all over the world, it has many different regional uses. In most regions where there are poisonous snakes, plantain is famous for its ability to heal and draw out the poison from snakebites. While this is not a problem we face in northern Canada, it's good to have plantain on hand if you are heading into snake country.

What we do have in the North are mosquitoes! A spit poultice or salve of plantain will take the itch out of a mosquito bite and help heal wounds, scrapes, cuts, and bruises.

In the North, plantain is also used for the lungs: taken as a tea, tincture, or in a cough syrup to help eliminate bronchial congestion,

Plantain

laryngitis, lung irritations, and coughs. It's also helpful during times of hay fever, runny noses, and excess mucus in the respiratory system.

Topically, plantain is used as a poultice, oil, or a salve for skin infections, leg ulcers, eczema, psoriasis, nettle and insect stings, burns, abscesses, cracked skin, cuts, abrasions, and hemorrhoids. As a styptic it helps staunch bleeding wounds and encourages the repair of damaged tissue.

For leg ulcers, gently heat the leaves and place over ulcerations to help draw out infections. Like comfrey and lungwort, plantain leaves contain allantoin and are therefore very good at healing skin and bones, and can be used as a poultice, oil, salve, or tea for broken bones, fractures, cuts, rashes, and abrasions.

A light, strained infusion made with distilled water can be used as eye drops, in a compress, or a wash for inflammation caused by pink eye and blepharitis.

Plantain is an active ingredient in commercial quit-smoking products. Ingesting plantain leaves will curb the desire to smoke and help clean the lungs. If you are trying to quit but are still smoking, try a cupful of tea before your ritual smoke—after you light up, you may experience the feeling of having over-smoked and may want to butt out!

If making tea isn't always convenient, consider making a quit-smoking tincture or spray. Making your own is simple (see Tincture Method, page 293). Once the tincture is complete, add 60 drops to a 30 mL bottle (2 tablespoons) of fresh water, shake (infusing your quit-smoking intention into it), and spray it under your tongue as needed throughout the day.

The digestive system benefits from the use of plantain as well. A tea infusion or tincture is helpful for diarrhea and irritable-bowel syndrome, and can help balance an overly acidic stomach by regulating stomach acids, to help heal ulcers.

The urinary system uses the active chemical constituent in plantain, aucubin, to help urinary-tract bleeding, bladder infections, cystitis, prostate inflammations, and infections of the urethra. Aucubin's actions also help increase the uric-acid excretion of the kidneys. Plantain may also be useful for gout because it's caused by high levels of uric acid in the body. The plant's diuretic effect helps with water retention and kidney infections.

For toothaches, chewing a little piece of the root and holding it in your cheek will help draw out infection and reduce pain and inflammation.

Plantain seeds are also useful. When gathered and soaked in water they can be drunk as a laxative to help relieve constipation. The seeds are mucilaginous and absorb eight to fourteen times their volume in water. Because of this it's important to consume more water than usual when using plantain seeds.

Plantain's cousin *P. ovata* is commercially grown and gathered for use in products such as Metamucil, and sold in bulk plant form in herb shops for use in lowering cholesterol.

> "One is tempted to say that the most human plants, after all, are the weeds."
>
> —John Burroughs, author, *Pepacton* (1881)

Part II: *Herbs*

ADDITIONAL RECIPES

Allergy-Aid Capsule, page 301

All-Purpose Botanical Ointment, page 313

Boreal Bitters, page 297

Boreal Skin Cream, page 317

Boreal Skin Oil, page 307

Cold-and-Flu Tincture, page 294

Cough and Cold Syrup, page 300

Good-for-Fever Infusion, page 282

Green Sauce, page 360

Hot Tomato, page 338

Lung-Tonic Infusion, page 283

Moisturizing Lip Balm, page 314

Plantain Poultice, page 319

Piquant Plantain, page 366

Rejuvenating Facial Steam, page 329

Styptic Powder, page 320

Food Uses

Fresh and full of enzymes, vitamins, and minerals, plantain can be used in a variety of ways in the kitchen: raw, chopped into salads, steamed as a side vegetable, cooked in soups, or sautéed in a bit of butter with nettle, dandelion leaf, and wild chives. No matter how you prepare it, the deep green leaves will provide a vitamin-rich alternative to other domestic greens.

I like to use plantain oil (see below) as a base for salad dressings, to drizzle on salads or other vegetables, meats, and meat alternatives, or spead on the body for massage, or use in the bath or shower. The oil can also be used as a base to make a green sauce or mayonnaise.

Nutritional Profile

Vitamins A, C, and K, potassium

Cautions

Plantain seed and husk powders can lower blood-sugar levels, so diabetics should be aware of this before using.

Plantain Herbal Healing Oil

This oil makes a great treatment for damaged skin. It's also good as a butter alternative or as a sauce for potatoes, salads, and pasta.

2 parts plantain leaves
3 parts olive oil
1 part hemp oil

Place plantain leaves in a blender. Cover with oil, blend, and pour into a jar with a lid and label. Shake daily. Strain after a month or so. Do you need your healing oil sooner? Make it quickly and easily with the Double-Boiler Method (see page 306).

RHODIOLA

Rhodiola integrifolia, Rhodiola Rosa

Other Names and Etymology
Roseroot, Arctic root, entire-leaved stonecrop, golden root. *Rhodiola* means "rose-like" in Greek. In 1725, Carl Linnaeus bestowed this name on the plant because when he cut into its roots he noticed that they gave off a subtle, rose-like scent. *Integrifolia* is from the Latin *integer*, meaning "entire" and *folium*, meaning "leaf."

Family
Crassulaceae (stonecrop family)

Botanical Description
Succulent dioecious herb, with a thick and much-branched, scaly, horizontal underground stem; erect stems numerous, 5–35 cm high, leafy. Leaves alternate, 1–4 cm long, hairless and glaucous, somewhat spoon-shaped, entire or dentate. Flowers in a dense terminal cluster; male flowers yellow; female flowers usually purple; dry fruit capsules reddish, erect.

Rhodiola plants are beautiful to look at. Their roots gives off an earthy, rose-like aroma. PL left, CA below

Part II: *Herbs*

Rhodiola integrifolia is the type of rhodiola found in the Yukon. It looks very similar to *R. rosea*, the eastern variety of rhodiola, and can be used the same way. It is inconclusive if they have similar medicinal actions, as more research is needed on *R. integrifolia*. The botanical illustration on the previous page originally appeared as *Rhodiola rosea L.* ssp. *Integrifolia* in *Flora of the Yukon Territory*, edited by William J. Cody.

Habitat and Range

Grows on moist, rocky alpine ledges, and gravelly beaches. Found throughout the northern boreal forest, including in Siberia, northern China, Alaska, and the Yukon, where it is common, and extending south through the mountains to California. *Rhodiola rosea* is found in the Prairies, northeastern North America, and in Scandinavia.

Plant Parts Used

Medicinally, the root and crown. The succulent leaves are good plucked directly from the plant and used to rehydrate you when you're out hiking.

Harvest Time

Greens: spring. Roots: autumn. Clean the roots, slice into thin strips, and dry. Dried roots will last up to three years if stored properly. (See Preserving Wild Plants, page 32.)

RHODIOLA INTRIGUES ME! So much so, that I decided to plant a hundred plugs in my garden four years ago. They will be ready for harvest soon after this book is published. This is exciting because rhodiola is indigenous to the Yukon and the circumpolar north, and needs the cold to germinate and fourteen hours of daylight to grow. While there are lots of wild stock, I don't want to risk contributing to its depletion. Sadly, that is happening to the wild Russian stocks, and has already happened to other much-loved medicinal herbs, such as goldenseal that now borders on extinction in its native environment. Rhodiola plants and seeds can be bought at www.richters.com.

Researchers in the former Soviet Union extensively studied the medicinal benefits of rhodiola following the Second World War. Their findings have exciting applications for the number one disease of our times: stress! The researchers had so much faith in rhodiola that Russian cosmonauts and Olympic athletes used it to reduce fatigue, stress, and to help boost energy.

But the Soviets were not the first to use rhodiola medicinally. Back in 77 A.D., the Greek physician Dioscorides documented *Rhodiola rosea* (then called *Rodia riza*) in his medical text *De Materia Medica*. The Vikings also used rhodiola to improve their endurance and stamina during long sojourns. In China the emperors would order expeditions to remote parts of the Siberian mountains to bring back the respected "golden root" for its healthful properties.

Rhodiola

Medicinal Actions
Adaptogen, anti-aging, anti-cancer, antidepressant, anti-diabetic anti-mutagenic, antioxidant, aphrodisiac, cardio protective, immune booster

Medicinal Preparations
Tea/decoction, tincture

Medicinal Uses
Enjoying a steamy decoction of wild rhodiola root is a subtle and aromatic experience: the pleasing, earthy, rose-like smell is soothing and restorative. The tea tastes as good as it smells, but also delivers some interesting results.

As an adaptogen, its main use is to aid the body in adapting to stress by helping to increase resistance to fatigue and depression, enhance mood and increase energy, which in turn promotes well-being and longevity. It brings relief to people who live in extreme climates and are challenged by seasonal affective disorder (SAD).

The bio-active ingredients in rhodiola roots, known as rosavins and salidrosides, also benefit athletes by enhancing physical performance and endurance, increasing energy, accelerating recovery time, returning the heart rate to normal more quickly, and decreasing pain caused by increased lactic acid buildup in the tissues.

Rhodiola root has also been found to help with asthenia, a condition that can cause mental or physical weakness and fatigue, and which can cause a decrease in quality of life, productivity, loss-of-work capacity, and sleeplessness, as is the case with chronic fatigue syndrome.

It's been noted that the heart benefits from regular use of rhodiola root. It appears to also strengthen the immune system, and the nervous and glandular systems (i.e., the

> "To be overcome by the fragrance of flowers is a delectable form of defeat."
> —Beverley Nichols, author, (1901-1983)

Rhodiola is found throughout the northern boreal forest. I have seen them growing in Finmark, Norway. PL

Part II: *Herbs*

A golden *root* opportunity?

Rhodiola could be a new and viable herb crop for northern herb growers and farmers who want to grow commercial medicinal herbs in the Yukon, NWT, Alaska, or northern British Columbia. The Alberta Natural Health Agricultural Network has an expanding network of growers planting rhodiola for its commercial value. (For more information visit, www.ANHAN.org.)

adrenal glands). When we are under stress our memory can suffer, and rhodiola is also known for its ability to improve our memory functions and increase mental alertness. This makes it an excellent tea when studying for exams or writing a book!

Rhodiola root has long been recommended to travellers as an antidote to altitude sickness. It can also act as an aphrodisiac, can help normalize menstruation, and has been used for symptoms of anemia. Studies have also shown it to help with cancer patients. It also helps speed up recovery from colds, flu, and poisonings.

Stress can age a person, and rhodiola root helps to bring the body back into a less-stressed state, therefore helping our bodies to act our biological ages and prevent premature aging.

A cup of rhodiola root decoction in the morning, or 10–20 drops of the tincture per day can help you get through life without the dangerous side effects of stress. Rhodiola has more of a stimulating effect in smaller amounts and a more sedative effect in larger amounts.

Rhodiola is best taken in the early part of the day.

Rhodiola tincture or tea can be swabbed on the gums or used as a gargle for treating pyorrhea, a gum disease that is a progression of gingivitis.

Food Uses

The succulent rhodiola leaves can be eaten as a trail snack or gathered and added to stir-fries or soups. I really like to add leftover decoction or tincture to my morning smoothie because it's a great way to start the day. The Norwegians use the root to make beer; I imagine that it must taste great!

Nutritional Profile

Leaves: Vitamins A and C, amino acids and trace minerals

Other Uses

"Rhodiola rosea flower essence is for issues surrounding ego, forgiveness and acceptance. It helps one de-personalize perceived injustice or acknowledgement of indifference as opposed to deliberate betrayal," says herbalist Robert Rogers.

Cautions

None known.

Rhodiola Decoction

Place a hearty pinch of the dried chopped-up root in a saucepan, cover with water, and bring to a simmer for 10 to 20 minutes. Remove from heat, let steep until warm, and enjoy your aromatic brew. (See Decoction section page 286 for more in-depth information on making decoctions.)

ADDITIONAL RECIPES

Athlete's-Aid Tincture, page 294

Athletes' Blend Smoothie, page 340

RIVER BEAUTY

Chamerion latifolia

Other Names and Etymology
Willowherb, dwarf fireweed, broad-leaf fireweed. The Inuit of Baffin Island call it "broad-leaved willow herb," "dwarf fireweed," or *paunnat*. River beauty was formerly known as *Epilobium latifolium*.

Family
Onagraceae (evening primrose family)

Botanical Description
Perennial herb with decumbent stems 15–40 cm high, tufted. Leaves opposite, 1–6 cm or more long, glaucous. Flowers larger than in fireweed; in small clusters, petals 15–30 mm long, pink to purple; seed capsules 3–10 cm long.

As its name suggests, river beauty grows near rivers and streams. Its flowers make a colourful addition to a summer salad. MC left, RF below

Wild Food and Medicine Plants of the North

Part II: *Herbs*

Habitat and Range
Frequent on gravel bars, stream banks, and reaching alpine elevations on scree slopes. Circumpolar, Arctic alpine. In North America from Newfoundland to Alaska, south to Colorado and California. It's widespread in the Yukon.

Plant Parts Used
Flowers and leaves

Harvest Time
Summer

RIVER BEAUTY IS a close relative of fireweed (page 91), a favourite Yukon plant. My best memory of eating river beauty is on the shores of Bennett Lake, Yukon. After many days without fresh greens it was so nice to have a crisp, nutrient-rich salad with lake trout for dinner. The flowers and leaves are tasty when eaten raw or steamed. I find frying up the leaves in a bit of butter, with garlic and hemp seeds, simply delicious. I save the bright magenta flowers and use them to garnish the cooked greens.

River beauty is the national flower of Greenland, where it's known as *nivaqsiaq*, meaning "little girl." PL

RIVER BEAUTY

Chamerion latifolia

Other Names and Etymology
Willowherb, dwarf fireweed, broad-leaf fireweed. The Inuit of Baffin Island call it "broad-leaved willow herb," "dwarf fireweed," or *paunnat*. River beauty was formerly known as *Epilobium latifolium*.

Family
Onagraceae (evening primrose family)

Botanical Description
Perennial herb with decumbent stems 15–40 cm high, tufted. Leaves opposite, 1–6 cm or more long, glaucous. Flowers larger than in fireweed; in small clusters, petals 15–30 mm long, pink to purple; seed capsules 3–10 cm long.

As its name suggests, river beauty grows near rivers and streams. Its flowers make a colourful addition to a summer salad. MC left, RF below

Part II: *Herbs*

Habitat and Range
Frequent on gravel bars, stream banks, and reaching alpine elevations on scree slopes. Circumpolar, Arctic alpine. In North America from Newfoundland to Alaska, south to Colorado and California. It's widespread in the Yukon.

Plant Parts Used
Flowers and leaves

Harvest Time
Summer

R IVER BEAUTY IS a close relative of fireweed (page 91), a favourite Yukon plant. My best memory of eating river beauty is on the shores of Bennett Lake, Yukon. After many days without fresh greens it was so nice to have a crisp, nutrient-rich salad with lake trout for dinner. The flowers and leaves are tasty when eaten raw or steamed. I find frying up the leaves in a bit of butter, with garlic and hemp seeds, simply delicious. I save the bright magenta flowers and use them to garnish the cooked greens.

River beauty is the national flower of Greenland, where it's known as *nivaqsiaq*, meaning "little girl." PL

River Beauty

Medicinal Actions
Antihistamine, anti-inflammatory, antioxidant, astringent, digestive

Medicinal Preparations
Oil infusion, tea

Medicinal Uses
Herbalist Robert Rogers says river beauty tops contain steroid compounds that act as gastrointestinal astringents to soothe the digestive tract. It also contains various flavonoids, including quercetin, which has powerful antioxidant properties. It's also a natural antihistamine and anti-inflammatory.

Food Uses
River beauty leaves and flowers are very tasty on their own or with other wild greens as a summer salad. River beauty's inner stems were a choice food of the Nuxalk (Bella Coola people). The Inuit of Baffin Island eat the leaves raw or mixed with fat; they mix the flowers with crowberries and oil.

Nutritional Profile
Vitamins A and C

Other Uses
Steve Johnson, founder of the Alaskan Flower Essence Project and author of *The Essence of Healing*, says that "River Beauty is used for emotional devastation; overwhelmed by grief, sadness, or a sense of loss; shock and trauma from emotional or sexual abuse." He adds, "The healing qualities of this essence help with emotional recovery, reorientation and regeneration; helps us start over after emotionally devastating experiences; empowers us to use adverse circumstances as incentive for cleansing and growth."

Cautions
None reported.

> "May all your weeds be wildflowers."
> —Author unknown

River Beauty Summer Salad
Enjoy the wild flavours and crunchy texture!

Gather river beauty leaves and flowers and tear up into a bowl. Using your hands, rub the wild salad greens with a bit of olive oil.

ROSE

Rosa acicularis

Other Names and Etymology
Prickly wild rose; the Van Tat Gwich'in name for wild rose and hips is *nichih*. *Rosa* comes from the Greek word *rodon*, meaning "red," in reference to crimson red roses. For us in the boreal forest we relate this to the colour of their fruit, the "hips." *Acicularis* means "needle shaped."

Family
Rosaceae (rose family)

Botanical Description
Low shrubs with stems covered in thorns and prickles. Leaves of 3–7 leaflets, 1.5–5 cm long, toothed, with small soft hairs. Fragrant flowers, usually solitary on the stem; petals 2–3 cm long, pink; many yellow stamens in the centre of the flowers. Sepals prominent and green with glandular hairs on the back. Fruits are fleshy red hips, elliptic or pear shaped with browned sepals still intact and erect.

From bud to flower to rosehip, wild rose is beautiful from spring through autumn. BG

Rose

Habitat and Range
Frequent on riverbanks and in woodland clearings or burns. Nearly circumpolar. In North America it grows from Quebec to Alaska, south to New Mexico. It's common throughout most of the Yukon.

Plant Parts Used
Roots, leaves, flowers, fruits

Harvest Time
Leaves, roots, and stems: early spring. Flowers: summer. Fruit: autumn, after the first frost. (Note: When harvesting rose petals do not over-pick from one bush. When taking the flowers, leave one petal behind to ensure that the flower is pollinated, which will let a hip develop.)

WE HAVE AN ABUNDANCE of wild roses growing right outside our door. They actually surround the whole house, and I feel so blessed to live among them with all their blatant and subtle energies.

The wild-rose bush offers a diverse range of healing properties. Its appearance is uplifting as each delicate flower bursts forth, and its aroma is gentle yet strong.

Rosehips are gathered in the autumn after the first frost and are generally dried for use as a tea that is high in nutrients, such as vitamin C. Jam and jelly can also be made with the hips.

Rose petals and hips are used in many therapeutic cosmetic preparations. I met a wonderful couple from The Hills Health Ranch, in British Columbia, who make and sell amazing wild-pressed rosehip oil that is highly energetic, emollient, and healing.

Medicinal Actions
Flowers: emollient. Leaves: astringent. Fruit: antimicrobial (flavonoids), antiscorbutic (vitamin C), antispasmodic (flavonoids), astringent (tannins), diuretic

Medicinal Preparations
Bush bandage, hydrosol, infusions in water and oil, jam, jelly, syrup, tea/tisane

Part II: *Herbs*

Medicinal Uses

Rose petals make an excellent backwoods "bush bandage" for cuts and scrapes, and they fit perfectly over fresh bug bites. The petals help take the heat out of the wound and stop inflammation.

On a hot day, drinking rose-petal sun tea not only keeps you hydrated but gives the body extra energy to combat the heat. A cooled petal infusion can also be made into an eyewash for irritated eyes. When making the eyewash, be sure to strain all the bits out of the water so they don't go into the eye and further irritate it.

Since rosehips are high in vitamin C and bioflavonoids, they help keep the human body healthy. Rosehip syrup is used by herbalists to help heal anemia because the hips are mineral rich with trace amounts of iron and B vitamins. It is believed that rosehips stimulate production of red blood cells.

A decoction of rosehips is helpful for menstrual cramping because it's antispasmodic. Its antibacterial properties lend a hand—or hip—in helping with bladder or kidney irritations.

The bioflavonoids in the rosehips make for an excellent heart tonic. The tonic affects the heart and circulatory system and helps strengthen the capillaries. This makes it a good remedy for hemorrhoids and varicose veins, and it may help regulate the blood circulation throughout the body, helping with cold hands and feet.

In spring, before the petals open, an infusion of the astringent leaves can be used as a blood-cleansing tonic. The roots can also be used for this purpose.

High in vitamins and minerals, especially vitamin C, the leaves contain bioflavonoids and tannins that are good for treating stress, infection, diarrhea, thirst, and gastritis.

Rose-petal water can also help on many levels to balance the hormones. Take one teaspoon (5 mL) in one cup of water (250 mL) every day for symptoms of PMS or menopause. (For preparation, see Hydrosols, page 321.)

Food Uses

Most of the rose plant is edible. We tend to focus on the flowers and hips, but the leaves, the peeled thorny stems, and the roots can also be used for nutritional purposes.

The flowers and leaves can be added to salads, jams, jellies, and they look great on top of a cake: they really brought my wedding cake to life!

The flower petals are lovely in a hot infusion, but are equally lovely in a sun tea.

Rose-petal jelly is always a treat. It's light and delicate, yet complex and divine. It can be made with fresh petals in the summer or with the dried petals any time of the year.

The highly nutritious fruit can be used in many ways in the kitchen, most commonly in making syrup, jam, and jelly. Try using the hips as a base for ice creams, sorbets, or even as a sauce for fish or other wild meats.

Vitamin C is an essential nutrient most famous for its ability to strengthen the capillaries and connective tissues. Three rosehips contain the same amount of vitamin C as one orange.

My daughter, Markie-May, loved eating rosehips when she was a little girl. This photograph was featured on the Northwestel $5.00 phone card. CA

Rose

Rosa acicularis has the highest vitamin C content of any rosehip in the world: 7.1%!

Rosehips are also full of bioflavonoids, making it truly a "heart-smart" food.

Nutritional Profile
Hips: Calcium, iron, magnesium, niacin, phosphorus, potassium, protein, riboflavin, selenium, sodium, vitamins A, E, and C, and zinc

Cosmetic Uses
Rose petals have a humectant effect and help retain moisture in and on the skin. This makes rose excellent for dry, mature, and dull skin. Cosmetically, many varieties of rose have been used as a prized beauty treatment throughout the ages.

Rose hydrosol can be bought commercially or made at home (see Hydrosols, page 321.) Rosewater controls and balances sebum production, making it useful for both dry and oily skin. It can balance and restore the skin's pH and helps tighten pores. Its antibacterial properties help fight acne, giving troubled skin a gentle, rather than a harsh treatment. It's reputed to be useful in the treatment of all sorts of dermatitis.

At Aroma Borealis we have been making a delicate rose-petal face cream for over a decade. One of our long-time wholesale customers was quoted in an Ottawa newspaper as saying it's "the champagne of creams."

Rose petals can be added to apple cider vinegar or vodka and used as a facial tonic that is astringent and cleansing. One of my favourite pampering treats in the summer is a rose-petal facial steam; when I'm done, I use the water to soak my feet.

Rose petals can also be infused in oil and used as a base for massage and body oils, creams, and bath preparations.

Other Uses
I make a Heart Chakra Vibrational Essence that I love! Wild roses are shaped like little hearts (another sign from nature) and the heart chakra is the middle (the fourth) chakra in a system of seven. It's related to love and is the integrator of opposites in the psyche: Mind and body, male and female, persona and shadow, ego and unity. A healthy heart chakra allows us to love deeply, feel compassion for ourselves and others, and have a deep sense of peace and centredness.

Steve Johnson, founder of the Alaskan Flower Essence Project and author of *The Essence of Healing*, says that prickly wild rose (*Rosa acicularis*) is used when "lacking trust and faith; feeling hopeless; apathetic and disinterested in life; unable to keep the heart open when involved in adverse circumstances." Of its healing qualities, he writes, it "helps us remain openhearted when we are faced with conflict and struggle; builds trust; encourages openness and a courageous interest in life."

Dr. Edward Bach, the father of flower essences, created a wild-rose essence for "Those who without apparently sufficient reason become resigned to all

Below: Rose essence being made in a rose quartz singing bowl on the summer solstice. BG

Bottom left: Aroma Borealis's Heart Chakra Vibrational Essence. BG

Bottom right: Wild roses are shaped like little hearts. BG

Part II: *Herbs*

ADDITIONAL RECIPES

Afternoon Floral Tisane, page 281

Boreal Healing Bath, page 325

Flower-Power-Peace Bath, page 325

High-Calcium Infusion, page 282

Rosehip Syrup, High-Iron Syrup, page 299

Living Fruit Leather, page 377

Nourishing Vinegar Tonic, page 296

Pink Drink (High-C Smoothie), page 340

Raspberry and Rosehip Vinaigrette, page 362

Rejuvenating Facial Steam, page 329

Revolutionary Raspberry Rosehip, page 335

Rose-Petal Jelly, page 351

Rosehip Decoction, page 286

Spirit-of-the-Boreal-Forest Tea, page 281

Sun Tea, page 288

Wild-Rose Petal Vinaigrette, page 361

Wild-Weed Salad, page 365

"I once had a rose named after me and I was very flattered. But I was not pleased to read the description in the catalogue: no good in a bed, but fine up against a wall."

—Eleanor Roosevelt (1884–1962)

BG

that happens, and just glide through life, taking it as it is, without any effort to improve things and find some joy. They have surrendered to the struggle of life without complaint."

Rose essential oil has been prized in aromatherapy circles since before it was called aromatherapy! It takes 5,000 pounds of fresh rose petals to make one pound of oil!

Rose is ruled by the planet Venus (named after the goddess of love and all things beautiful). Venus rules the sun sign of Taurus, and those born under this sign are said to express their earthiness through their deep love of the planet Earth. (If you haven't guessed by now, I am a Taurus!)

Cautions

The seed hairs of the rosehips can be irritating to the digestive tract, so try not to ingest too many whole. That said, in the past the seeds were used to get rid of intestinal parasites. In many northern communities rosehips are often called "itchy bums" because that's what happens when you eat too many of them!

Wild-Rose Petal Healing Ointment

1 cup (250 mL) wild-rose petals
1½ cups (375 mL) almond oil or sunflower oil
½ cup (125 mL) jojoba oil
1 teaspoon (5 mL) vitamin E
1 to 2 oz. (30 mL to 60 mL) beeswax, depending on desired consistency

Place rose petals and oil in a double boiler. Warm slowly on a medium heat. Let simmer 20 to 40 minutes. Stir often. Strain petals and wipe the pot clean, so that there are no petals left. Add beeswax to pot and let melt. Add strained oil. Once blended, pour into a jar. Cap jar only after the ointment has cooled down and solidified.

WILD SAGE

Artemisia frigida
Artemisia tilesii

Other Names and Etymology
Artemisia frigida: wormwood, prairie sagewort, pasture sage, sage, mountain sage, female sage. *Artemisia tilesii*: The Van Tat Gwich'in of Old Crow, Yukon, call *A. tilesii gyùutsanh*. It is commonly called stinkgrass, stinkweed, mugwort, mountain wormwood, or mountain sagewort. The genus name *Artemisia* may be derived from the moon goddess Artemis, or it may be named after Artemisia, a botanist and medical researcher who died in 350 B.C. In Greek the word *frigida* means "cold bath."

Family
Asteraceae (aster, daisy, or sunflower family)

Wild sage has a pungent aroma and taste. Its silvery green leaves are unmistakable. BG

Part II: *Herbs*

Wild sage leaves help relieve tension in sore muscles—add them directly to a hot bath or infuse them to make an aromatic massage oil. CA

"(*Salvia boreale*) The Yukon wild sage which grows so profusely on the rocky, sunny hillside above Dawson is a diminutive member of the salvia or mint family.* Gathered and dried it is better for seasoning than the tinned powdered sage."

—Martha Louise Black, *Yukon Wild Flowers* (1940)

Botanical Descriptions

Mat-forming, strongly aromatic perennial herb, with a stout, fibrous and much-branched crown. Leaves crowded, highly dissected, with fine linear divisions, silvery-grey; upper stem leaves smaller and less dissected. Flowering stems usually arising from prostrate or ascending branches, 10–40 cm or more high, whitish-grey hairs. Flower clusters not showy, the bracts surrounding the flowers being the same colour as the leaves and stems.

Artemisia tilesii is much taller than *A. frigida*. The leaves are aromatic, sage-green in colour, and their shape is irregular and coarsely toothed, with a silvery underside.

Habitat and Range

Steep, open slopes and sandy river terraces. Central Asia and North America; it grows from southern Manitoba to British Columbia, and in the eastern part of Alaska and northeast to the Arctic coast in western NWT, and south to Arizona, New Mexico, and Texas. In the Yukon it's found north to the Dawson area, and also near the Porcupine River. *Artemesia tilesii* is found throughout the Yukon.

*Because wild sage is used like common sage, Black assumed it was also a member of the *salvia* or mint family (Lamiaceae). Interestingly, I thought the same thing when I first met this plant. As it turns out, it's actually a member of the Asteraceae family.

Wild Sage

Plant Parts Used
Flowers and leaves

Harvest Time
Spring through late summer

AS MENTIONED earlier, *Artemisia* may be derived from the name Artemis, the Greek goddess of the moon. She was also the goddess of feminine energy, herbalists, midwives, and the hunt, and was called upon to protect women in labour, small children, and wild animals.

Wild sage is often called "wormwood" and gets this common name from the fact that, traditionally, it was drunk as a warm tea to help expel worms (such as pinworms and roundworms) from the body. It's useful to note that this is not the same kind of wormwood as *Artemisia absinthium*, the plant used to make absinthe, a powerful alcoholic drink.

Wild sage is silvery in colour and pungent in aroma and taste. *Artemisia frigida* is traditionally used by First Nations for ceremonial purposes and for smudging. Wafts of smoke from the burning plant are said to clear negative energy. It's also one of the herbs added to kinnikinnick, a smoking mixture that includes bearberry and Labrador tea.

Artemisia tilesii, known as stink grass, is used for colds, muscle and joint pain, and as a wash for sores.

The potent and volatile oils of the *Artemisias* are used to repel mosquitoes. For most of us in the North, mosquitoes are an unavoidable part of the summer season. In fact, some refer to the mosquito as the Yukon's territorial bird! To give yourself some respite from the biting bugs, try rubbing the leaves on your neck, wrists, and ankles, or if you are sitting outside, burn sage in a bowl.

At my store, Aroma Borealis, we include an infusion of wild sage and yarrow in our all-natural insect repellent called Skeet-Addle.

Artemisia tilesii leaves are aromatic and have a silvery underside. CA

Part II: *Herbs*

> Sage and Wild Roses
>
> Take my hand I want to walk with you
>
> through sage and wild roses
>
> Aspens whisper love is new, trembling
>
> sage and wild roses
>
> —Nicole Edwards, *Sage and Wild Roses* (CD)

Medicinal Actions
Anthelmintic, antibilious, antimicrobial, aromatic, bitter, emmenagogue, hepatic, stimulant, tonic, vulnerary, vermifuge

Medicinal Preparations
Bath, gargle, hydrosol, massage oil, oil infusion, poultice, salve, tea, tincture

Medicinal Uses
As a cool tea, the bitter wild sage acts as a digestive tonic and is a potent agent to expel worms from the digestive system. A warm tea helps with painful or delayed menstruation.

Its antimicrobial actions make it a good choice for gargling to heal a sore throat and to wash wounds.

A light tea of the leaves and stalks can be drunk in moderation to treat sore throats or to help a sluggish digestive system.

Wild sage can be applied topically in a salve, poultice, or compress to aid the healing of cuts and wounds, and on other skin problems such as acne, boils, and blisters.

The leaves help relieve tension in sore muscles: Add to a hot bath or infuse in oil to make an aromatic massage oil.

According to Susun Weed, herbalist and author of *Healing Wise*, "*Artemisias* produce hundreds of different aromatic oils, including camphor and thujone. This keeps them bug-free and provides us with medicines. Aromatic oils in fresh *Artemisias* may be extracted into vinegar, vodka, or fat such as olive oil, and used externally to counter bacterial and fungal infections, and internally to prevent or cure digestive parasites in wo/man and beast. Small doses are said to improve appetite and digestion."

Wild Sage

Food Uses
Wild sage is a nice herb to add to campfire feasts. A wonderful northern spice will not only add a unique flavour to your meal, but it's also a great tonic for the digestive system.

I like to add wild sage and wild cranberries to turkey stuffing for a distinct northern flavour.

Nutritional Profile
Protein, crude fibre, traces of vitamin A

Other Uses
Steve Johnson, founder of the Alaskan Flower Essence Project and author of *The Essence of Healing*, says that mountain wormwood, *Artemisia tilesii*, is used in cases of "unresolved anger and resentment; cannot easily forgive the self or others for past actions regardless of the intent behind them." He adds that it "stimulates the healing of old wounds and the release of resentment; supports us in surrendering unforgiven areas within ourselves and in our relationships with others."

A sachet made of the leaves adds a lovely scent to clothing drawers and closets.

You can chew a tiny bit of the fresh leaves as a breath freshener.

Cautions
Use in moderation and small doses. High doses can cause stomach upset and cramping. Not recommended for young children, pregnant women, or people with severe allergies to the plants in the Asteraceae family.

The Hiker's Herb

How many times have you been out hiking and your feet have gotten sore? A simple remedy is to line the bottom of your shoes with some wild sage leaves. The result: Relief for aching arches and hiking boots you don't mind having in the tent!

> "There is hope if people will begin to awaken that spiritual part of themselves, that heartfelt knowledge that we are caretakers of this planet."
>
> —Brooke Medicine Eagle, Earth Wisdom Teacher and author of *Buffalo Woman Come Singing*

ADDITIONAL RECIPES

All-Purpose Botanical Ointment, page 313

Aromatic Herbal Balm, page 314

Cold-and-Flu Bath, page 326

Skin-Disinfectant Spray, page 295

SHEPHERD'S PURSE

Capsella bursa-pastoris

Other Names and Etymology
Poor man's pepper, lady's purse. The genus name *Capsella* means "little box" and the species name *bursa-pastoris* means "pastoral purse," in reference to the shape of the seed pods.

Family
Brassicaceae (mustard or cabbage family)

Botanical Description
Winter annual, with a thick central root; stems 10–50 cm high. Basal rosette of leaves 3–20 cm long; stem leaves alternate, small, clasping the stem. Loose flower clusters, sepals greenish or whitish, about half the length of the petals; petals four, 2–4 mm long, white or pinkish. Dry fruits with heart-shaped cases, lobes split off to release seeds; 4–8 mm long.

Above: Shepherd's purse leaves are delicious on their own or in a mixed salad. BG

Right: The plant has small, heart-shaped, purse-like seed pods. BG

Shepherd's Purse

Habitat and Range
Introduced plant that grows in disturbed soils, like gardens and along roadsides. Cosmopolitan in North America, it grows from Newfoundland to Alaska, south to Central America. In the NWT and Yukon it's scattered north to Dawson.

Plant Parts Used
Aeriel parts, leaves, roots, seeds

Harvest Time
Leaves: before flowering. Seeds: in late summer to snow fall. Note: Shepherd's purse is a winter annual, i.e., a plant that germinates in the autumn, lives over winter, produces its seed the following spring, and then dies.

CA

I USED SHEPHERD'S PURSE tincture almost immediately after the home birth of my youngest child. A midwife friend told me it was good to have shepherd's purse on hand because its styptic qualities can help control any hemorrhaging that may occur. So, in between birthing the baby and placenta, I ingested some. I knew if my body needed it, it would utilize it. The tincture is a good addition to any midwife's birthing kit and to your home medicine chest.

I have personally used the herb since and have also recommended it as part of a blend to my clients who have excessive and prolonged menstruation. It blends well with yarrow flowers, raspberry leaf, crampbark, and nettle leaf.

Medicinal Actions
Alterative, anti-inflammatory, astringent, diuretic, haemostatic, styptic, uterine stimulant, vulnerary

Medicinal Preparations
Infusion, ointment, poultice, spit poultice, tea, tincture, vinegar, wash

Medicinal Uses
For medicinal usage, the whole plant in flower (the aerial parts) is used. Dried shepherd's purse loses its quality within six months, so it's better to use it fresh or to tincture it while fresh.

Shepherd's purse acts as a mild diuretic and can be used to treat water retention due to mild kidney problems and to help to cleanse the kidneys. It acts as a gentle tonic for strengthening the body.

One of its most impressive attributes is that it's extremely high in vitamin K, an

Part II: *Herbs*

> Shepherd's purse seeds are reported to be toxic to mosquito larvae when put in water. So you may wish to add some seeds to your garden water barrel to stop it from becoming a mosquito breeding ground.

effective blood clotter, and that makes the plant excellent for halting internal and external bleeding and hemorrhages.

Shepherd's purse is indicated for helping to stimulate menstruation, while also being used to decrease excess flow or flooding. It can be used while menstruating for menorrhagia (excessive bleeding), endometriosis, is effective for postpartum bleeding, and can be used throughout menopause.

Shepherd's purse is recommended for topical use on cuts, abrasions, and for nosebleeds. Because it acts as an astringent it's good to help stop excessive diarrhea. It's also noted in a few references that it may help lower blood pressure and act to regulate blood pressure.

Food Uses
The basal green leaves are not only nutritious but have a taste reminiscent of spinach. They are good added to a salad, in a sandwich or steamed as a side green. Shepherd's purse is also known as "poor-man's pepper" because the seed pods have a peppery taste that lends itself nicely as a cooking spice. The dry ground-up pods can then be used in cooking.

Nutritional Profile
The leaves are very high in beta carotene (A), vitamin K, niacin, iron, and rutin, thiamine (B1), choline, inositol, and fumaric acid. They are a good source of ascorbic acid (C), riboflavin (B2), calcium, potassium, and phosphorus.

Cosmetic Uses
Shepherd's purse can be used to help balance oily skin. Either steam or wash the face with a regular infusion. For long-term use you can make up a batch of Shepherd's-purse facial tonic by infusing the herb in witch hazel for four weeks. After four weeks, strain, bottle, and label the liquid as a facial astringent and cleaner.

Other Uses
According to Herbalist Michael Tierra in his book *The Spirit of Herbs: A Guide to the Herbal Tarot*, the spiritual properties of shepherd's purse help to, "invigorate the life force energy within and helps one to recognize and implement talents and abilities that have been hidden. It strengthens the will and spiritual tenacity." He adds, "A talisman of the seeds can be worn or the flower essence can be taken to inspire hope and opportunity."

Cautions
This plant should not be used during pregnancy or labour (until after the baby is born).

Shepherd's Purse Tincture
Add fresh shepherd's purse herb (cut up the entire aerial part of the plant) to a clean jar, cover with vodka and shake every day for 4 to 6 weeks. Strain, bottle and label. (See Tincture Method, page 293.)

MOUNTAIN SORREL

Oxyria digyna

Other Names and Etymology
Alpine sorrel, scurvy-grass. The name is from the Greek *oxys*, meaning "sour or acid," *di* refers to the two partitions of the fruit, and *gyna* is derived from *gyne*, meaning "female organ system" or "woman."

Family
Polygonaceae (buckwheat family)

Botanical Description
Perennial herb with most leaves basal and flowering stem mostly leafless, 10–30 cm high. Leaves alternate, somewhat succulent, kidney-shaped, long-petioled, 0.5–5 cm long. Flowers tiny (1.5–2.5 mm long), numerous, in loose clusters 2–20 cm long. Dry one-seeded fruits 3–6 mm wide, flattened, wing-margined.

Habitat and Range
Moist alpine and tundra situations, often around late snow beds. Circumpolar, Arctic alpine. In North America it's found from Newfoundland to Alaska south to Colorado. In the Yukon it's found in mountain and tundra habitats, north to the Arctic coast.

The sour taste of mountain sorrel leaves indicates the presence of vitamin C, a powerful antioxidant. L to R: CA, BL

Wild Food and Medicine Plants of the North

Part II: *Herbs*

The fleshy leaves of mountain sorrel can be eaten right off the plant. CA

Plant Part Used
Leaves

Harvest Time
Leaves: late spring through late summer

GATHERING MOUNTAIN SORREL leaves in early summer is always a treat. The sour taste indicates the presence of vitamin C that acts as an antioxidant in the body. Eating wild foods feeds the body on so many levels. Just think: it's pure nourishment from the planet—food that hasn't been touched by any other human hand before yours and was grown solely by Mother Earth. The energy is pure, wild, and free, and it feeds us on so many levels.

Mountain Sorrel

Medicinal Actions
Antiscorbutic, astringent

Medicinal Preparations
Oil, poultice, vinegar, wash

Medicinal Uses
As an astringent herb, sorrel can be used to reduce or balance secretions and discharge from the body, such as diarrhea. The leaves are high in vitamin C and have been used to prevent scurvy.

Topically it can be used to help alleviate the itch and rashes of bug bites by simply making a spit poultice and placing it directly on the bite.

Food Uses
The fleshy leaves and young stems can be eaten raw, right off the plant, as a trail snack or gathered and brought home to make various dishes, like vegetarian lasagna, wild-weed soup, stir-fries, or sandwiches. The zesty flavour adds a unique and interesting taste to salad oils and dressings.

A traditional food plant, the Inuit fermented mountain sorrel into a sauerkraut-like food to eat in winter.

Nutritional Profile
Vitamins A, B, C, and trace minerals including iron

Cautions
Sorrel leaves contain oxalates, an acid that is also found in spinach, Swiss chard, potatoes, and rhubarb. Cooking and freezing the plants will help break down the acids. (For more information on oxalic acid, see page 27).

> "The more clearly we can focus our attention on the wonders and realities of the universe about us, the less taste we shall have for destruction."
>
> —Rachel Carson, biologist, environmentalist, and author of *Silent Spring* (1907–1964)

ADDITIONAL RECIPES

Green Sauce, page 360

Wild-Weed Dressing, page 361

Mountain Sorrel Iced Tea

In a large teapot, infuse a handful of mountain sorrel leaves in hot water for up to fifteen minutes or until cooled, strain, sweeten with honey or birch syrup, and refrigerate. Once cold, pour into a glass and enjoy!

SHEEP SORREL

Rumex acetosella

Other Names and Etymology
Common sheep sorrel, field sorrel, red sorrel. *Rumex* is from Latin meaning "a lance" in reference to the shape of the leaves. In Greek, *acetosella* means, "vinegar salts," and this refers to the acidity of the leaves.

Family
Polygonaceae (buckwheat family)

Botanical Description
Dioecious perennial; stems succulent, erect, 10–60 cm high. Stem leaves with short petioles; basal leaves long-petiolate. Leaf shape variable but at least some arrowhead-shaped, 1–8 cm long. Flowers small, numerous, in loose clusters. Dry one-seeded fruits 1–2 mm long.

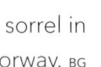

Right: Sheep sorrel in northern Norway. BG

Below: Sheep sorrel can be eaten raw or cooked. BG

Sheep Sorrel

Habitat and Range
Likes to make its home in disturbed soils. Naturalized from Europe and grows across Canada. It grows along the Yukon–Alaska border near Fraser. It's been reported throughout the Southern Yukon and along the Alaska coast.

Plant Parts Used
Leaves

Harvest Time
Late spring through late summer

SHEEP SORREL'S CLAIM TO FAME is its blood cleansing, detoxifying, and cell-regenerating properties. It's one of the key ingredients in the Canadian herbal cancer remedy Essiac Tea, which has its roots in Ojibwa medicine. Essiac Tea is a blend of herbs created by the late Rene Caisse, a Canadian nurse, who came up with the product name by spelling her last name backward. The "tea" contains not only sheep sorrel but other wonderful herbs such as burdock root, slippery elm bark, and Indian rhubarb root.

Yukon's famed birch-syrup guy, Berwyn Larson, says that sheep sorrel is one of his favourite edible wild plants, "It tastes like crab apples. I eat it straight up. I often find it in the mountains by a creek—stop and get a drink, and hey, there it is!" He advises when harvesting it, "Try to get the crown—breaks nicely off the root, chew it, eat the whole thing, it tastes lemony. It's super nutritious. There are so few wild vegetables that give you calories and a few nutrients."

Medicinal Actions
Anti-inflammatory, antioxidant, antiseptic, astringent, diuretic, hepatic, laxative, vermifuge

Medicinal Preparations
Mouthwash, poultice, tea, tincture

Medicinal Uses
Infusions of the leaf can be used as an astringent mouth rinse for sore gums and cankers. The herb also contains several anthraquinones that are effective antioxidants and radical scavengers. A tea made from the leaves acts as a diuretic and also as a vermifuge that will help to expel worms from the digestive system. Intestinal worms are thought to have no resistance to the properties of this herb.

> In the 1868 *Canadian Pharmacy Journal*, the leaves of sheep sorrel were included in the list of Canadian medicinal plants.

Part II: *Herbs*

> "You have to leave the city of your comfort and go into the wilderness of your intuition. What you'll discover will be wonderful. What you'll discover is yourself."
>
> —Alan Alda, actor

Considered a very useful herb for women, in China raw sheep sorrel is given as a tea after birthing to "cool" the reproductive area and prevent infection. It's said to contain significant levels of phytoestrogens that have estrogen-receptor binding activities, similar to those found in red clover, licorice, and soy.

Sheep sorrel is also considered a good tea remedy for profuse menstruation because of its high tannin content.

When taken in low doses, the astringent action of sorrel tea is useful for treating diarrhea. At higher doses it's good to relieve constipation. A tart, sorrel-leaf infused tea is best drunk cold. It will cool you on a hot summer day, or if you are feverish.

Because sheep sorrel enduces sweating, it has traditionally been used by herbalists to help cool the body and promote mild detoxification through the skin. Its diuretic properties help to maintain healthy kidney and urinary functions. It's also thought to nourish the glandular system.

Sheep sorrel is considered the most active herb in the Essiac Tea formula for stimulating cellular regeneration, detoxification, and cleansing.

Nicholas Culpeper recommended that the leaves be used for their diuretic property and taken as an infusion with a dosage of 1 oz. (28 g) to 1 pint (568 mL) of boiling water, in doses of 2 fluid oz. (56 mL). He recommended the leaf juice as a tonic for the kidneys and urinary tract taken in doses of ½ to 1 teaspoon.

The finely chopped leaves can be used as a face mask or in a steam for decongesting sinuses.

Food Uses

Once used to prevent scurvy, today sheep-sorrel leaves can be chewed as a breath freshener. Sheep sorrel's tangy, lemon-like flavour makes it a delectable herb to eat as a green vegetable either raw or cooked.

The Sami used the acidic effect of sorrel leaves as a rennet substitute in cheese making. In Europe, sheep sorrel is used to curdle the whey for ricotta cheese.

Sheep sorrel was considered a delicacy in the early twentieth century, and was used in making dill pickles.

Nutritional Profile

Vitamins A, B, C, D, E, and K, calcium, beta carotene, phosphorus, potassium, manganese, copper, iron

Cautions

Sheep sorrel contains oxalates, an acid that is also found in spinach, Swiss chard, potatoes, and rhubarb. Cooking and freezing the plants will help break down the acids. (For more information on oxalic acid, see page 27.)

Sheep Sorrel

BG

Borealis Green-Goddess Dressing

2 cups (500 mL) of fresh sheep sorrel leaves
1 cup (250 mL) hemp oil
1 cup (250 mL) of olive oil
2 tablespoon (30 mL) apple cider vinegar
Handful of wild chives

Blend ingredients together. Store in covered jar or bottle in the refrigerator. Lasts for up to three weeks. Shake before using.

ADDITIONAL RECIPE

Wild Dock Sorrel-Potato Soup, page 90

STRAWBERRY BLITE

Chenopodium capitatum

Other Names and Etymology
Strawberry spinach, strawberry goosefoot. The Dene (Slavey) call it *tsa dzhi* meaning "beaver berry." *Chenopodium* is from the Greek *chen*, meaning "goose," and *podos*, meaning "foot." *Capitatum* is derived from the Latin word *caput* meaning "head" or *capitate* meaning "head shaped."

Family
Chenopodiaceae (goosefoot family)

Right: Strawberry blite fields forever! PL

Below: In Victorian times, strawberry blite was prized as an ornamental food plant. CA

Strawberry Blite

Botanical Description
Annual, with stems erect or ascending, simple or branched, up to 60 cm high. Leaves up to 10 cm long, shorter above, triangular to arrowhead shaped, sometimes toothed, somewhat fleshy, long-petioled toward base. Unusual-looking flower clusters yellowish-green, juicy and globose in shape; becoming more succulent and bright red.

Habitat and Range
Clearings, roadsides, and disturbed soil situations. In North America, from Nova Scotia to central Alaska, south in the western United States; in the Yukon, north to the Dempster Highway.

Plant Parts Used
Leaves and seeds

Harvest Time
Leaves: summer. Red flower/seed heads: blooming

STRAWBERRY BLITE is an amphi-Atlantic plant, meaning it is native to both sides of the Atlantic Ocean, Europe and North America. In the early 1800s it was valued and used as an ornamental food plant, which it is!

As is obvious when you look at its leaves, strawberry blite is in the same family as lamb's quarters and can be used in much the same way.

While there isn't very much information available on strawberry blite's medicinal uses, the plant's bright red flowers taste like a juicy berry and are great on their own as a snack or added to summer salads.

BG

Medicinal Preparations
Decoction, tea/infusion, mouth rinse

Medicinal Uses
Strawberry blite is useful in cases of nutritional deficiencies. When decocted, the whole plant is used for healing mouth and throat ulcers; or cooled and used as a wash or compress for hemorrhoids.

Part II: *Herbs*

> "What is a weed? A plant whose virtues have not yet been discovered."
>
> —Ralph Waldo Emerson, author (1803–1882)

Strawberry blite's bright red flowers taste like juicy berries. BG

Food Uses

The seedy red flowers are sweet and tasty raw or cooked.

I have heard that a bright red jelly can be made with flower heads; I invite you to try making it by following any of the flower-jelly recipes in this book (see Jam, Jellies, Chutney, and Topping Syrups, page 348).

For information on how to use strawberry blite's leaves as food, see lamb's quarters, page 111.

Other Uses

Strawberry blite is loaded with bright red flower/seed heads that stain red, and were used by First Nations for decorating baskets, porcupine quills, marking beadwork patterns on moose hide, and rubbing on the body for ceremonial purpose.

In fact, the Thompson of the Interior of British Columbia called strawberry blite "smeared blood top plant," an apt description of its staining ability.

Spruce roots, used for baskets, were also dyed red with the strawberry blite fruits.

Cautions

Strawberry blite leaves contain oxalates, an acid that is also found in spinach, Swiss chard, potatoes, and rhubarb. Cooking and freezing the plants will help break down the acids. (For more information on oxalic acid, see page 27.)

Strawberry Spinach Smoothie

1 cup (250 mL) red juicy strawberry blite flowers
1 cup (250 mL) yogourt
½ cup (125 mL) milk (or rice, soy, almond, or hemp beverage)

Gather fresh strawberry blite flowers. Add all ingredients to blender and blend until smooth.

Makes enough for 2 people.

SWEETGRASS

Anthoxanthum hirtum
Anthoxanthum monticola

Other Names and Etymology
Vanilla grass, holy grass, alpine holy grass, sacred grass.

Anthoxanthum is from the Greek *anthos*, meaning "flower" and *xanthos* meaning "yellow." Its former name, *Hierochloë* is from the Greek *hieros*, meaning "sacred," and *chloa*, meaning "grass."

Family
Poaceae (grass family)

Botanical Descriptions
Anthoxanthum hirtum: A perennial grass with stems reaching 20–60 cm high, arising from a creeping rhizome. The sweet fragrance of grasses in this genus distinguishes them from other grasses. Basal leaves narrow, tapering at the tips. Sheaths at the base of leaves not conspicuously purplish. *Anthoxanthum monticola*: Also known as alpine holy grass, sweet and fragrant. Stems reaching 10-30 cm high. Basal leaves narrow, curling upwards at their margins, and with sheaths at the base of leaves conspicuously purplish. Stem leaves short, wider. Flowering clusters 2–5 cm long.

Habitat and Range
Anthoxanthum monticola: Usually found on dry tundra and in rocky alpine habitats. *Anthoxanthum hirtum:* Found along sandy stream banks, river edges and lakeshores. Both are circumpolar plants. In North America they grow from Newfoundland to Alaska and British Columbia, and throughout the Yukon.

Sweetgrass's medicine lies in its spiritual healing qualities. BG

Wild Food and Medicine Plants of the North | 167

Part II: *Herbs*

The sweet fragrance and the telltale pinkish sheaths of sweetgrass distinguish it from other grasses. BG

Plant Parts Used
Aerial parts of the plant

Harvest Time
Throughout the summer

WHEN OUT IN THE WILDS, it is usually the scent of sweetgrass that catches me first and then I look for the plant with its telltale purplish base. I then give a blade of the grass a rub; if it releases a distinctive, sweet smell I know I've found the sweetgrass.

I vividly remember the first time the smoke from a sweetgrass smudge wafted past my nose. I was a young girl, maybe thirteen years old, attending a First Nations gathering in Ontario. The sweet, vanilla-like fragrance intrigued me, and I had a realization then of just how old this tradition was, that it had been used by indigenous cultures around the world. Since then, I have been honoured by being invited to many sweat-lodge ceremonies, an ancient ritual where each participant is cleansed with a sacred sweetgrass smudge before entering the lodge.

Sweetgrass grows along the shores of Bennett Lake, along the Porcupine River and the Yukon River. If you're visiting the Canadian historical site of Hootalinqua, along the Yukon River, note the rising smell of the coumarin from the sweetgrass as you pull ashore. It's a very pleasant and welcoming scent, almost as if you are being cleansed before entering the site.

Medicinal Actions
Astringent

Medicinal Preparations
Cream, hydrosol, oil, salve, smudge, tea

Medicinal Uses
Sweetgrass's medicine truly lies in its spiritual healing qualities. However, it has physical healing properties as well.

As an astringent herb, when it's ingested or used topically, it can reduce secretions and discharges. Sweetgrass contains tannins that in turn cleanse the body and assist in detoxifying by discharging excessive morbid (stagnant) matter and help achieve firmness of tissue.

A sweetgrass infusion can help with a sore throat when drunk or used as a gargle. When applied topically in a cream it can relieve chapped skin that is caused by harsh,

Sweetgrass

dry temperatures and wind. The First Nations of the Prairies use sweetgrass as a cleansing smudge and for infections, post-partum bleeding, and as a wash for eye irritations.

Food Uses
Sweetgrass can be used as a tea or to flavour blended teas that can be added to delicate cakes for a subtle, vanilla-like flavour; use the tea as a replacement for water in a recipe.

Other Uses
First Nations from all over Canada use sweetgrass for smudging, a sacred washing and cleansing of oneself in smoke. A smudge is used by individuals, in sacred ceremonies, and to purify a space. When smudging, the dry herb is placed in a bowl, often a shell, and set alight. The flame is put out and the herb starts to smolder. The smoke then rises and is fanned towards the person being cleansed.

Personally, I like to smudge every day. I keep a smudge bowl in front of my wood stove to have my spiritual shower. On some days I use sweetgrass, and on others wild sage, cedar, white sage, or copal that I brought back from Mexico.

The First Nations tradition of braiding the sweetgrass has deep and significant teachings attached to it, each piece of grass that is used has meaning, as does each section of the three braided strands.

> "Death must be so beautiful. To lie in the soft brown earth, with the grasses waving above one's head, and listen to silence. To have no yesterday, and no tomorrow. To forget time, to forgive life, to be at peace."
>
> —Oscar Wilde, author,
> *The Canterville Ghost* (1887)

Sweetgrass is used to make cleansing smudges. BG

Part II: *Herbs*

My friend Steve Williams makes a sweetgrass hydrosol that is used to cleanse and create sacred space. The hydrosol is great when in places where you can't burn smudge: such as in hospitals, workplaces, and confined spaces.

Steve Johnson, founder of the Alaskan Flower Essence Project and author of *The Essence of Healing*, says sweetgrass (*Hierochloë odorata*) is used for "energy blockages in the etheric body; low-energy flow during the day; difficulty bringing a healing process to final completion." He adds that its healing quality "cleanses and rejuvenates the etheric body; brings lessons and experiences to completion on the etheric level; removes disharmonious energies from our home or work environments."

Sweetgrass is used in basket weaving and other art forms.

Cautions
Sweetgrass contains coumarin-like compounds that may accentuate effects of blood thinners, such as Aspirin or warfarin.

BG

Sacred-Spirit Sweetgrass Cream
A wonderful healing cream that can be used from head to toe.

1½ cup (375 mL) sweetgrass-infused sunflower or grapeseed oil (see Medicinal Oils, page 304)
⅔ cup (160 mL) of coconut oil
1 oz. (28 g) beeswax

Water mixture
1⅓ cup (325 mL) distilled water or sweetgrass hydrosol
⅔ cup (160 mL) aloe vera gel
15 drops lavender essential oil (optional)

Place sweetgrass infusion, coconut oil, and beeswax in a double boiler over low heat.

Heat ingredients just enough to melt them.

Pour into the blender, and let cool to room temperature, intermittently turning on the blender to mix the ingredients.

With the blender mixing, slowly pour the water mixture into the blender. It's done when the water-and-oil blend looks creamy.

Pour into jars, let cool before putting the lids on, and label.

ADDITIONAL RECIPES

Aromatic Herbal Balm, page 314

Flower-Power-Peace Bath, page 325

Spirit-of-the-Boreal-Forest Tea, page 281

TWINFLOWER

Linnaea borealis

Other Names and Etymology
The Dena'ina call it *k'ela tl'lia*, meaning "mouse's rope." *Linnaea* is named after the Swedish botanist Carl Linnaeus. Twinflower was one of Linnaeus's favourite plants as a young man. He used the twinflower as his own personal symbol and included it in his coat of arms when he was raised to the Swedish nobility in 1757. Twinflower is also the provincial emblem of Linnaeus's home province, Småland. *Borealis* is from Latin, meaning "of the north" or "Nordic."

L. borealis

Family
Caprifoliaceae (honeysuckle family)

Botanical Description
Dwarf shrub; stems freely branching and trailing, rooting at nodes. Leaves oval with a few rounded teeth, almost stalkless. Pink, funnel-shaped flowers nodding in pairs from a stalk, sweet-scented, 1.5–6.5 cm high; fused petals forming funnel 6–12 mm long. Pairs of dry fruits replace the nodding flowers on the stalks; 1.5–3 mm long, hairy, one-seeded.

Habitat and Range
In open woods, or mossy and turfy openings in thickets. Twinflower is a circumboreal/circumpolar plant. Two subspecies have been delineated, based on subtle differences in the shape of the funnel-flowers. Subspecies *americana* grows from Labrador to Alaska (north to about the treeline), south to California; it's found throughout much of the Yukon territory. Subspecies *borealis* grows in Eurasia and Alaska, and has been found growing in Dawson City and along the Dempster Highway.

The Doctrine of Signatures hints at twinflower's medicinal uses. CA

Part II: *Herbs*

> **T**winflower fruits are tiny stick burrs that spread themselves by attaching to animals.

Plant Parts Used
Aerial parts of the plant; leaves, flowers

Harvest Time
Summer, as they flower. As they are very delicate to gather, I like to use scissors to cut them from the base of the stem so as not to pull up the root system so the twinflowers can continue to flourish and grow. This way you get the leaves and flowers together.

> This photograph was taken by Fritz Mueller, a Yukon photographer, the day his wife, Teresa Earle, went into labour. They had twin girls. FM

THE DELICATE PAIR of pink, trumpet-like twinflowers are so lovely that when you lie down and look at them it feels as though you have entered some magical fairy land. At first glance, the leaves look like those of wild cranberries, but unlike the wild cranberries that grow in clusters, you will notice the twinflower's leathery leaves are attached to little runners that form a mat-like community.

Twinflower

Medicinal Actions
Galactagogue, astringent

Medicinal Preparations
Tea/infusion

Medicinal Uses
The Doctrine of Signatures gives us hints as to what twinflower's medicinal uses are. Traditional usage seems to be indicated for organs or body parts that come in pairs. It could be a coincidence, but twinflower tea is good for coughs and colds (the lungs), increasing breast milk (the breasts), and for sore limbs (legs or arms).

Other Uses
Steve Johnson, founder of the Alaskan Flower Essence Project and author of *The Essence of Healing*, says that twinflower is indicated for: "defensive or reactive communication; unable to clearly communicate from different aspects of the self; has difficulty understanding what others are saying." He adds that twinflower "promotes balance in communication; helps us learn to listen and speak to others from a place of inner calm and focused neutrality."

> "Earth laughs in flowers..."
>
> —Ralph Waldo Emerson, poet,
>
> "Hamatreya" (1845)

Twinflower Vibrational Essence

Fill a glass bowl with pure spring water. Choose an area where flowers are growing abundantly. When you are feeling attuned to the spirit and vibration of the flowers, carefully snip the blossoms with scissors, allowing them to fall randomly onto the surface of the water. Make sure your hands don't touch the flowers or water as to not interrupt the energetic flow, or contaminate the water.

Set the bowl down in front of the plants you have picked from and allow the sun to penetrate the water and flowers for about three hours, or until the blossoms begin to look faded or spent. Gently remove flowers with tongs. Measure the flower water, pour into a jar, and add equal amounts of brandy.

This preparation is called the "mother essence." The mother essence is used to make smaller stock bottles that you can share with others. These are prepared by adding 2 to 4 drops of mother essence to 30-mL (2 tablespoons) dropper bottles filled with brandy.

To use the flower essence, add 2 to 4 drops to 1 cup (250 mL) of water and drink. Flower essences are a wonderful way to shift gears emotionally, mentally, physically, and spiritually.

VALERIAN

Valeriana capitata
Valeriana dioica

Other Names and Etymology
Tobacco root, wild heliotrope, marsh valerian, little root. *Valerian* is derived from the Latin *valor*, meaning "courage," or the Latin *valeo*, "to be strong" or *valere*, "to be healthy." *Capitata* comes from the Latin *capitatus*, meaning "having a head," in reference to the flowers, while *dioica* in Latin means "two-leafed."

Family
Valerianaceae (valerian family)

Botanical Description
Perennial herb, stems 30–100 cm or more high, leafy. Lowermost stem leaves and those of sterile shoots oval or heart-shaped with slender petioles; middle or upper stem leaves cut into narrow segments, stalkless.

Valerian can be used to help treat insomnia. L to R: BL, RF

Valerian

Flower clusters in dense heads, opening into a diffuse cluster of fruits; petals 4–7 mm long, pink to white. Clusters of dry, single-seeded fruit 3–4 mm long.

Valerian is derived from the Latin *valor*, meaning "courage," or the Latin *valeo*, "to be strong" or *valere*, "to be healthy." PL

Habitat and Range

Moist, turfy woodland bogs, tundra, and river flats. From Siberia to Alaska, and the Yukon, extending eastward to the east slopes of the Mackenzie and Richardson mountains of the Northwest Territories. In Yukon territory, found mostly west of longitude 135° west and north to the Arctic coast.

Bruce Bennett, a Yukon biologist, explains that *V. dioica* is found in bogs, lakeshores, and riverbanks, but becomes "replaced by *V. sitchensis* in the subalpine and throughout the mountainous regions of the south."

Plant Parts Used

Roots

Harvest Time

Spring or autumn

I LOVE VALERIAN SO MUCH, I always have it growing in my garden. What I love most about it is that it likes to spread. I planted the domesticated European variety of valerian (*Valeriana officinalis*), but over the years it has adapted quite well to our cold northern climate. This gives me an ample supply of root to use in my remedies.

Even though *V. officinalis* is the "official" medicinal valerian, the other variations that grow in the North are also just as effective as medicinal plants.

The first time I dug the wild roots, I was amazed at how pungent and sweet smelling they were. The aroma hit my olfactory system and I felt calm, yet alert. Since that day I have used valerian in teas, tincture (my preferred method), and as an essential oil to add to blends that require the sedative qualities that valerian offers.

Interestingly, my cat also has an affinity for the smell of the plants and loves to roll around in my valerian patch. What I find fascinating about this is that herbalist Robert Rogers mentions that various First Nations traditionally used valerian root in traps for cougars, mountain lions, and lynx—big cats that are attracted by valerian's scent.

Medicinal Actions

Analgesic, anti-inflammatory, antispasmodic, aromatic, carminative, hypotensive, hypnotic, nervine, sedative, stimulant, tonic

Part II: *Herbs*

> "Valerian" is said to mean "an accommodating disposition."
>
> —Margaret Pickston, author, *The Language of Flowers* (1968)

Medicinal Preparations
Capsule, essential oil, syrup, tea/decoction, tincture

Medicinal Uses
Valerian is monographed in some of the most renowned pharmacopoeias in the world, including those of Austria, Brazil, Egypt, France, Germany, Greece, Hungary, Italy, the Netherlands, Norway, Romania, Russia, Switzerland, and Yugoslavia.

Valerian's small yet sweet-and-pungent smelling roots help tone and strengthen the nerves. This makes valerian a good candidate for alleviating stress, tension, anxiety, nervous excitement, nervous sleeplessness, and nervous palpitations.

It's effective for people who suffer from insomnia, and can help create a calming, restful, and healing sleep. For sleeplessness, valerian is best taken before bedtime.

Valerian also acts as a pain reliever that is useful when there is pain associated with tension—like migraine headaches and rheumatic pain. Its antispasmodic properties can help relieve uterine cramping during painful menstruation. Its carminative properties make valerian useful for intestinal colic and gas.

For treating insomnia, valerian combines well with wild chamomile. For cramps it works well with crampbark, raspberry leaf, and nettle.

Dosage is an issue with valerian root, and more is not always better. If you use an excessive amount, the results can be the opposite of your intention: For instance, you may be awake all night instead of getting a good night's sleep. So to start, it's recommended to try a small dosage.

For an infusion of valerian roots: Pour a cup of boiling water over 1 teaspoon of root and let infuse for 10 to 15 minutes. Drink as needed. (Note: Although the general rule with roots is to decoct them, with valerian root an infusion works better.)

To use the tincture: Use up to 2 to 4 mL (½ teaspoon to 1 teaspoon) of tincture a day.

Because valerian affects different people in different ways, it's important to observe how you react to it. If you get drowsy soon after ingestion, it's important not to operate machinery or drive after use. Just go to bed and have a great sleep!

> Herbalist, physician and astrologer Nicholas Culpeper (1616–1654) wrote in his book *Culpeper's Complete Herbal* that valerian is under the influence of Mercury.

Valerian

Cosmetic Uses

Adding a cup of valerian root infusion to a bath at the end of a long day helps sooth frayed nerves or sore muscles. It can also be mixed with other herbs, like wild chamomile and rose, and added to a cotton bag and infused in the bath.

When added to shampoo, valerian-root tincture can help lift dandruff from the scalp and is reputed to darken hair. Acne-prone complexions can also benefit by washing with a light valerian infusion or dabbing it directly on pimples.

Other Uses

Steve Johnson, founder of the Alaskan Flower Essence Project and author of *The Essence of Healing*, says that *Valeriana officinalis* helps us "slow down in order to gain perspective on our priorities, especially when we feel under pressure to do or decide; promotes harmony in relationships; helps groups find a peaceful common ground."

Cautions

Do not mix with pharmaceutical antidepressants or prescription sedatives. High doses of valerian can cause headaches. Not for prolonged use. It is noted in many field guides that wild valerian root should be fully dried before being used medicinally.

> Juliette de Bairacli Levy (1912–2009), a pioneer of modern herbal medicine, said that "excessive dependence on valerian causes headaches, mental agitation, much restlessness, and severe cases of delusion."

Valerian-Root Tincture

1 part dried valerian root
2 parts vodka

Put roots in a jar. Cover with vodka. Shake every day for 4–6 weeks. Strain, bottle, and label. (See Tinctures, page 293.)

ADDITIONAL RECIPES

Aromatic Herbal Balm, page 314

Good-Sleep Infusion, page 283

Pain-Aid Capsule, page 301

Pain-Aid Tincture, page 294

Valerian's small yet sweet- and pungent-smelling roots help alleviative stress, tension, and sleeplessness. The roots should be fully dried before being used medicinally. BG

YARROW

Achillea millefolium

Other Names and Etymology
Wound wort, staunchweed, herbe militaris, milfoil. The Van Tat Gwich'in name for yarrow is *ch'at àn dagàii*. *Achillea* (Latin) may be named after the Greek doctor who recorded the medicinal usage of yarrow, or after Achilles, who used yarrow to heal his soldiers' wounds in battle. *Millefolium* is derived from Latin, meaning "thousand-leafed." Yarrow is derived from the Dutch *yerw*, meaning "to repair."

Family
Asteraceae (aster, daisy, or sunflower family)

Botanical Description
Stems simple or somewhat forked above, 20–60 cm high, with long soft hairs. Leaves in basal rosettes and alternate on stems, 3–15 cm long, highly dissected and fern-like in appearance, fresh green in colour. Basal leaves with long petioles, lower stem leaves with petioles, but middle and upper stem leaves stalkless, reduced upwards. Fragrant flowering heads numerous, in flat or round-topped clusters; petals white or cream coloured.

A circumpolar plant, yarrow is easily identified by its fern-like leaves and clusters of fragrant, white flowers. L to R: BG, CA

Yarrow

Habitat and Range
Meadows, sandy slopes, dry areas; also common in garden and waste areas. Circumpolar. Two subspecies grow in the Yukon: subspecies *borealis* occurs across Canada from the Maritime provinces to British Columbia, reaching the Arctic coast in northwestern Northwest Territories and the Yukon, where it's widespread.

Plant Parts Used
Flowers, leaves, roots, stalk

Harvest Time
Summer, as soon as they are in flower. I gather yarrow in bunches using scissors to cut the stems and then tie them up at the base and hang to dry. Then, after harvesting season when I have more time, I remove the leaves and flowers for medicinal use.

During the excavation of a forty- to sixty-thousand-year-old Neanderthal tomb, pollen from yarrow (among other herbs) was found.

IN GREEK MYTHOLOGY, it's said that when Achilles was a young child his mother dipped his body in a vat of yarrow tea to protect him from the dangers of war. Many years later, when he was a famous warrior, Achilles died from an arrow wound that pierced his heel, where his mother had held him while dunking him in the vat.

We use yarrow in our popular Aroma Borealis all-natural mosquito repellent, Skeet-Addle. It works in the battle against the abundant pesky boreal mosquito. This plant has come to my rescue many times, not only for its ability to repel insects, but for its ability to stop bleeding.

Yarrow makes great bush medicine! Prepared in a tea, the leaves and flowers of this plant have the ability to stop internal bleeding.

Accidents on northern trails can happen and this plant may make a difference if you have a long way back to amenities. I consider yarrow one of nature's best band-aids. For cuts and scrapes, chew up a few leaves to make a spit poultice and place over the affected area. A bleeding nose can be remedied by putting a gently chewed leaf up your nostril.

Medicinal Actions
Analgesic, antibacterial, anti-catarrhal, anti-inflammatory, antiseptic, astringent, bitter, carminative, diaphoretic, digestive, diuretic, emmenagogue, expectorant, febrifuge, hepatic, hypoglycemic, hypotensive, stimulant, styptic, tonic, vulnerary

Medicinal Preparations
Bath, cream, essential oil, hydrosol, oil infusion, poultice, salve, spit poultice, tea/infusion, tincture, wash

Part II: *Herbs*

Medicinal Uses

Yarrow's diaphoretic properties make it one of the best herbal teas to aid with fever, especially when the tea is hot. It also helps eliminate toxins and promotes perspiration, thereby helping break up colds and flu.

Yarrow is an immune stimulant and acts as a mild expectorant helping get rid of excess phlegm. Ingest yarrow tea the minute you feel a cold coming on and it may help to stop the cold from taking hold. Gargling with yarrow helps prevent or heal a sore throat, inflamed gums, or mouth infections.

If your mouth needs attention right away, try chewing up a few yarrow leaves and stuffing them into your cheek over the infected tooth or gum—it will start to work immediately. Just be sure to change the leaves every couple of hours.

Herbalist Robert Rogers notes that fresh yarrow root can also relieve toothache pain for up to eight hours—making it handy to have in the bush!

When prepared as a tea or taken as a tincture, yarrow has the ability to help lower blood pressure and to assist with good blood flow through the body.

Yarrow is an aromatic bitter that aids digestion and is used for mild stomach indigestion, bloating, flatulence, and nausea. Take ten drops of tincture before, or after a heavy meal.

If you have cold extremities, yarrow tea promotes warmth and circulation throughout the body and will help warm up your hands and feet.

Yarrow is a powerful topical wound healer and can stop bleeding immediately. Its analgesic properties will help with pain relief, and its antiseptic, antibacterial, anti-inflammatory properties help keep the wound free of infection.

It can also be used topically for eruptive conditions like measles, chickenpox, or insect bites.

It's best to use yarrow fresh, but keeping dried, powdered yarrow in your herbal first-aid kit or pantry is a good idea.

I like to infuse the flowering tops and fern-like leaves in carrier oil; my preference is an olive-and-sunflower oil combination. I like to add this oil combo to skin-healing salves and creams, and I use this for animals as well.

A Yukon musher once approached me to make a preparation for sled dogs' pads and paws. The preparation needed to do three things: heal the wounds quickly, provide lubrication to prevent snow buildup between toes, and, equally important, repel licking (we added cayenne pepper and bitter almond oil to the mixture for this). The yarrow in the mixture stopped the bleeding of the dogs' injured paws and helped speed up healing.

Yarrow is an especially good women's herb. It can slow down excessive bleeding during heavy menstruation, relieve pelvic congestion, reduce cramps, and help cleanse the liver so hormones like progesterone and estrogen are processed efficiently in the body.

> "Yarrow or Milfoil (*Achillea borealis*) frequently called old man's pepper, probably because of the pungent odour and taste...The plant must be of ancient heritage, as it was from the bruised leaves Achilles made the ointment to heal his wounds after the battle of Troy."
>
> —Martha Louise Black, *Yukon Wild Flowers* (1940)

Yarrow can also help clear up bladder infections and help with incontinence. Its diuretic properties help with premenstrual water retention.

Yarrow is an emmenagogue and helps bring on normal menstruation. It can help with amenorrhea, a lack of menstrual cycle, while also being a good herb for dysmenorrhea—painful menstruation—because of its astringent and vulnerary properties.

Herbalist Matthew Wood, author of *The Earthwise Herbal*, says that yarrow is called "master of the blood" because of its ability for clotting and unclotting. "Thus, it is cooling and warming, fluid-generating and controlling. Remedies with contradictory but complementary properties are often of great utility since they are able to normalize opposing conditions. This is true for yarrow."

For postpartum treatment, a yarrow infusion in a sitz bath can help heal tearing of the perineum, or an episiotomy, and bleeding hemorrhoids. Make up a spray bottle with a strong yarrow infusion to use after urination between sitz baths.

Twelfth-century nun and herbalist Hildegard von Bingen touted yarrow as an antimetastasis beverage. She recommended taking three pinches of powdered yarrow in fennel tea followed by three pinches of powdered yarrow in warm wine. She also recommended using yarrow before surgery or after internal injury and to continue use for at least eight days.

Centuries after her death, German physicians familiar with von Bingen's work suggest that yarrow also protects against radiation damage—something that wouldn't have been discussed during her lifetime!

Food Uses

The best way to benefit nutritionally from yarrow is either to make an infusion of the leaves and flowers, or make nutritive vinegar (see page 296). The vinegar can be used for salad dressing, or any recipe that calls for vinegar. A tablespoon of vinegar infusion can also be added to a cup of hot water and sipped. The leaves can be crushed and added to spice blends.

Enjoying yarrow plants in the southern Yukon. CA

Part II: *Herbs*

An immune stimulant, yarrow also acts as a mild expectorant, helping to get rid of excess phlegm. PL

Nutritional Profile
Calcium, chromium, cobalt, iron, magnesium, manganese, niacin, phosphorus, potassium, protein, riboflavin, selenium, silicon, sodium, thiamine, tin, vitamins A and C, zinc

Cosmetic Uses
Because of yarrow's anti-inflammatory and astringent properties it's an excellent plant to use for inflamed skin. Yarrow can be added to a facial steam to reduce inflammation and unclog congested pores. A facial tonic can also be made by infusing the flowers and leaves in witch hazel for a few weeks, straining, bottling, and using on a cotton pad to clean facial skin.

Other Uses
In aromatherapy, yarrow's indigo-coloured essential oil is considered antispasmodic, cicatrizant, febrifugal, and vulnerary. It's an excellent topical remedy for healing skin because of its high content of azulene, an anti-inflammatory.

The hydrosol that is made in the distilling process is also very powerful. Suzanne Catty, author of *Hydrosols: The Next Aromatherapy*, says of the yarrow hydrosol: "It is a good digestive aid and is significantly detoxifying but in a gentle manner." She notes that, "A three-week course will improve digestion, increase elimination and calm gastric spasms and rumbles and is recommended as part of a weight-loss program." She also adds she believes it improves digestion of fatty foods because it has bile-releasing properties, since it can quickly relieve the indigestion and heartburn caused by overindulgence.

Steve Johnson, founder of the Alaskan Flower Essence Project and author of *The Essence of Healing*, says that yarrow (*Achillea borealis*) is useful if one is "oversensitive to

Yarrow

the environment; looking for protection from outside rather than from within the self," or if "integrity of the aura has been compromised by injury or trauma in this or another lifetime." He adds that yarrow "seals energy breaks in the aura; strengthens the overall integrity of the energy field; helps us know and be the source of our own protection."

I made a Third-Eye Chakra Vibrational Essence with the yarrow flowers for balancing the ajna chakra (third eye; sixth chakra). It really speaks to the sacred light that we all have within us. Yarrow has the ability to help us expand our own inner psychic abilities and heighten our perceptions.

A United States Army study showed yarrow tincture to be more effective than DEET at repelling ticks, mosquitoes, and sand flies. (I sometimes carry a little spray bottle of yarrow tincture with me when I'm outside, and wet my skin every half-hour or so.)

Cautions

Not recommended for people with severe allergies to plants in the Asteraceae family. Not recommended during pregnancy.

Yarrow Oily Skin Treatment

This naturally astringent formula will help balance your skin. It is also great added to the bath to cleanse the whole body physically, emotionally, and energetically.

- 1 teaspoon (5 mL) yarrow
- 1 cup (250 mL) water

Pour boiling water over dried yarrow flowers. Cover and let steep for at least 15 minutes. Strain.

Pour onto a clean cotton cloth and gently wash your face.

If you have any left over you can use it later the same day, but it should be made fresh every day.

ADDITIONAL RECIPES

All-Purpose Botanical Ointment, page 313

Boreal Skin Cream, page 317

Boreal Skin Oil, page 307

Cold-and-Flu Tincture, page 294

Cold-and-Flu Bath, page 326

Cough-and-Cold Syrup, page 300

Good-for-Fever Infusion, page 282

Lung-Tonic Infusion, page 283

Styptic Powder, page 320

Urinary-Tract Tincture, page 294

BERRIES

BERRY PROFILES AND RECIPES

Blueberry 187
   The Purple Drink 190
   Blueberry Beauty RX (Blueberry Fruit Mask) 190

Cloudberry 191
   Cloudberry-Buckwheat Pancakes 194

Cranberry 195
   Cranberry Chutney 198
   Gratitude Gathering Cranberry Sauce 199

Highbush Cranberry 200
   Moon-Cycle Nourishing Vinegar 203

Currants 204
   Currant Juice 207

Juniper 208
   Detoxifying Juniper Sea-Salt Scrub 213
   Juniper-Berry Cleanse 213

Mossberry 214
   Mossberry Syrup 216

Raspberry 217
   Raspberry Jam 220

Saskatoon Berry 221
   Saskatoon-Berry Pie 224

Soapberry 225
   Soapberry Hair Shampoo 228

Strawberry 229
   Wild-Strawberry Exfoliating Face Mask 231

Previous page: An offering of ripe blueberries. FM

BLUEBERRY

Vaccinium uliginosum
Vaccinium ovalifolium

Other Names and Etymology
Alpine bilberry, bog bilberry, bog blueberry, wild blueberry, huckleberry. The Van Tat Gwich'in word for blueberry is *jàk zraii*. The common name "blueberry" comes from the Scottish *blae*, "bluish," or from the Danish *bollebar*, "dark berry." The genus name probably comes from *Baccinium*, from the Latin *bacca*, for "berry." However, some linguists believe *Vaccinium* comes from the Latin *vacca* or *vaccines*, meaning "cow." *Uliginosum* means "of the marshes."

Family
Ericaceae (heath family)

Botanical Description
Low, much-branched, erect or depressed shrubs, forming extensive colonies, 20–60 cm high; twigs hairless or with fine fuzz of hairs. Leaves deciduous, small, 0.4–2 cm long, oval or variously shaped, firm, entire-margined, dull green above, glaucous beneath. Flowers one to several from scaly buds; pedicels short; pink petals fused into a drooping bell shape, 3–4 mm long; berries 5–8 mm in diameter, slightly elongated, blue to blackish with a bloom.

Habitat and Range
Vaccinium uliginosum: Common in acid soil, muskegs, swamps, woodlands, and heath on alpine slopes. Circumpolar; in North America from Labrador to Alaska, south to California; common in the Yukon, north to the Arctic coast.

Vaccinium ovalifolium: Found at high elevations, such as near Fraser on the Alaska–Yukon border.

Left: Blueberries near Old Crow, northern Yukon, ripen about a month before those in the southern part of the territory. BG

Below: In autumn, blueberry bushes add splashes of vibrant colour to the forest. BK

Part II: *Berries*

Plant Parts Used
Berries and leaves

Harvest Time
Late summer, when berries are ripe. (This always seems to be weather dependent.) Leaves gathered before the plant produces berries.

ANYONE WHO HAS SPENT TIME with our family knows our famous morning "purple drink," made with greens and wild blueberries. Drinking this nutrient-packed smoothie first thing in the morning is a great way to start the day off right. Before I started adding wild blueberries to the drink, it was green, and the girls would not touch it. So one day I decided to add wild blueberries and—voilà—the purple drink was born. In the summer, when we have some left over, we just pour it into popsicle moulds to create healthy frozen treats.

Berry picking is an annual event for many people around the circumpolar north. Top: BK; above: PL

Medicinal Actions
Anti-inflammatory, antioxidant

Medicinal Preparations
Concentrates, such as fruit leather, tea, whole raw berry

Medicinal Uses
Considered an antioxidant super-fruit, wild blueberries have some very favourable health benefits, including claims of fighting oxidative stress that can lead to chronic illness and premature aging. The antioxidant activities of the wild blue fruit are said to be higher than those of many other fruits including apples, raspberries, red grapes, and strawberries.

Antioxidants are beneficial for heart health because they help control the bad cholesterol that contributes to cardiovascular disease and stroke. They may also help regulate blood pressure and fight atherosclerosis, a plaque buildup inside the arteries.

Blueberries are also touted as having excellent anti-inflammatory properties because they contain polyphenols and anthocyanins that help reduce chronic inflammation in the body and have been linked to the prevention of cancer.

The Highbush Blueberry Council of the United States reported that including blueberries in your diet on a regular basis may improve motor skills and help reverse the short-term memory loss that comes with aging and age-related diseases such as Alzheimer's.

Laboratory studies have provided evidence that consuming bilberry (wild blueberry) may help with eye fatigue and slow down or even reverse eye disorders, such as macular degeneration and cataracts.

Blueberry

Blueberries, like cranberries, contain chemical compounds that prevent the bacteria responsible for urinary-tract infections from attaching to the bladder wall.

Traditionally, blueberry leaves were used to treat gastrointestinal ailments, such as diarrhea and upset stomachs. All parts, including leaf and stem, may be useful for lowering blood sugar in Type-2 diabetes. Topically, an infusion of blueberries can be used to prevent skin infections.

Food Uses

I love raw blueberry fruit leather! It's one of those foods in which you can taste the nutrient goodness in each flavour-packed bite. Raw foods contain more enzymes and preserve more of the vitamins and minerals than cooked foods. Most fruit-leather recipes call for cooking the fruit, adding sugar, and straining the skins off, so many of the nutrients are cooked away. (See fruit leather recipes, page 376.)

> "During my early winters in the Klondyke my wild huckleberry jam was particularly popular among the men folk. In fact they used to say I was 'famous' for it, but you see I had little competition!"
>
> —Martha Louise Black, *Yukon Wild Flowers* (1940)

Martha Louise Black, First Lady of the Yukon

Martha Louise Black, a naturalist who is quoted frequently in this book, first arrived in the Yukon under rather unusual circumstances. At the height of the 1898 Klondike gold rush, she had set off for the goldfields with her first husband, a railroad tycoon, but along the way he decided to go to Hawaii and she decided to continue north without him. She crossed the Chilkoot Pass, while pregnant, and wearing long skirts, a corset, and bloomers.

In 1900, after travelling south to visit her family, she returned to Dawson City and set up a sawmill and a stamp mill for assaying ore. She also staked a claim at Excelsior Creek, which proved to be very rich.

In 1901, she divorced her first husband. Three years later, she married George Black, a lawyer who

went on to become a long-serving Yukon politician. During the First World War, the Blacks travelled to Europe, where George served as an infantry captain and Martha gave over 400 illustrated lectures on the Yukon. These talks led her to be inducted as a fellow of the Royal Geographic Society in 1917. In 1935, when George was too ill to defend his seat as a Member of Parliament, Martha campaigned in his place, becoming the second woman elected to the Canadian House of Commons.

In the 1940s, Martha wrote *Yukon Wild Flowers*. In this rare little book she shared her passion for the wild flowers of the Yukon, writing in the foreword, "I would like to take you with me in spirit on a Wild Flower quest, along the banks of our golden creeks, which were a mecca for fortune seekers during the mad gold rush days of '98." Her little flower book has over 100 illustrations and photos taken by George Black.

Part II: *Berries*

BK

Of course, there is so much that can be done with wild blueberries. You can eat them au naturel, or add them to yogourt, cereal, porridge, muffins, breads, scones, jams, jellies, chutneys, juice, or tea.

This super-food is very valuable to have on hand at all times, so spending a day picking them is a great investment. Sure you can buy some at the grocery store, but nothing will compare to your own wild, hand-picked berries.

Nutritional Profile
Vitamins A, C, E, K, riboflavin, folate (vitamin B9), calcium, iron, magnesium, phosphorus, potassium, sodium

Cautions
Unripe berry juice may be psychotropic to some and may vivify dreams.

ADDITIONAL RECIPES

Athletes' Blend Smoothie, page 340

Wild-Blueberry Chutney, page 353

Wild-Blueberry Jam, page 349

Blueberry-Mossberry Juice, page 336

Blueberry Muffins, page 342

Herb-and-Berry Sweet Summer Days Syrup, page 355

Living Fruit Leather, page 377

Wild-Blueberry Fruitsicles, page 373

The Purple Drink

1 cup (250 mL) wild blueberries (frozen)
1 cup (250 mL) yogourt
1 cup (250 mL) rice, almond, or soy milk
1 tablespoon (15 mL) spirulina or herb of your choice (i.e., nettle, dandelion or fireweed greens, chickweed, echinacea, and/or yarrow)

Put ingredients in a blender. Mix until smooth. Serve and enjoy!

Blueberry Beauty RX (Blueberry Fruit Mask)
When used once a week, this highly antioxidant and hydrating mask leads to a glowing complexion.

⅔ cup (160 mL) wild blueberries
Juice of 1 lemon
1 tablespoon (15 mL) full-fat yogourt
1 tablespoon (15 mL) honey

Put ingredients, in order, into a blender and mix. Blend till you have a medium-thick paste. Transfer to a small bowl. Chill in refrigerator for no more than 20 minutes. Spread mixture evenly onto the face and neck and let penetrate for up to 20 minutes. Gently rinse off with warm water.

CLOUDBERRY

Rubus chamaemorus

Other Names and Etymology
Salmonberry, baked apple, baked-apple berry, bake apple, yellowberry. In Scotland it's called avron. The Van Tat Gwich'in name for salmonberry is *nakàl*. The Inuvialuit name is *akpiks*. The Latin *Rubus* means "red," while *chamaemorus* is derived from the Greek *chamai*, meaning "brown," and *morus*, meaning "bramble."

Family
Rosaceae (rose family)

Botanical Description
Dioecious glabrous herb, arising from extensively creeping and branching rootstocks; stems simple, erect, 10–20 cm high, bearing one to three leaves. Leaves mostly five-lobed, toothed, somewhat leathery. Flowers solitary; petals 6–14 mm long, white, showy; fruit raspberry-like, at first bright red, becoming salmon coloured.

Left: Plump cloudberries are hard to resist. L to R: RF, PL

Below: The delicate shape of a cloudberry flower. BL

Wild Food and Medicine Plants of the North

Part II: Berries

Habitat and Range
Moist, peaty wetlands and turfy grassy places. Circumpolar. Widespread in Scandinavia. In North America it grows from Newfoundland to British Columbia and Alaska; found throughout the NWT and Yukon, north to the Arctic coast.

Plant Parts Used
Leaves and berries

Harvest Time
Mid-summer

I LOVE THE NAME "CLOUDBERRY." The flavour and texture of the fruit evokes exactly that: clouds of tasty berries. I hope that's where I end up after this lifetime.

Medicinal Actions
Antioxidant, astringent

Once picked, cloudberry berries can be stored in a cool place for months without extra preservation. BK

Medicinal Preparations
Leaves: compress, poultice, tea-infusion, tincture. Berries: juice, syrup, tea

Medicinal Uses
Cloudberry and raspberry both belong to the rose family, and their leaves have similar healing properties.

The large maple-leaf-shaped leaves are astringent and can be infused in boiling water to drink as a tea for menstrual cramping and diarrhea. Topically, the leaves can also be used as a compress or poultice for weeping wounds.

The berries are high in water-soluble vitamin C and are rich in antioxidants and anthocyanins. (Anthocyanins are the plant chemicals responsible for changing the colours of leaves from green to red, orange, purple, yellow, etc.) In humans, anthocyanins have the ability to fight free radicals in the body, and help fight disease caused by oxidative stress, which can cause premature aging.

Food Uses
Cloudberries can be enjoyed in so many ways—raw, frozen, as juice, syrup, fruit leather, jam, jelly, etc.

Cloudberry

The Finnish two-Euro coin is embossed with cloudberry fruit and leaves.

"Thank God, they cannot cut down the clouds!"

—Henry David Thoreau, author (1817–1862)

The ripe, salmon-coloured fruits are tart, soft, and juicy—making them irresistible to eat while picking. But if you hit a eureka patch and have good berry-picking discipline, you may make it home with some. If you do, there are a variety of ways to preserve this late-summer beauty.

Cloudberries are rich in benzoic acid, a natural preservative. This allows the berry to be stored fresh for many months without extra preservation. Not surprisingly, this made cloudberries a winter staple for aboriginal people before refrigeration.

Because the berries contain a natural preservative, they can be kept in a cache or any cool place and used when needed. Enjoy the berries by adding them to your morning yogourt, nutritional smoothie, in pancakes, or in muffins. For dessert, try cloudberries with a dollop of fresh whipped cream on top. Cloudberry jam in mid-winter is a tasty treat and an enticing reminder of the long days of summer. The syrup has many uses, including as a nice drizzle on top of cheesecake, ice cream, or pancakes. A liqueur can also be made with the berries.

A eureka patch of cloudberries can lead to jam, syrup, or liqueur. RH

Nutritional Profile
Calcium, iron, protein, vitamin C

Cosmetic Uses
Cloudberry leaves can be used in a facial steam or in a tonic for oily skin.

Other Uses
Steve Johnson, founder of the Alaskan Flower Essence Project and author of *The Essence of Healing*, says the cloudberry "helps us recognize the light of purity deep within ourselves; for replacing low self-esteem with an awareness of inner value; helps us open to the true source of our being and reflect this outward for others to see."

Cautions
It's important to note that when the cloudberry leaves are drying, they go through an important chemical process. Like all other leaves in the rose family they should only be used fresh or fully dried, but not when wilted.

Blueberries and cloudberries, Old Crow, Yukon. BG

Cloudberry-Buckwheat Pancakes
A dark and hearty pancake!

½ cup (125 mL) spelt flour
¼ cup (60 mL) buckwheat flower
1 teaspoon (5 mL) baking powder
¼ teaspoon (1 mL) sea salt
1 organic egg, beaten
¾ cup (175 mL) milk
1 tablespoon (15 mL) honey
1 tablespoon (15 mL) molasses
1 tablespoon (15 mL) butter or oil
Handful of cloudberries

Mix the dry ingredients together in a bowl. In another bowl beat the egg; add milk, honey, molasses, and butter or oil. Stir until well-blended. Stir wet ingredients into dry ingredients. Gently fold in cloudberries. Cook in a hot and oiled frying pan.

CRANBERRY

Vaccinium vitis-idaea

Other Names and Etymology
Mountain cranberry, lingonberry, low-bush cranberry, cowberry, mountain bilberry, partridgeberry, alpine cranberry. The Van Tat Gwich'in of Old Crow, Yukon, name for cranberry is *natl'at*. The word "cranberry" came from "crane-berry," an early American name given to the plant because its flower bud looks like a crane. The genus name *Vaccinium* may be derived from the Latin *vacca*, meaning "cow," or it may come from the *baccinium*, that comes from the Latin *bacca*, meaning "berry." *Vitis-idaea* means "the vine of Mount Ida," after a mountain in Crete.

V. vitis-idaea

Family
Ericaceae (heath family)

Left: Vivid red cranberries are easy to spot. CA

Below: Most northerners gather wild cranberries after the first frost when they are sweeter. PL

Part II: Berries

Tangy wild cranberries are smaller than commercial cranberries—and pack a greater taste! FM

Botanical Description
Low, mat-forming shrub; stems decumbant-ascending and 5–20 cm high; branchlets mostly rounded in cross-section with fuzzy hairs. Leaves leathery, evergreen, nearly stalkless, 6–15 mm long, narrowly elliptic, blunt-tipped, margins curling under, glossy above, pale below, dotted with brownish glandular hairs. Flowers in short terminal clusters, few; petals white, cup-shaped, about 5 mm long. Juicy red berries 6–10 mm in diameter, often persisting over the winter.

Habitat and Range
Open and acidic soils, turfs, and bogs; pine stands; rocky tundra; and lichen woodlands. Circumpolar and subarctic. In North America from Labrador to Alaska, south to Minnesota and north to the Arctic coast. It's found throughout Alaska, the Yukon, NWT, Nunavut, and Greenland.

Plant Parts Used
Berries and leaves

Harvest Time
Most northerners gather wild cranberries in early autumn. After the first frost is the best time to gather wild cranberries because it softens the berries and makes them more flavourful (but it's okay to pick them earlier, too). When washing your wild cranberries in a vat of water, the damaged ones will generally sink to the bottom.

Bursting with colour and flavour, wild cranberries, or lingonberries as they are also well known, are one of the most recognized berries in the North. Wild cranberries are smaller and tangier than commercial cranberries, and pack a greater taste! Every autumn people go out in droves to their sacred spots to gather enough berries to get them through the winter months.

Cranberries were and are still valued by First Nations who use the berries to keep themselves healthy and nourished throughout the long boreal winters.

Cranberries can be dried, frozen, or kept in a cool cache over the winter months. The array of uses for cranberry as food and medicine is amazing. It's now the star product of many nutraceutical companies for its lead role in many products touted to help with chronic bladder infections.

Medicinal Actions
Anti-inflammatory, anti-oxidant, antiseptic, astringent, diuretic, refrigerant, tonic, urinary antiseptic

Cranberry

Medicinal Preparations
Capsule, juice, poultice, tea, tincture

Medicinal Uses
Cranberry contains high concentrations of flavonoids such as quercitin and anthocyanins. These can help lower blood-sugar levels and reduce symptoms of allergies such as hay fever. These chemicals are also found in grapes and red wine, both famous for being high in antioxidants.

Cranberries also contain antioxidant polyphenols that benefit the cardiovascular and immune systems. They have the potential—though this isn't well-researched, yet—to prevent tumour growth.

Cranberry's antiseptic properties make the berry good for preventing and treating urinary-tract infections. Wild cranberry juice is touted for its ability to prevent bacteria such as E. coli from binding to the wall of the bladder and for creating an inhospitable environment for infection. Ingesting the juice will also help reduce the odour of ammonia associated with urine. The juice also helps stop kidney stones from developing.

Cranberry stimulates the production of digestive enzymes, so it's good to have a handful before a meal. You can also add cranberry sauce to your meats to help with heartburn or indigestion.

Pregnant women and new mothers may find that eating cooked or raw cranberries, or making them into a tea, will help curb nausea and reduce the pain of uterine cramping during and after childbirth.

Some of the chemical constituents in the cranberry can also help prevent tooth decay and reduce plaque formation on the teeth, reducing the bacteria in the mouth that causes gingivitis.

Cranberries are related to the wild blueberry (bilberry) and are said to be good for eyesight.

Like the leaves of its relative, the bearberry (*A. uva-ursi*), cranberry leaves help fight urinary-tract infections (UTI) because they also contain the chemical arbutin. Depending on the UTI imbalance (too alkaline or too acidic), cranberry leaves alkalize urine like bearberry leaves. You can test your urine with pH paper.

In Finland, lingonberry-leaf tea is routinely prescribed for nerve pain such as sciatica. And throughout Scandinavia, cranberries have traditionally been used both as food and as a medicine to treat inflammatory diseases and wounds.

When using cranberry juice medicinally, or nutritionally for that matter, it's important not to confuse the pure juice with the commercial cranberry cocktails: the latter contain tons of white sugar and a little bit of juice concentrate. It's better to pick your own berries and make your own juice or tea.

Cranberries can be topically used in a poultice to soothe and heal cuts, scrapes, and abrasions.

> "Fable says that cranberry is a cure for heartache but far more prosaically the sauce wards off scurvy."
>
> —Martha Louise Black, *Yukon Wild Flowers* (1940)

Part II: *Berries*

Cranberries stimulate the production of digestive enzymes, so it's good to eat a handful of them before a meal. BG

Food Uses

Christmas and Thanksgiving seem to be the feast days when cranberries get the most attention. After all, what's a turkey dinner without tart-tasting cranberry sauce? I always make my sugarless scarlet-red sauce at the same time we cook the bird, and serve it warm as a bright red gravy. I also include cranberries in my stuffing.

Rich in pectin, cranberries are wonderful to make into jams, jelly, syrups, and conserves such as cranberry chutney.

The flavourful, bright red berries look and taste wonderful in pancakes, scones, muffins, breads, and strudels. They are terrific in pies, either on their own or mixed with apples or other berries.

The Yukon Brewing Company brews a Cranberry Wheat Ale that is very popular because of its tartness. Cranberries also make a nice wine or liqueur.

If you gather lots of cranberries at once, don't worry if you can't get to preparing them as soon as you hoped: wild cranberries keep really well as long as they have a nice, cool home. This is due to their benzoic acid content that acts as a natural preservative.

If you want to freeze your cranberries for the long term, spread them on a cookie sheet (this stops them from freezing in a clump) and place them in the freezer. When they are fully frozen, put them into a freezer container or freezer bag. Freezing also sweetens the berries.

Nutritional Profile

Calcium, iron, niacin, phosphorus, potassium, small amounts of protein, sodium, vitamins A and C. The seeds are rich in omega-3 fatty acids

Cautions

Use in moderation if you are prone to calcium-oxalate kidney stones.

Cranberry Chutney

This chutney goes great with East Indian food, curries, chicken, and vegetarian dishes. Makes 2 cups (500 mL).

1 cup (250 mL) water
½ cup (125 mL) brown sugar, date sugar, or organic cane sugar
3½ cups (875 mL) fresh or frozen wild cranberries
½ cup (125 mL) organic raisins
½ cup (125 mL) organic pecan or walnut pieces
1 cup (250 mL) dried organic apricots, diced
½ teaspoon (2 mL) nutmeg

Cranberry

½ teaspoon (2 mL) cinnamon
½ teaspoon (2 mL) cardamom
1 teaspoon (5 mL) fresh, organic ginger, grated
2 tablespoons (30 mL) brandy

Place sugar and water in a large pot and bring to a boil. Add cranberries and bring to boil again, stirring occasionally. Reduce heat and let simmer until the cranberries start to pop open. Add the rest of the ingredients and simmer for 10–15 minutes, stirring occasionally. Stir in the brandy and simmer another 1–5 minutes. Remove from heat and let cool before serving.

The chutney should be stored in a jar in the fridge, where it can keep for a couple of months.

Gratitude Gathering Cranberry Sauce

6 cups (1.5 L) of water
4 cups (1 L) cranberries

Bring water to a boil, add the cranberries and simmer on a low heat for hours (while the turkey is cooking), stirring occasionally, until it cooks down to a tart-tasting sauce.

Makes enough for a turkey dinner with friends and for leftovers.

ADDITIONAL RECIPES

Cranberry-Mint Muffins, page 343

Cool Cranberry Juice, page 335

Living Fruit Leather, page 377

Pink Drink (High-C Smoothie), page 340

Urinary-Tract Tincture, page 294

Tart wild cranberry sauce is a welcome addition to any poultry dish. BG

HIGHBUSH CRANBERRY (CRAMPBARK)

Viburnum edule

Other Names and Etymology

Squashberry, low-bush cranberry, bush cranberry, arrowwood, crampbark, mooseberry. The word *Viburnum* comes from the Latin *vieo*, which means, "to tie," referring to the flexibility of the stems for tying bundles, and also can mean "wayfaring tree." The Latin *edule* means "edible." Cranberry is derived from "crane-berry," an early American name given to the plant because its flower bud looks like a crane. (Note: The common name of this plant has caused some confusion for foragers. In the North we commonly call it "highbush cranberry," but in fact *V. edule* has been called "low-bush cranberry" by some botanists and *V. opulus*, which does not grow in the Yukon, is known as "highbush cranberry." The name "cranberry" is also deceiving as crampbark actually belongs to the honeysuckle family, whereas the true cranberry belongs to the heath family.)

Family
Caprifoliaceae (honeysuckle family)

Left: Rich in vitamin C, highbush cranberries are great for making juice, medicinal syrup, and jelly. CA

Right: Highbush cranberry bark should be harvested before, or after, the plant goes into berry. CA

Highbush Cranberry

Botanical Description
Erect, branching shrub, up to 2 m high; bark smooth, dark grey. Branching opposite, leaves opposite. Leaves petioled; vaguely maple-leaf shaped: rounded, palmately three- to five-veined, often three-lobed tips, coarsely toothed. Flowers in axillary compound corymbose cymes; calyx less than 1 mm long, five-lobed; corolla 4–7 mm wide, five-lobed, with lobes longer than tube, milky-white; fruit a bright red one-seeded edible drupe.

Habitat and Range
Woods, riverbanks, streams, and thickets. In North America it's found from Newfoundland to Alaska, south to Pennsylvania, Minnesota, Colorado, and Oregon. In the Yukon it's found north to the Porcupine River.

Plant Parts Used
Bark, inner bark, berries

Harvest Time
Bark: before the plant goes into berry, or after. Berries: the berries are hard and sour until after the first frost in early autumn when they become ripe and soft. Some people like to collect the berries before frost to make preserves, while others like to pick them after—it's up to you!

The bright red autumn foliage of *V. edule*. RF

IF MUSCLE RECALL hasn't set in and you're feeling the miles you have put in on the trail, you may want to keep an eye out for the highbush cranberry.

Also known as crampbark, for its virtues as an antispasmodic, the bark of the highbush cranberry helps stop cramping when used topically or internally. To prepare, collect a few branches along the trail, whittle the bark off into a pot, and simmer it to make a tea or poultice. It's a perfect campfire activity!

Medicinal Actions
Bark: alterative, antispasmodic, astringent, diuretic, nervine, sedative

Medicinal Preparations
Bark: bath, compress, cream, decoction, liniment, oil, poultice, salve, tea tincture. Berries: syrup

Part II: *Berries*

As the name "crampbark" suggests, the bark can be used for treating cramping muscles. CA

Medicinal Uses

As the name "crampbark" suggests, the bark can be decocted or tinctured to make an antispasmodic which can help relieve cramping of the uterus, bladder, and stomach. I recommend making the tincture in advance and having it on hand, as this is far more convenient than making the tea. It can also be used for uterine pain caused by endometriosis and pain in the testes.

Many midwives and herbalists recommend the inner bark of *V. edule* for cramping of the uterus after childbirth. It has also been used as a herbal treatment to prevent miscarriages in the third trimester: it slows down and potentially stops rhythmic uterine contractions.

In combination with yarrow, crampbark tincture can help menopausal women with muscle pain, excessive bleeding, and cramping.

For spasmodic coughing, the tincture of crampbark on its own, or blended with coltsfoot and plantain, can help ease and soothe the lungs. An inner-bark decoction can also help heal strep infections of the skin or throat.

Including inner bark in topical preparations such as salves and liniments for sore muscle pain, cramping, and spasms—those caused by restless-leg syndrome—is very valuable. The smell is fresh and the results are amazing!

The bark can be used as a gargle for sore throats, and as a rinse for gingivitis and loose teeth. Its mild diuretic properties help cleanse the kidneys as well.

Dr. William Cook, one of the great American eclectic physicians who used botanical remedies in the late nineteenth and early twentieth centuries, praised the highbush cranberry. Because of its nervine and antispasmodic properties, Dr. Cook recommended an infusion of the branches and twigs for bronchial irritation and coughs. He also recommended it for various forms of indigestion, irritable stomach, and in cases of "uterine sensitiveness."

"With leucorrhea and nervous irritability," he said, "I am confident it will be found of significant benefit."

Food Uses

The aromatic and musty-smelling autumn fruits of highbush cranberry are stunning. They light up the forest like little ruby-red, twinkling jewels.

While the berry is tasty, removing its small flat seed is a bit of a pain, so using highbush cranberries in baked goods takes effort. This said, these vitamin C-rich berries are great for making juice, medicinal syrup, and jelly.

The juice can be used as a liquid replacement in baking, when appropriate, and a dollop of jelly is really nice on thumbprint cookies, in tarts, and as a condiment for wild game and fowl.

Highbush Cranberry

Nutritional Profile
Bark: Calcium, chromium, cobalt, iron, magnesium, manganese, phosphorus, potassium, selenium, tin, zinc. Berries: Vitamins C, K

Other Uses
Highbush cranberry can be used as a pink dye plant.

Cautions
None known.

Some berries will last through the winter to the following spring. PL

Moon-Cycle Nourishing Vinegar
A nourishing remedy to use before and during menstruation.

2 parts raspberry leaf
2 parts crampbark
1 part peppermint leaf
1 part oatstraw (optional)
8 parts apple cider vinegar

The herbs in this preparation can be used fresh or dried.

Chop herbs, put in a jar and cover with apple cider vinegar.

Infuse for 4–6 weeks, shaking daily. (See Tincture Method, page 293, for more details.)

Add 1 tablespoon (15 mL) to 1 cup (250 mL) of warm water and drink or use it in a salad dressing.

ADDITIONAL RECIPES

Detoxifying Infusion, page 283

Boreal Bitters, page 297

Cough-and-Cold Syrup, page 300

Highbush Cranberry Applesauce, page 358

Moon-Time Infusion, page 284

Muscle-and-Pain Relief Oil, page 307

Muscle-Ease Cream, page 317

Pain-Aid Capsule, page 301

Pain-Aid Tincture, page 294

CURRANTS

Ribes triste (swamp red currant)
Ribes hudsonianum (northern black currant)
Ribes oxyacanthoides (Canada gooseberry)

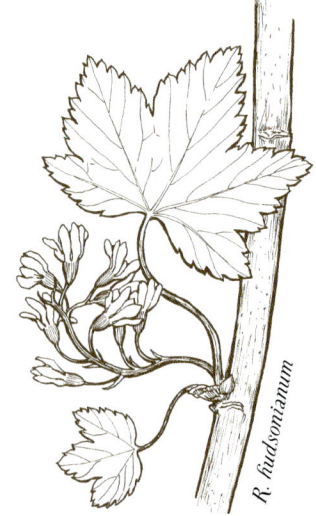

Other Names and Etymology

Ribes is derived from the Danish word *ribs*, for "red currant." The common name, "currant," may be derived from Corinth, the location where the plant is said to have first been reported. Van Tat Gwich'in word for black currant is *neeyùu zraii*

Family

Grossulariaceae (currant family). Previously Saxifragaceae (saxifrage family)

Botanical Descriptions

Ribes triste (red currant): Woody shrub, branches trailing with ascending tips, pubescent in youth, rooting at nodes; bark exfoliating without spines. Leaves thin, hairless, or thinly pubescent below in youth. Flower clusters drooping, six to fifteen flowered; sepals and petals reddish or purplish. Berries bright red, oval, 6–10 mm in diameter, hairless. *Ribes*

It's easy to distinguish northern black currant (left) from wild red currant because of the berry colour. L to R: BG, PL

Currants

hudsonianum (northern black currant): Low shrub with branches erect-ascending, about 1 m high, smooth, no spines. Leaves three-lobed, 5–7 cm in diameter; lower surface pale, scattered with resin-dots and long soft hairs on veins. Flower clusters ascending, eight to ten flowered; petals whitish, 4 mm long. Berries 5–12 mm long, black, smooth to somewhat glandular. *Ribes oxyacanthoides* (wild gooseberry): Woody shrub with branches prostrate to ascending or erect, with spines at nodes and bristles at internodes, and soft hairs. Leaves deeply five-lobed, with glandular hairs below. Flower clusters one to two flowered; sepals and petals green-yellow. Berries blue-purple, hairless.

Gooseberries CA

Habitat and Range

Ribes triste (red currant): Moist woods and clearings. Boreal North America; from Labrador to Alaska, south to Oregon; found throughout the Yukon, north to the Arctic coast. Also found in central and eastern Russia. *Ribes hudsonianum* (northern black currant): Moist woods. In North America, from northwestern Quebec to Alaska, south to California; in the Yukon, north to the Porcupine River valley. *Ribes oxyacanthoides* ssp. *oxyacanthoides* (wild gooseberry): Gravel banks, rocky slopes, and borders of woodlands. In North America, from James Bay to Alaska, south to Minnesota, Wyoming, and Nebraska; in the Yukon, north to the Peel Watershed.

Plant Parts Used

Leaves, berries

Harvest Time

Leaves: spring and summer before the plant goes into berry. Fruit: as it ripens, generally mid- to late summer.

CURRANTS ARE VERY COMMON PLANTS in the North, with predominantly red and black berries that favour moist and open woods. Old homesteads often have currant bushes growing nearby. The Robinson Roadhouse in the Carcross Valley has a huge currant bush. I love going there to eat and gather berries.

Medicinal Actions

Anti-inflammatory, antioxidant, antiseptic, astringent, digestive, diuretic, laxative

Medicinal Preparations

Berries: Cooked, raw, syrup. Leaves: Infused as a tea

Part II: *Berries*

> "Soothsayers tell us that to dream of currants denotes happiness in life, success in one's undertakings. While to the farmer and tradesman, riches."
>
> —Martha Louise Black, *Yukon Wild Flowers* (1940)

Medicinal Uses

Eating the berries raw or drinking the diluted juice of *Ribes* species helps treat yeast infections, specifically *Candida albicans*, which can cause a person to feel run down all the time.

Raw berries contain anthocyanosides that are beneficial antioxidants that inhibit inflammation, and can, in turn, help heal and keep the body healthy and free from rheumatic pain and gout. The leaves, used in the form of a tea or tincture, help with pain relief.

Black-currant seed oil is very rich in essential fatty acids including gamma linolenic acid (GLA). It's marketed commercially for use for migraine headaches, skin irritations (eczema), psoriasis, acne, arthritis, and helps relieve symptoms of premenstrual syndrome.

Black-currant juice is beneficial drunk hot at the beginning of a cold or flu, and can be blended with yarrow to support the immune system.

Red-currant juice or tea can help reduce fever and induce sweating.

As with the berries, a tea infusion made from dried red-currant leaves is said to ease the symptoms of gout and rheumatism, and can be used as a gargle for mouth infections. The leaves are useful as a compress or poultice for wounds that are healing slowly.

> A tea infusion made from dried red-currant leaves is said to ease the symptoms of gout and rheumatism, and can be used as a gargle for treating mouth infections. CA

Food uses

Currants are said to help keep the body in an alkaline state. When the body is alkaline, it's less likely that disease will set in.

Ribes species make excellent jams, jellies, syrups, and juice. These not only taste good, but preserve the berries' medicinal properties.

Currants

Of all the currants, I think that red currant makes the best jam. A tart gooseberry sauce compliments poultry, and the astringent nature of black currants brings out the flavour in many sauces, meat dishes, and desserts.

The berries can be dried and eaten or made into a pulp and dried as fruit leather.

In Scandinavia, a favourite first course of a meal is a hot or chilled soup made with gooseberries, wine or chicken stock, and thickened with cornstarch or potato, and topped with sour or whipped cream.

Black currants are famous as the main ingredient in the French liquor crème de cassis. The berries are also a favourite ingredient among northern home vintners.

Nutritional Profile
Calcium, (high in) fibre, iron, magnesium, manganese, phosphorus, potassium, vitamins A, B1 (thiamine), B2 (riboflavin), B3 (niacin), B5, C

Other Uses
Herbalist Robert Rogers says the flower essence of black currant is for "getting in touch with the deep instinctual feminine forces. It deals with the cycles of conception, menstruation, and birthing. This essence helps us attune with the heartbeat of the Earth, which may help strengthen feelings of self worth and adequacy."

Cautions
Eating copious amounts of currants can cause diarrhea.

Currant Juice

    7 cups (1.75 L) berries
    1 cup (250 mL) water

Put berries and water in a saucepan. Simmer over low heat, stirring, and mashing. When fully integrated, remove from heat and strain through cheesecloth or jelly bag into a large measuring cup with a spout (for easy pouring).

The juice will not keep for very long in the refrigerator. I recommend freezing the juice in ice cube trays, then transferring the cubes to a container for use when needed.

Add a currant-juice ice cube to fizzy water for an excellent natural pop.

JUNIPER

Juniperus communis (common juniper)
Juniperus horizontalis (creeping juniper)

Other Names and Etymology
Juniper communis is also known as prickly juniper. The Van Tat Gwich'in word for juniper is *deetru' jak*. In Old Crow it is also sometimes called "sharp tree." Juniper is derived from the Celtic word *juneprus*, meaning "acrid, biting or rough." *Juniperus* means "youth producing" and *communis* means "common."

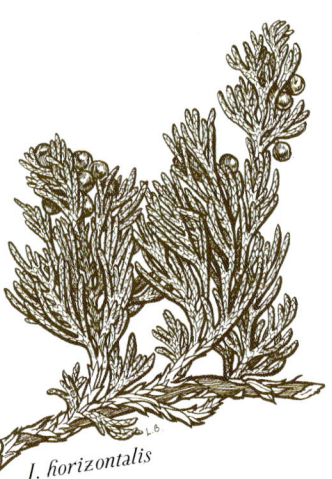

Family
Cupressaceae (cypress family)

Botanical Descriptions
Juniperus communis (common juniper): An evergreen shrub that grows close to the ground. Needle leaves are awl-shaped, 5–20 mm long, and crowded in whorls of three along the branch. Bands of glaucous stomata along the green needles give it a striped appearance. The mature seed cones are berry-like in appearance, resinous, fragrant and dark blue with a powdery coating. The "berries" require three summers to fully mature (turn dark blue). In the Yukon there are two subspecies of common juniper, ssp. *alpina* and ssp. *depressa*.

When eaten raw for the first time, many people are surprised to discover that juniper berries taste like gin. The berries are often coated with a dusty white powder. L to R: BG, PL

Juniper

Juniperus horizontalis (creeping juniper): A prostrate and mat-forming evergreen shrub, creeping juniper has long-trailing and free-rooting main branches. The scale-like leaves are blue-green, partially folded and pressed against and covering the stem like shingles. Very young shrubs have needles like common juniper. Berry-like cones are very similar to those in common juniper.

Habitat and Range
Juniperus communis (common juniper): Circumpolar, growing on dry, sunny slopes. In North America, north to, and in some places slightly beyond, the treeline. Subspecies *alpina* grows in alpine and exposed situations, whereas ssp. *depressa* is found in clearings and wooded areas, usually in lowlands. *Juniperus horizontalis* (creeping juniper): Grows on bluffs, alluvial fans, terraces, and rocky slopes. In North America it grows from Newfoundland and southern Labrador to southwest Alaska south to the northern United States. In the Yukon, creeping juniper grows north to about latitude 65° north.

Plant Parts Used
Berries, needles

Harvest Time
Juniper's blue, resinous fruits take three years to mature and are gathered when they are ripe. The needles can be gathered anytime.

JUNIPER'S SPICY BERRIES make a wonderful trail nibbler. Most people who taste juniper berries for the first time find that they taste like gin; this is because juniper berries are what Dutch gin is made from.

Interestingly, "gin" is a short form of the Dutch word *jenever*, or the French word *geneve*. The city of Geneva was named after the juniper forest that once grew there.

I really like to use juniper topically in an infused oil rub for sore muscles, aches, and pains. It's wonderfully soothing after a long day of hiking, gardening, or kayaking.

Medicinal Actions
Alterative, analgesic, antibacterial, anti-inflammatory, antimicrobial, anti-rheumatic, antiseptic, aromatic, carminative, diaphoretic, digestive aid, disinfectant, diuretic, emmenagogue, rubefacient, stimulant, tonic, urinary antiseptic, uterine stimulant

Medicinal Preparations
Bath, cream, essential oil, hydrosol, infusion, infused oil, liniment, poultice, salve, syrup, tea, tincture, wash

Part II: *Berries*

Northerners use juniper berries as a spice when cooking wild game. The warm and aromatic qualities of the berries aid digestion. CA

Medicinal Uses

Warming, spicy, and antiseptic juniper berries can be brewed as a tea for urinary-tract problems such as cystitis, urethritis, prostatitis, and vaginitis.

Juniper berry is a diuretic and promotes increased urine flow, which in turn helps to clear the bladder, prostate, gallbladder, and kidneys of excess wastes like uric acid, which causes gout.

Juniper's anti-inflammatory properties can help to ease pain of rheumatic conditions, sore painful joints, arthritis, sore muscles, gout, and nerve pain caused by sciatica. In the latter case, the berries can be used as a tea, but for prolonged use it's best applied topically as a poultice, in a cream, salve, or bath.

Juniper berries' warm and aromatic qualities aid digestion, expel gas, ease stomach cramps, indigestion, and stimulate the appetite. In laboratory testing juniper has been shown to lower blood-sugar levels.

Juniper needles have long been used as a favourite in traditional saunas because of their antimicrobial actions. Juniper berry is strengthening to the immune system and is good for preventing colds and flus. If you have a cold, the berry tea can be a healing ally for the lungs. Its expectorant properties help clear excess phlegm, aiding with conditions like bronchitis and sinusitis. It's also a good sore-throat gargle.

Juniper berry is considered a purifier of the blood and an overall system cleanser. By removing acid and toxic wastes from the body, it helps to reduce overall susceptibility to disease.

Northern First Nations have used juniper as a survival food and know juniper is "good medicine," and that it has very powerful healing properties, not only for cleansing the body but the spirit as well. They use juniper to keep away infection, treat arthritis, and stomach aches. The Van Tat Gwich'in use it as a tea for colds and flu and to help increase the appetite.

Since juniper can be irritating to the kidneys, combining it with demulcent local herbs like plantain and/or coltsfoot, or non-local herbs like slippery elm or marshmallow root, will help to minimize irritation. It also blends well with bearberry leaves and bedstraw.

Food Uses

Move over Beano! The volatile oil in juniper berries helps to eliminate gas and expel intestinal flatulence, assisting in the digestion of gas-producing foods like beans and cabbage. Adding hard, dried juniper berries to a peppermill is an easy way to include juniper as a seasoning—especially to dishes that may cause gas. Adding a few berries to your homemade sauerkraut before you ferment the cabbage helps prevent farting after eating it. If you don't want to add the berries to your meal, just eat a few dried berries before you eat.

One of my favourite ways to use juniper in my kitchen is to add a very light dusting of ground juniper berries to salmon fillets, which I then serve with a couple of lemon wedges.

Many northerners use juniper berries when cooking wild game such as moose, caribou, bison, and fowl, like grouse. The needles and berries are also often used to smoke fish.

Juniper

An abundance of wild harvested juniper berries at the Aroma Borealis Herb Shop. BG

To add a sharp, clear flavour to duck, chicken, rabbit, goose, fish or lamb, experiment by adding crushed juniper berries to your favourite marinades. Juniper can also be added to soups and stews. In Sweden, juniper berries are made into a conserve and used as a condiment for meats.

Juniper berries have a light coating of yeast over the skin and can be used to make breads rise. To do this, make a standard sourdough starter from flour and water, and add a small handful of berries to the mix. The microorganisms on the berries will feed on the flour and water, multiply, and take up residence in the starter. When the starter is bubbly and smells yeasty, you can discard the berries and continue bread making.

Juniper berries may be infused into vodka to flavour it. Juniper beer is held in high esteem in Norway, Finland, and Sweden, where it's valued for its antiviral, antibiotic, and antifungal properties.

Nutritional Profile
Chromium, cobalt, iron, magnesium, manganese, niacin, phosphorus, potassium, protein, riboflavin, selenium, silicon, sodium, thiamine, tin, vitamins A and C

Cosmetic Use
A decoction of the needles and branches can be used as a wash for dry, dandruff-prone scalps; as a facial astringent to help clear up acne; and used in the bath, or added to a scrub, to help diminish cellulite.

Part II: *Berries*

Other Uses

Juniper berry oil has a warm, woody, peppery, pine smell that is uplifting and strengthening. It's highly antiseptic and is astringent for the skin. The oil is chemically similar to turpentine and tea-tree oil.

Using antimicrobial juniper essential oil in an aromatherapy diffuser can help combat a wide range of airborne bacteria, and helps ease respiratory complaints.

Energetically, the oil strengthens and uplifts the spirit during times of low energy, anxiety, and weakness, and helps us clear emotional overload and impurities.

Diluted juniper essential oil (or the crushed berries) acts as a topical antiseptic. It stimulates circulation and is a tonic for oily or congested skin. It's also helpful for treating eczema, dermatitis, acne, wounds, and can be added to general detoxifying blends.

Juniper berry or needle essential oil should not be used directly on the skin, but instead should always be used with a carrier such as oil, sea salt, or witch hazel.

The Aroma Borealis Chakra Vibrational Essence includes juniper berries to energetically support the crown chakra (seventh chakra). It's clearing and cleansing to the mind, and helps one have clear thoughts.

A juniper hydrosol has many beneficial uses as well. In her book *Hydrosols: The Next Aromatherapy* author Suzanne Catty writes, "Highly energetic, juniper berry is one of the best for vibrational work." She adds, "Apply one drop to the palm of each hand, rub them together, and wipe them through your auric field to cleanse yourself. Very protective; use it with intention to create an effective, 'light shield' that both blocks the outward and inward movement of inappropriate energies."

Juniper branches can also be burned as a smudge to cleanse and detoxify, and to energetically clear a space.

> "Juniper berries are used in making gin but I never heard of them being used for that purpose in the Yukon, though we have been known to make 'home brew.'"
> —Martha Louise Black, *Yukon Wild Flowers* (1940)

Cautions

Berries can be irritating to the kidneys and should be avoided by those who have kidney disease or problems. Not for use in pregnancy. Use in moderation.

> In the Middle Ages, juniper branches were used as a fumigant to disinfect houses and hospitals. Its smoke was also used to clear away evil spirits.

Juniper

Detoxifying Juniper Sea-Salt Scrub

A circulatory stimulant that removes dead surface-skin cells, leaving your skin feeling as soft as a baby's bottom! This scrub is nice to do once a week.

- 2 cups (500 mL) sea salt
- 2 cups (500 mL) juniper berries, ground
- 4 cups (1 L) almond, olive, or grapeseed oil
- 1 cup (250 mL) water

Mix ingredients together in a bowl and pour into a jar for use.

In the shower or bath, take a dab of the combo in the palm of your hand and scrub from the chest all the way to the tip of your toes. Do not use on your face.

Be aware that the surface of your shower or tub may become slippery with oil so be careful stepping out. (I recommend using baking soda to clean up afterward so the next person doesn't slip either.)

Juniper-Berry Cleanse

Traditional recipe by 19th century German Doctor, Sebastian Kneipp. Use only blue, ripe, dried berries from a non-polluted area.

Caution — This is a very stimulating cleanse and should not be used by pregnant or nursing mothers or those with compromised health and/or kidney disease. Consult your health practitioner before starting cleanse.

Day 1 Chew 4 berries, either all at once or over the course of the day (at the beginning of the treatment, either way is possible).

Days 2–12 Chew 1 berry more each day than you did the previous day, until the daily dose totals 15 berries. Note: The more berries you take each day, the more important it is to distribute them over the course of the day. It's advisable to divide the berries into the four daily doses, drinking at least one full glass of water with each dose.

Days 13–24 Once you have reached a daily total of 15 berries (on day 12), reduce the amount by one berry per day until you finally reach the initial dose of 4 berries again (on day 24).

ADDITIONAL RECIPES

Athletes' Blend Smoothie, page 340

Cold-and-Flu Bath, page 326

Juniper Butter, page 357

Juniper-Willow Liniment, page 317

Muscle-Ease Cream, page 317

Muscle-and-Pain Relief Oil, page 307

Oh-My-Aching-Bones Bath, page 326

Respiratory Steam, page 329

MOSSBERRY

Empetrum nigrum

Other Names and Etymology
Blackberry, crowberry, black crowberry, curlewberry. The Van Tat Gwich'in word for it is *dineech'ùh*. *Empetrum* is derived from Greek, meaning, "upon a rock" referring to where they grow. *Nigrum* means "black" referring to the colour of the berry. As a common name, "mossberry" refers to the berry's tendency to grow in mossy areas, while "crowberry" comes from crows eating them, and from the glossy, black colour of the fruit.

Family
Empetraceae (crowberry family)

Botanical Description
Matted, freely branching and trailing evergreen shrub; branches minutely glandular. Leaves alternate or whorled, linear to narrowly elliptical, blunt-tipped, about 5 mm long, spreading and becoming reflexed. Flowers early in spring; flowers inconspicuous, solitary in leaf axils; both perfect and unisexual flowers on plants of ssp. *hermaphroditum*. Purple-black berries, 6–9 mm wide, shiny, very juicy, sweet.

Habitat and Range
Tundra, heathlands, swamps, and bogs. Circumpolar, wide-ranging. In North America, it's found from Newfoundland to Alaska, south to California. It's found throughout most of the NWT and Yukon.

Plant Parts Used
Berries, branches

Harvest Time
Ripe berries: mid-summer to autumn. Branches: ideally before berry, but anytime is okay

> Mossberries are also known as "crowberries" because of their glossy black colour and because crows eat them. CA

Mossberry

I FIND MOSSBERRIES very interesting looking. They are so perfectly round and firm, and the same midnight black as a raven's wing. It doesn't seem to matter what time of year it is, when out walking in the woods mossberries can often be spotted with little effort.

Mossberry has always been an important food and medicine staple for northern First Nations. Eating a small handful while in the bush is a great way to quench your thirst, but they have a very unique flavour, so don't compare the taste to other round northern berries, such as blueberries and cranberries. They work synergistically with other berries to create flavourful northern delights.

Medicinal Actions
Berries: Rich in antioxidants and anthocyanins. Branches: Antifungal, antibacterial

Medicinal Preparations
Berries: raw, syrup. Branches: decoction, tea

Medicinal Uses
A mossberry-bark infusion has been traditionally brewed by northern First Nations for colds and flus, diarrhea, and stomach problems. Decoctions of the root and bark have also been used for treating sore eyes and cataracts.

It was also traditionally used by First Nations to combat tuberculosis. University of British Columbia researcher Dr. Allison McCutcheon and her colleagues studied the effect of *Empetrum nigrum* and found that when ingested it does inhibit the growth of *Mycobacterium tuberculosis*. They also found that the branches exhibited strong antifungal and antibacterial properties.

Mossberries have the highest levels of anthocyanins of any of the northern berries, according to research conducted at the University of Kuopia in Finland. This makes them extremely antioxidant-rich, so eating the berries will help combat oxidative stress caused by free radicals in the body, assisting in the prevention of disease.

Mossberry also contains the compound quercetin that acts as a powerful antihistamine and can help relieve the symptoms of allergies; it also has anti-inflammatory properties. The quercetin can help to reduce inflammation and provide pain relief for aches

Mossberry has always been an important food and medicine staple for northern First Nations. PL

Part II: *Berries*

> "The clearest way into the Universe is through a forest wilderness."
>
> —John Muir, naturalist, *Travels in Alaska* (1915)

and pains, symptoms of arthritis, and can reduce inflammation in the prostate gland. Quercetin may also help combat fatigue and anxiety.

Food Uses
The flavour of mossberries seems to sweeten up with frost, freezing, and cooking. They can be used on their own or in combination with other berries, like blueberries, to make pies, muffins, pancakes, syrup, sauce, or fruit leather. The jam is a wonderful addition to meat, fish, fowl, homemade ice cream, or jam cookies.

Juice can be made by cooking the berries, straining them though a cheesecloth and then freezing the juice in ice cube trays. The cubes can then be added to boost your morning smoothie, a stew, or just as an ice cube in a herbal iced tea. More adventurous people can even try making mossberry wine!

Nutritional Profile
Vitamin C

Other Uses
Steve Johnson, founder of the Alaskan Flower Essence Project and author of *The Essence of Healing*, says *Empetrum nigrum* "stimulates awareness of cycles of light and darkness, internally and externally; enables us to hold these variations with respect and gratitude, rather than with attachment or aversion."

The berries create a black dye.

Cautions
None known.

ADDITIONAL RECIPES

Allergy-Aid Capsule, page 301

Blueberry-Mossberry Juice, page 336

Cold-and-Flu Tincture, page 294

Herb-and-Berry Antioxidant Smoothie, page 340

Herb-and-Berry Sweet Summer Days Syrup, page 355

Skin-Disinfectant Spray, page 295

Mossberry Syrup
A quick and easy-to-make topping for pancakes, ice cream, and cakes.

1 cup (250 mL) of mossberry juice (see page 334)
2 cups (500 mL) of cane sugar or honey
1 teaspoon (5 mL) of lemon juice

Place ingredients in a pot, simmer but do not boil. Stir and cook until you like the consistency. Store in the fridge, it will keep for months.

RASPBERRY

Rubus idaeus

Other Names and Etymology
Wild red raspberry. *Rubus* means "red" in Latin and *idaeus* is from "Mount Ida," in honour of the mountain in Crete where raspberries are said to have originated.

Family
Rosaceae (rose family)

Botanical Description
Erect woody shrub, to 1.5 m high, with branches covered in prickles; canes biennial, freely branching. Leaves with three or five leaflets, double-toothed, green above, densely white, woolly hairs below. Flowers in small clusters, axillary or terminal; sepals 4–10 mm long, finely pubescent and with gland-tipped hairs; petals about as long as the sepals, white. Fruit red.

Habitat and Range
Woodland clearings. Circumpolar. In North America, found from Labrador to Alaska, south to California and northern Mexico.

In the Yukon, it's found north to the Porcupine River valley.

Left: There is nothing like eating fresh wild raspberries right off the bush! CA

Right: Raspberry leaves are famous for helping women with various reproductive issues. PL

Part II: *Berries*

Plant Parts Used
Leaves and berries. The top first-year growth leaves are the best to use as medicine. Harvesting these leaves also helps the plant produce more fruit the following year.

Harvest Time
Leaves: spring/early summer before the berries start to form. Berries: when juicy, red, and ripe

THE RASPBERRY PLANT invokes a childlike wonder in me. It's such a familiar plant that most people don't need a field guide to identify it, and its fragrant aroma reaches us even before we see it.

I love the smell of raspberry leaves drying in my home. The light aroma wafts off the racks and permeates the whole house.

Medicinal Actions
Antidiarrheal, antiemetic, antioxidant, antiseptic, antispasmodic, astringent, cardiac tonic, emmenagogue, febrifuge, galactagogue, refrigerant, stimulant, styptic, tonic

Medicinal Preparations
Leaf: Capsules, glycerite, infusion, tea, tincture, vinegar. Berry: Syrup, tea

Medicinal Uses
Raspberry fruit acts as a blood tonic and is high in vitamins C and B, and minerals like magnesium, calcium, iron, and phosphorus.

The dried leaves are famous for their use as a women's tonic for all the stages of reproduction, from the first menstrual cycle to the last, and beyond.

The leaves contain fragarine, an alkaloid that helps tone and strengthen the muscles of the pelvic region, including the uterus. Fragarine can also help regulate the menstrual cycle and prepare the uterus for conception.

For menstrual cramping and excessive bleeding, raspberry-leaf infusion or tincture can help. Raspberry tea can also help with the nausea that some women experience during menstruation and ovulation.

A leaf infusion can be used in combination with red clover for aiding both male and female fertility and conception.

During pregnancy, the nutrient properties keep the mom-to-be nourished and supported and can ease morning sickness and can help prevent miscarriage and haemorrhaging.

Like many women before me, I used raspberry-leaf infusions for both my pregnancies to support the uterus. My midwives told me to drink one cup per day in the first trimester, two cups per day in the second trimester, and three cups per day in the third trimester.

> Raspberry-leaf tea can also be given to pregnant animals to support the birthing process.

Raspberry

Yukon midwife Heather Ashthorn says, "The teachings around raspberry leaf that have been passed down from one midwife to the next has been that raspberry leaf is a beneficial herb for most women during pregnancy, but like anything we use medicinally, we treat it with respect and listen to what it is telling us." She adds, "If you have any adverse reactions, whether repelled by the taste or feel crampy after a cup or two it may not be the herb for you." She urges women who are pregnant and want to use herbs to call a local midwife.

Raspberry leaf is known for its relaxant and pain-relief properties, and for its ability to help speed the recovery process after childbirth. In preparation for my second childbirth, I froze raspberry-leaf tea in ice cube trays, crushed up the ice cubes, and sucked on bits of them between contractions during labour.

After giving birth, the tea helps with cramping, blood loss, and restoring the uterus to pre-birth size.

Due to its high vitamin and mineral content, raspberry-leaf tea is a beneficial breastfeeding tonic that helps produce plenty of nourishing milk. Though, by this time, some women may be tired of the astringent taste of raspberry leaf and may want to opt for blending the tea with other teas or juices.

For children, raspberry leaf can help calm an upset tummy. (See Dosages, page 38.)

A leaf infusion makes an effective mouthwash for inflammations such as ulcers, cankers, bleeding gums, and can be used as a gargle for sore throats. The tea can help clear up diarrhea. A poultice can be used for cuts, scrapes, and wounds.

Raspberries have the highest percentage of ellagic acid of any the red berries. Ellagic acid is reported to have antioxidant, anti-mutagen, and anti-carcinogenic properties that actively help treat cancers of the breast, esophagus, skin, colon, prostate, and pancreas.

> "Tea is instant wisdom—just add water!"
>
> —Astrid Alauda

Tasty raspberries are high in vitamins C and B, and minerals like magnesium, calcium, iron, and phosphorus. CA

Food Uses

Who doesn't love the sweet, juicy taste of a bright red, wild raspberry? Nutritionally, raspberries are full of healthy compounds like flavonoids and anthocyanins that support and benefit human health and help prevent disease.

There are so many ways to use raspberries in the kitchen. In the late summer, I love topping off a garden salad with fresh berries. In combination with raspberry vinaigrette they create an amazing taste sensation!

Raspberries are high in pectin and citric acid, so raspberries are great for making jam. Raspberry jam is a favourite in my household and a staple we use not just on toast, but in plain yogourt, cakes, and in our Christmas almond-raspberry thumbprint cookies.

Raspberry syrup can be used on its own as a tasty topping for desserts like cheesecake and pancakes, or combined with other berries to create a berry medley extravaganza! The syrup can also be used medicinally in nutritious tonics and blends for supporting the blood and whole system.

Part II: *Berries*

ADDITIONAL RECIPES

Afternoon Floral Tisane, page 281

Good-for-Fever Infusion, page 282

Good-for-Gout Infusion, page 284

Good-Sleep Infusion, page 283

Herb-and-Berry Antioxidant Smoothie, page 340

High-Calcium Infusion, page 282

Moon-Cycle Nourishing Vinegar, page 203

Moon-Time Infusion, page 284

Nourishing Vinegar Tonic, page 296

Raspberry-and-Rosehip Vinaigrette, page 362

Revolutionary Raspberry Rosehip, page 335

Tea-for-Two Pregnancy Infusion, page 284

Nutritional Profile

Leaves: Calcium, chromium, cobalt, iron, magnesium, manganese, niacin, phosphorus, potassium, protein, riboflavin, selenium, silicon, sodium, thiamine, tin, aluminum, ash, vitamins A and C, and trace amounts of zinc. Berries: High in phytochemicals (such as the antioxidant quercetin), vitamins A, C, E and K, folate, choline, calcium, iron, magnesium, phosphorus, potassium and dietary fibre.

Cosmetic Uses

An infusion of raspberry leaves in witch hazel can make an excellent facial tonic. Steaming with the leaves unclogs facial pores.

Other Uses

Herbalist Michael Tierra, author of *The Spirit of Herbs*, notes that, "Raspberry leaf helps one tune into the child within and allows natural enthusiasm for life to be manifested. It will soften the heart and allow for greater receptivity and intuition."

Herbalist Robert Rogers says, "Raspberry flower essence helps people to get back on track with their own life path and journey. It helps one turn the corner and get on with life. It is for people who have unfulfilled dreams eating away at them. It is an essence for facilitating endings and opening the door for new beginnings."

I developed Aroma Borealis's Root Chakra Vibrational Essence with raspberry fruit and leaves to help balance the root chakra (first chakra).

Cautions

The small seeds of raspberry berry may irritate the colon of those with diverticulosis. It's important to note that when the leaves are drying, they go through an important chemical process. They should only be used fresh or dried, but not when wilted.

Raspberry Jam

Here is a quick-and-easy raw alternative to homemade jam making. It can be made as needed from fresh or frozen berries. You can make as little or as much as you like!

1 cup (250 mL) of fresh berries (if you are using frozen, let thaw)
1 teaspoon (5 mL) of honey (optional)
Juice of half a lemon (optional)

Place berries and other optional ingredients in a blender on low, or blend by hand with a potato masher.

Use immediately, leftovers can be stored in a sealed container in the fridge or dehydrated and made into fruit leather. It can be used on toast, muffins, or in yogourt or as a dessert topping for cheesecake. Use your imagination!

SASKATOON BERRY

Amelanchier alnifolia

Other Names and Etymology
Serviceberry, juneberry. *Alni* is Latin for "alnus" or "alder," and *folia* means "leaf." As the scientific name indicates, Saskatoon berry does indeed have an alder-shaped leaf. "Saskatoon" originates from the Cree name *misask-wa-toomina*. The city is also named after the tree.

A. alnifolia

Family
Rosaceae (rose family)

Botanical Description
In the northern part of its range, it is a tree-like shrub rarely more than 1 m high, with underground horizontal stems forming small colonies or thickets; bark grey or dark brown, smooth. Leaves simple, alternate; elliptic with rounded tips, sharply toothed margins, hairless at least when mature. Flowers in clusters; petals five, 5–14 mm long, white. Berries juicy, purplish-black when ripe.

Habitat and Range
Open woodlands and slopes. In North America it grows from Hudson Bay and western Ontario through to British Columbia, south to California and north to southeastern Alaska. In the NWT and in the Yukon, grows north to Dawson City.

Saskatoon berries become purply black when ripe. They have anti-inflammatory and antioxidant properties. CA

Part II: *Berries*

Plant Parts Used
Leaves, bark, pomes (berries)

Harvest Time
Leaves: gathered in the spring/early summer before the berries start to form. Berries: gathered when they are juicy, red and ripe (unless treating diarrhea, in which case they should be harvested prior to ripening).

Saskatoon flower essence can be taken to connect the inner self and the higher spirit. CA

I HAVE AN AFFINITY for the Saskatoon berry, simply because my mom, Deanna, was born in Saskatoon. During a recent visit to the city, I was impressed with the array of Saskatoon-berry products on sale at the airport. It was good to see so many value-added products lining the shelves that visitors would be able to take home and enjoy.

Medicinal Actions
Anti-inflammatory, antioxidant, antiviral, astringent

Medicinal Preparations
Cooked syrup, raw berries, tea/infusion

Medicinal Uses
Saskatoon berries can be eaten raw or made into a decoction before the berries ripen to help with diarrhea but—be careful—after they are ripe, eating an excess amount can cause diarrhea.

Being in the rose family, Saskatoon leaves contain tannins similar to those in the leaves of other plants in the same botanical family, such as raspberry and strawberry. The leaves' astringent nature helps to firm tissues and reduce excess mucus in the respiratory system.

Both Saskatoon-fruit pits and leaves contain cyanide compounds that give them an awesome dried almond flavour. This said, eating large amounts of raw berries may cause stomach upset; cooking destroys the compounds in both the leaves and berries.

The berry is an anti-inflammatory and antioxidant and helps reduce oxidative stress and inflammations in the body.

Ethnopharmacological research into traditional medicines at the University of British Columbia indicates that the Saskatoon plant extract acts as an antiviral agent.

Saskatoon Berry

First Nations have a long history of using Saskatoon as both food and medicine. They boiled the berries for juice to help relieve digestive disturbances, and then sometimes mixed this with the plant's inner bark and roots to help prevent miscarriages, stop excess menstrual bleeding and cramping, and to help women with transition into menopause. The juice and inner bark were also used after birthing for pain relief and as a blood tonic. The berries were also used to make eye- and ear-drop treatments.

Herbalist Robert Rogers states that it's important to know, "when using the bark, root or berry in a decoction to use the tea warm in the first trimester of pregnancy to prevent miscarriage, and to use cool tea for hemorrhaging and for use in menopause."

> It's the great, broad land 'way up yonder,
>
> It's the forests where silence has lease;
>
> It's the beauty that thrills me with wonder,
>
> It's the stillness that fills me with peace.
>
> —Robert Service, poet, "The Spell of the Yukon" (1916)

Food Uses

Saskatoon berries are sweet and have many uses in the kitchen. An excellent plant-based source of protein and fat, they are good added to your morning porridge for an extra protein boost. Saskatoon pie seems to be a autumn favourite, and preserves such as jam, jelly, and syrup line the panty shelves of many a northern kitchen.

The berries can also be dried whole and eaten like raisins or puréed and made into fruit leather.

Nutritional Profile

Vitamin C, calcium, copper, iron, magnesium, manganese, potassium, phosphorus, sulfur. High in protein, fat, and fibre.

Compared with other berries, like blueberries, strawberries, and raspberries, Saskatoons provide higher amounts of usable energy, protein, carbohydrates, fibre, vitamin C, iron, and potassium. They contain three times as much iron and copper in the same weight as raisins.

Other Uses

Herbalist Robert Rogers says, "Saskatoon flower essence is related to the soul qualities of giving thanks." He adds, "It is difficult for many to think of prayer as other than a religious ritual. Saskatoon essence can be taken to connect the inner self and the higher spirit, in a way that is free of dogma."

Cautions

Eating too many raw, ripe Saskatoon berries can lead to stomach upset or diarrhea. After they are cooked they should be fine.

Robert Serviceberry Jam?

When transplanted from the wild, Saskatoons take very well in the North. The shrubs are also readily available at local garden centres.

I see a lot of potential for commercial growing and processing of Saskatoons in the North. At the moment it's the second-largest commercial fruit crop on the Canadian Prairies (after strawberries) and studies indicate that there is a worldwide demand for Saskatoon-berry products.

Since the Saskatoon berry's other common name is "serviceberry," I could see someone in the Dawson City area of the Yukon creating a whole product line of serviceberry jams, syrups and teas, and tying them into a Robert Service storytelling theme. I think it would sell really well!

Part II: *Berries*

MN

ADDITIONAL RECIPE

Toon-Town Smoothie, page 337

Saskatoon-Berry Pie*

Filling
3½ cups (875 mL) Saskatoon berries
½ cup (125 mL) water
¾ cup (175 mL) sugar
3 tablespoons (45 mL) cornstarch
2 teaspoons (10 mL) lemon juice
1 tablespoon (15 mL) butter
¼ teaspoon (1 mL) almond extract

Pastry
2 cups (500 mL) pastry flour
1 teaspoon (5 mL) sea salt
½ cup (125 mL) butter
¼ cup (60 mL) ice-cold water

The filling: Place Saskatoon berries in water, cook until boiling. Add mixed cornstarch and sugar, boil until clear, stirring constantly. Remove from heat. Add lemon juice, butter, and almond extract. Cool. Use as a filling for pie, tarts, or as a cheesecake topping.

The crust: Using a pastry blender or the tips of your fingers, cut half of the butter into the flour-salt mixture until it becomes the consistency of cornmeal. Cut the remaining half into the dough until you have corn-nibblet-sized pieces. Mix in the cold water. Form into two balls.

Roll out one ball and place in pie dish. Fill pie crust with filling. Roll out second ball for the top crust, place it on top of the pie and puncture with a fork to let excess steam escape.

Bake at 425° F (220° C) for 15 to 20 minutes.

---

*This recipe was provided by Wendy Cooper of Saskatoon. Find her blog "The Cooking Blog" at www.thecookingblog.blogspot.com for more recipes.

SOAPBERRY

Shepherdia canadensis

Other Names and Etymology
Buffaloberry, mooseberry, soopolallie. The Latin *Shepherdia* is named after English botanist John Shepherd, and *canadensis* means "from Canada."

Family
Elaeagnaceae (oleaster family)

Botanical Description
Erect-ascending dioecious shrubs, to 1 m tall (to 2 m in southern range); branches with coppery scales. Leaves elliptical to oval, densely covered with silvery hairs and coppery scales below, green with few hairs above. Flowers in early spring (when crocuses are flowering); flowers yellow, small, in clusters (or with pistillate flowers single), with a slight fragrance. Berries 4–7 mm long, red with scattered coppery scales; juicy. Only the female plants produce fruit.

Left: Only female soapberry plants produce fruits—they seem to help with digestion. PL

Below: Soapberry flowers give off a slight fragrance. PL

Wild Food and Medicine Plants of the North

Part II: *Berries*

Soapberry flowers in early spring, usually in stride with the crocuses. BG

Habitat and Range
Riverbanks; clearings in spruce woods; dry, south-facing, well-drained alpine and subalpine slopes. In North America, it grows from Newfoundland to Alaska, south to Oregon. It's common throughout the NWT and Yukon, north to the British and Richardson mountains.

Plant Parts Used
Berries, branches

Harvest Time
Late summer, early autumn. The berries sweeten up after the first frost.

The plant's common name comes from the berry's high saponin content—the chemical compound that makes the berries foam up when whipped or shaken with water. When squished between your fingers, the berries feel slippery and a bit oily.

Soapberries are abundant in the Yukon. The area where I live is no exception—the open spruce and aspen forest floor is covered with soapberry shrubs. I love the colour of the berry and how it contrasts with its deep green leaves. I also like the taste of the berries. They have a layered flavour: First sweet and then bitter—they make my mouth water and get my digestive juices flowing.

When on the trail, it's fun to pick a few berries and rub them into your hands to feel the slippery soap-like saponins working. Add a little spit or water and your hands will get nice and clean.

Medicinal Actions
Cathartic, haemostatic, hypotensive, laxative, stomachic

Medicinal Preparations
Decoction, infusion, poultice, tea, wash

Medicinal Uses
Modern herbalism gives few uses for soapberry, but First Nations people have used this plant extensively as both food and medicine.

The fruits, with their bitter nature, seem to improve digestion. I've noticed eating too many can cause loose stools, leading me to believe that they are good for constipation. I've heard from other people that they've used soapberries to help stop diarrhea as well.

Soapberry

The fruit juice has been drunk as a treatment of digestive disorders and the berries have been eaten to treat high blood pressure. The berries and the juice can be applied externally in the treatment of acne and boils.

The roots are anti-haemorrhagic and cathartic. It's been reported that an infusion of the roots has been used in the treatment of tuberculosis and the coughing up of blood.

A decoction of the stems and/or inner bark has been used as a stomach tonic and also in the treatment of constipation and high blood pressure.

I've heard that a decoction of the stems and leaves is useful as a wash for treating sores, cuts, and swelling tissue.

A decoction of the plant, berries, stems, and roots have been used externally as a wash and rub for sore, aching limbs, arthritic joints, and skin sores. I imagine it would also be nice to add these plant parts to a bath or foot soak.

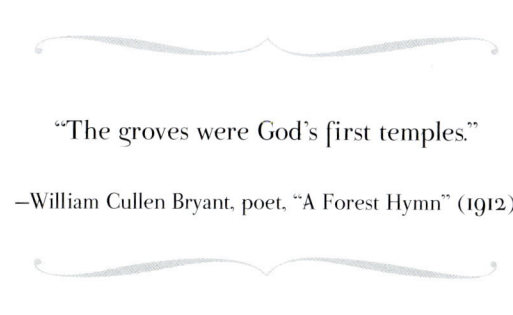

"The groves were God's first temples."

—William Cullen Bryant, poet, "A Forest Hymn" (1912)

Food Uses

The soapberry flavour is tart but pleasant, with a strong aftertaste. The berries become sweeter after frosts.

Many of my First Nations friends talk about eating "Indian ice cream" as kids. It sounded so neat that I had to try it! With enough sugar added the results are yummy—it makes a good topping on a vanilla or chocolate cake or as a condiment with dry meat. To make, add 1 cup (250 mL) soapberries to ¼ cup (60 mL) water and add 6 tablespoons (90 mL) white or brown sugar. Mix or beat ingredients until you have a stiff and pink foam.

Herbalist Robert Rogers suggests that soapberry was commonly called "buffaloberry" because it was used to spice up the buffalo meat eaten by the voyageurs. The ripe berries may also have been a sign that it was time for the buffalo hunt.

Soapberries can be made into a jelly, sauce, and even added to ice cream. Michele Genest, a Yukon chef and the author of *The Boreal Gourmet*, has a Soapberry Jelly Ice Cream in her book that is mouthwatering, and truly inspired.

Soapberries can also be added to high-iron tonics along with strawberry leaves and nettle.

It's been reported that to add nutrition to their meals during the long, cold, and dark Yukon winters, First Nations made soapberry patties that they dried over a fire, and then stored.

Nutritional Profile

Iron, vitamin C

Cosmetic Uses

The mid-summer branches of the soapberry shrub can be used as a brown hair dye. Simply boil the branches for two to three hours in water until it looks really brown. Then strain and bottle the liquid for use.

Part II: *Berries*

Other Uses

Steve Johnson, founder of the Alaskan Flower Essence Project and author of *The Essence of Healing*, says that soapberry is indicated for "fear of the power of nature; fear of one's own power; using one's power in irresponsible, inappropriate, or unbalanced ways." He adds that the essence "stimulates the release of tension from the heart associated with a fear of nature; helps us move through fear with an open heart; supports us in channeling the expression of power through our hearts."

The fruit of the soapberry produces a red dye.

Cautions

Use fruit in moderation.

Soapberry Hair Shampoo
Completely biodegradable! An environmentally friendly alternative to store-bought shampoos.

1 cup (250 mL) fresh soapberries
1 cup (250 mL) hot water

Mash or blend soapberries and water together until completely integrated. Strain through cheesecloth, ringing every last drop into a cup.

Massage desired amount into wet hair. Rinse.

This is a great shampoo to make out in the bush. It doesn't lather like commercial shampoos, but it also leaves no residue. After you've mastered this one, try adding other plants to it like mint, horsetail, or wild sage. It will keep for a couple of days.

Many helping hands make light work. BG

STRAWBERRY
Fragaria virginiana

Other Names and Etymology
Virginia strawberry, wild strawberry. The Northern Chipewyan call the berry *idziaze* or "little heart." *Fragaria* is from the Latin *fragrans*, meaning "fragrant," and *frango*, meaning, "fragile." The common name is derived from the Anglo Saxon name *strauberige* or *streowberie*, which may have referred to the runners that strawberries send out on the ground. It could also come from straw being laid in between rows of garden strawberries to keep the fruit clean.

Family
Rosaceae (rose family)

Botanical Description
Perennial herb with scaly rhizome and with long, slender nodally rooting stolons. Leaves all basal, long petioled, with three leaflets; leaflets coarsely toothed, with silky hairs below, hairless above. Flowers in small clusters; petals five, 6–8 mm long, showy, white. Fruit somewhat oval, red when ripe, seeds imbedded on surface of the fruit.

Habitat and Range
Clearings in woods and open slopes. Often in disturbed areas.

Plant Parts Used
Leaves and berries

Harvest Time
Leaves: spring/early summer before the berries start to form. Berries: when they are juicy, red, and ripe

It's hard not to get excited by a patch of delicious, wild strawberries.

Part II: *Berries*

WHEN OUT WALKING in the backcountry you may be surprised by the sudden appearance of wild strawberries. Their runners weave a web across the forest floor, but it's their bright crimson fruits that really captivate us and draw us to ecstasy with their stunning flavour.

Because strawberries are one of the sweeter boreal berries, I wanted to have some nearby, so I transplanted some into a little garden beside my kitchen. On a hot day, the subtle aroma of the leaves and berries waft through the air. When the berries are ripe and at their juiciest, I pick a handful to add to my breakfast porridge or pancakes, and invite the children to pick them before the birds do.

Small, sweet, and juicy, this miniature strawberry is the mother of the larger species we see in grocery stores today. (At some time in the past they were hybridized with a large South American variety.)

Medicinal Actions
Antidiarrheal, antioxidant, antiseptic, astringent, tonic

Medicinal Preparations
Leaves: syrup, tea/infusion, tincture, vinegar. Berries: juice, raw, syrup, tea

Medicinal Uses
The berries and the leaves both have medicinal value, as was discovered in the 1700s by botanist Carl Linnaeus, who used strawberry to treat his gout. I'm not sure how he used it, but I assume he would have eaten the berries and perhaps drunk the leaves in an infusion.

Wild strawberries are not only delectable, but also packed with phytonutrients that are beneficial to our overall heath and wellness, such as flavonoids, anthocyanins, and ellagic acid. Ellagic acid has been reported as having antioxidant, anti-mutagen, and anti-carcinogenic properties that actively help treat cancers of the breast, esophagus, skin, colon, prostate, and pancreas.

Because they are high in vitamin C, strawberry leaves—on their own or in combination with other herbs—can help keep colds and flus at bay.

In a tea infusion, the dried leaves help regulate menstruation, calm morning sickness, promote abundant breast-milk production, and can act as a mild nerve tonic.

As a mouthwash, the leaves and fruits can help alleviate toothache and heal ulcers of the gums. A poultice made from fresh leaves can be used on open wounds, eczema, and psoriasis to accelerate healing.

The iron-rich leaves can be added to teas, tinctures, tonics, or elixirs to help prevent anemia.

Wild strawberry flowers are a promise of good things to come. PL

Strawberry

Food Uses
Strawberries are usually eaten raw, but they can also be frozen, dried, and made into jams, jellies, fruit leather, sauces, and syrup.

A wild-strawberry smoothie is a great way to boost energy throughout the day. The berries can also be added to breads, muffins, cakes or, better yet, made into a wild-strawberry sauce to accompany desserts, rice, and poultry dishes. And summer just wouldn't be complete without eating strawberry shortcake. I love infusing and blending the berries in apple cider vinegar to use in a salad dressing; it's great on a strawberry pasta salad.

Nutritional Profile
Calcium, copper, fibre, folate, iodine, iron, magnesium, manganese, omega-3 fatty acids, potassium, vitamins A, C, B2, B5, B6, K

Cosmetic Uses
The leaves are astringent and when used in a facial steam, help reduce excess sebum (which causes oily skin). The leaves can be transformed into a facial toner.

Eating raw strawberries also helps whiten the teeth without harming or removing enamel.

Other Uses
For many aboriginal people, the wild strawberry is considered the first fruit of the womb of Turtle Mother, and is a spiritual and healing entity. You will often see the image of the womb-shaped strawberry, symbolic of fertility and sacred renewal, in First Nations' baskets, quill work, moose-hair tufting, beadwork, and wood carvings.

Cautions
When strawberry leaves are drying, they go through an important chemical process. They should only be used fresh or dried, but not when wilted.

> "I had rather have one pint of wild strawberries than a gallon of tame ones."
>
> —Perry Medsger Oliver, author, *Edible Wild Plants* (2007)

Wild-Strawberry Exfoliating Face Mask
Strawberries are purported to have the same pH as that of human skin, making them nourishing and balancing for our skin. You could just mash up a handful of wild berries and use them to exfoliate your face right away, or you could make this sweet-smelling strawberry exfoliating mask.

A handful of wild strawberries
Juice of 1 organic lemon
½ cup (125 mL) yogourt

In a bowl, squeeze lemon juice over berries, add yogourt, and mash together. Chill for twenty minutes. Apply generously to face and neck, then sit back in a bath and feel the berries work their magic. This mask helps reduce inflammation and pulls the heat out of irritated skin caused by acne, sunburn, or scarring.

ADDITIONAL RECIPES

High-Iron Syrup, page 299

Nourishing Vinegar Tonic, page 296

Strawberry Fields Forever, page 335

Pine-Forest Smoothie, page 338

TREE PROFILES AND RECIPES

Alder **235**
   Alder Poultice *237*
Trembling Aspen **238**
   Aspen Toothache Remedy *240*
Birch **241**
   Birch-Syrup Granola *246*
Subalpine Fir **247**
   Sweet-Balsamic Body Oil *250*
Lodgepole Pine **251**
   Pine Tree Ache-and-Pain Liniment *254*
   Pine Household Cleaner *254*
   Pine Bark Bread (Crackers) *255*
Balsam Poplar **256**
   Boreal Healing Oil *258*
Spruce **259**
   Spruce Winter Salve *264*
Tamarack **265**
   Tamarack Tea *268*
Willow **269**
   Willow Inner-Bark Tincture *273*

ALDER

Alnus species

Other Names and Etymology
Speckled alder, red willow, mountain alder. The Van Tat Gwich'in word for alder is *k'oh*. Many Yukon First Nations people call the alder "red willow" because of its red sap which is high in tannic acid. *Alnus* is from the Anglo Saxon *alr* or the Old English *alor*, that in turn are from the Old German *elawer* or *elo* meaning "reddish-yellow." Over time it progressed to "aler," then "aller" or "aldir" before becoming "alder." The name is in reference to the wood's characteristic change of colour after felling. In Ogham, the Celtic tree alphabet, alder is called *Fearn*, which means "strength."

Family
Betulaceae (birch family)

Botanical Description
Large shrubs or small trees; bark is reddish or greyish brown. smooth, in age the bark can break open into irregular plates. Leaves are oval to broadly ovate and serrated—they are greyish green above, paler beneath. Flowers develop before leaves in the early spring; cones short peduncled and nutlets are wingless. One of the two species of alder in Yukon, the speckled alder, reaches tree size.

The illustration we are using for *A. viridus* originally appeared as *A. crispa* in *The Flora of the Yukon*, edited by William J. Cody. Yukon biologist Bruce Bennett says that *A. crispa* is now considered to be the subspecies *A. veridus* ssp. *fruiticosa*.

Alder catkins are high in protein. CA

Part II: Trees

Habitat and Range
Common on riverbanks and lakeshores. Forming thickets on stream banks, mountain slopes, in woods, and on tundra. Grows throughout boreal North America.

Plant Parts Used
Catkins, leaves, dried inner bark

Harvest Time
Inner bark: in the spring or autumn. Leaves: spring to early autumn. Catkins: in spring

IN FOLKLORE THE ALDER tree is known as the "King of the Waters," and the willow tree as its queen. You know you are at an alder patch if it's close to water. The roots of alder like a wet habitat and they have nitrogen-fixing nodules that help improve soil conditions.

Medicinal Actions
Anti-inflammatory, astringent, diuretic, hemostatic

Medicinal Preparations
Bath, compress, decoction, facial steam, foot bath, oil infusion, poultice, salve, tea, wash

Medicinal Uses
The leaves and dried inner bark can be used in tea as a bitter medicinal tonic that aids the digestive system.

Like willow bark, alder bark contains the anti-inflammatory salicin and can be used for stomach ache, to reduce inflammation, and in a compress, wash, or poultice for hemorrhoids, arthritis, sore muscles, and general pains. As a skin-healing remedy, alder can help to take the itch out of skin irritations like chicken pox, bug bites, and weepy wounds.

On its own, or in combination with other plants like juniper or wild chamomile, alder leaves can be used in a face steam to help cleanse and tighten facial pores. A cotton ball dipped in alder tea can help to stop a fresh-squeezed pimple from weeping.

Herbalist Janice Schofield says, "alder decoctions can be sipped in emergencies for internal bleeding, when no doctor is available." She adds, "Alder's tonic properties make it helpful for strengthening and toning the whole system. For a more palatable tonic tea, blend with nettle leaves, dandelion, and devil's club roots."

Alder leaves and the dried inner bark can be used to help digestion. Always be sure to dry the inner bark of the alder tree before using it as medicine. PL

Food Uses
The alder catkins are a good survival food because they are high in protein. They can be

Alder

dried and used in soups and stews as well. It's always nice to find plant-based proteins.

Many Yukon First Nations chip up the alder and use it for smoking fish and wild meats. The flavour is earthy and delicious; there is no other taste like it!

Nutritional Profile
Protein

Other Uses
Steve Johnson, founder of the Alaskan Flower Essence Project and author of *The Essence of Healing*, says alder essence is indicated for "taking life at surface value; for those who are unable to see what one senses to be true." He adds that the healing qualities promote clarity of perception on all levels, and help us integrate seeing with knowing so that we can recognize our highest truth in each life experience.

Alder was traditionally used for tanning hides.

In her book, *Discovering Wild Plants*, Janice Schofield notes that "an ancient remedy for ridding your house of fleas is to place dew-laden alder leaves on the floor of your home. The fleas are said to be attracted to the moist alder, so one can easily gather up the flea-covered leaves."

Alder wood makes beautiful music! Fender uses alder for building electric guitar bodies because the resonance of alder creates a full, well-rounded sound.

Cautions
The fresh inner bark can induce vomiting. Dry the inner bark before using it.

> "In all creation, trees, plants, animals, and gem stones, there are hidden secret powers which no person can know of unless they are revealed by God."
>
> —Hildegard von Bingen, nun and herbalist (1098–1179)

Alder Poultice

An alder-leaf poultice is used to help regenerate, to soothe, and to heal tissue. It can also help stimulate circulation, and relax and warm muscles. Use it for taking the itch out of skin irritations such as chicken pox, bug bites, and weepy wounds.

Place ½ cup (125 mL) of alder leaves in a saucepan with enough water to cover. Heat and simmer for 15 minutes.

Let the herbs steep until the preparation has reached room temperature. Place warm herbs in a couple of layers of cheesecloth over the affected area.

Repeat if necessary

TREMBLING ASPEN

Populus tremuloides

Other Names and Etymology
Quaking aspen, white poplar. *Populus* comes from the Latin name for poplar, which means "tree of the people." *Tremuloides* is also Latin meaning "trembling." In Ogham, the Celtic tree alphabet, aspen's name is *Eadha* and means "endurance."

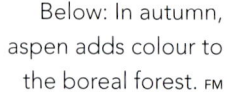

Below: In autumn, aspen adds colour to the boreal forest. FM

Below right: Aspen bark is covered with a white powder that can be used as a mild sunscreen. CA

Family
Salicaceae (willow family)

Botanical Description
Slender trees to 35 m high in favourable situations, bark smooth, white-greyish with a greenish tinge, sometimes becoming furrowed with age. Mature leaves oval to ovate; petioles long and slender, flattened laterally, causing the leaves to tremble in the slightest breeze.

Trembling Aspen

Habitat and Range
Common forest species on dry, burnt-over slopes and ridges; in the autumn the leaves turn golden yellow, in sharp contrast to the dark green of a spruce forest. Found throughout boreal North America, from Newfoundland to Alaska, south to Virginia and Mexico. In the Yukon, found north to the Porcupine River.

Plant Parts Used
Buds, leaves, catkins, inner bark, aspen dust

Harvest Time
Leaves, catkins, dust: spring and summer. Bark: in spring, from pruned branches.

IN THE SUMMER, I like to garden late into the evening and sometimes into the early morning. It always feels like a time of grace. I love the silence and the mystical feeling. Aspen contribute to this energy with their spontaneous dancing, creating music throughout the whole forest. This quote by Nathaniel LeTonnerre really sums up what an aspen grove feels like to me: "Breeze is the conductor, trees the musicians, leaves the instruments."

Aspen trees move and make music with the wind. Top: BG, Above: CA

Medicinal Actions
Analgesic, anodyne, anticoagulant, anti-inflammatory, anti-rheumatic, antiseptic, astringent, bitter, cholagogue, diuretic, febrifuge, (mild) sedative, tonic

Medicinal Preparations
Bath, cream, infusion, liniment, oil, poultice, salve, soak, steam, tea, tincture

Medicinal Uses
The bark of the aspen tree is covered with a white powder that when rubbed on the skin can be used as a mild sunscreen. Adding it to jojoba oil will increase the protection factor. It's good only for moderate sunlight.

Being in the willow family, the leaves and bark contain salicin, an Aspirin-like substance that acts as an anti-inflammatory, analgesic, and as a mild sedative. Because of this, it can be used for pain relief.

A tea infusion, topical liniment, or oil infusion of the bark can help treat inflammation, rheumatoid arthritis, aches, and pains. A wash, poultice, salve, or cream can be made from the bark or leaves for itchy rashes, cuts, abrasions, scratches, wounds, and burns.

A light tea is good for use with intermittent fevers caused by the common cold, flu, for treating coughs, or as a gargle for sore throats. Taste is bitter and odourless

Part II: *Trees*

> "The trees are whispering to me, reminding me of my roots, and my reach... shhhhhh... can you hear them? Selflessly sharing their subtle song."
>
> —Jeb Dickerson, blogger

Aspen's diuretic, anti-inflammatory, and antiseptic properties help with cystitis (an inflammation of the urinary bladder and ureter) and with benign prostatic hypertrophy (BPH).

An infusion of the bark can also be used for chronic diarrhea, dysentery, indigestion, heartburn, nausea, flatulence, general stomach upset, and can help support the gallbladder to stimulate the secretion of bile.

Food Uses

Like balsam poplar, aspen's catkins can be eaten raw and are high in vitamin C. The catkins can also be added to a campfire soup or stew. The cambium—the thin layer between the outer bark and the inner sapwood—is a very nourishing survival food.

Herbalist Robert Rogers says the white powder on the bark is a yeast that can be used to rise bannock and other breads.

Nutritional Profile

Catkins: High in vitamin C

Other Uses

Woodland Essences, a lovely little company in upstate New York, created a quaking-aspen-tree essence that is used for "Mother Courage and a strong heart." This is good for people who are in the throes of doing what needs to be done and for people who are "breaking the path, the ice and new beginnings."

It also helps with "Fearlessness in the face of the unknown and understanding of and an easy connection with subtle planes." I used this essence a lot when I first opened my store Aroma Borealis Herb Shop.

Dr. Edward Bach, the creator of the Bach remedies, recommended aspen for "Vague unknown fears, for which there can be given no explanation, no reason. It is a terror that something awful is going to happen even though it is unclear what exactly. These vague inexplicable fears may haunt by night or day. Sufferers may often be afraid to tell their trouble to others."

Cautions

Aspen can cause an allergic reaction in those with tree allergies.

Aspen Toothache Remedy

Chew up an aspen leaf and pack around the sore tooth. Let the juices of your saliva mix with the astringent, medicinal juice of the leaf; this will help relieve pain and draw out infection.

BIRCH

Betula species

Other Names and Etymology

Betula neoalaskana: Alaska paper birch, paper birch, resin birch, Yukon white birch. *Betula glandulosa*: Ground birch. The Van Tat Gwich'in word for both the *B. neoalaskana* tree and *B. glandulosa* shrub is *aat'oo*, and the bark is *k'ii*. *Betula* might come from the Sanskrit word *burga* that means "a tree whose bark is for writing on," or it could be derived from the Gallic word *betu* that translates as "heart." *Neoalaskana* is from Latin, and refers to the fact that the tree is found in Alaska. *Glandulosa* is from Latin and means "with glands." In Greek, *papyrifera* means "paper bearing." In Anglo Saxon, "birch" is derived from, *birk* or *birce* meaning, "white or shining." In Ogham, the ancient Celtic tree alphabet, birch is called *beithe,* which means "beginnings."

Left: The outer bark of the birch tree was traditionally used to make berry-picking baskets. CA

Right: Birch leaves, as well as the inner bark of the tree, help the digestive system. PL

Part II: *Trees*

In the early spring, birch trees around the world are tapped to make syrup. BL

Family
Betulaceae (birch family)

Botanical Descriptions
There are six types of birch found in the Yukon, four are shrubs including *B. glandulosa*, dwarf or ground birch. *B. neoalaskana* and *B. papyrifera* both have thin exfoliating papery bark that is white, reddish, or grey-brown. Twigs are densely resinous and glandular.

Betula neoalaskana: Leaves are deltoid in shape, broadly triangular in outline, dentate margins with scattered fine glands below. *Betula papyrifera*: Leaves are ovate. The leaf margins are double-serrate.

Betula glandulosa: Is a shrub with one to several main stems, 0.3–2 m, or more, high, often forming thickets; twigs densely covered with resin glands, leaves orbicular or circular. Grows in low Arctic tundra, woodland muskegs, and peat bogs.

Habitat and Range
Betula neoalaskana: Grows in acid and peaty not-too-wet situations. In North America, from Manitoba and the southwest District of Keewatin through Saskatchewan to the District of Mackenzie, northern British Columbia and Yukon to central Alaska; the common paper birch in the Yukon territory found north to latitude 68° north. *Betula papyrifera*: Grows in open woodland at lower altitudes. Throughout boreal North America, from Labrador to the Alaskan Panhandle, south to Pennsylvania, New York, Minnesota, Nebraska, Colorado, Montana, and Washington; in the Yukon territory, the Mackenzie drainage including the Liard and Peel rivers, all other reports appear to be *B. neoalaskana*. *Betula glandulosa*: Boreal and subarctic North America, from southwest Greenland to Alaska, south to Nova Scotia, New Brunswick, Minnesota, Colorado, and northern California; found throughout all of the Yukon territory.

Plant Parts Used
Bark, buds, catkins, leaves, sap

Harvest Time
Sap: early spring, when the sap makes it upward journey to the tips of the branches. Branches and leaf buds: late spring, before the buds open. Leaves: when they open, or anytime throughout the summer, preferably in the morning before the sun's heat releases their volatile essential oils. Inner bark: from small branches harvested in the spring, or in the autumn, when the leaves just start to turn, and before the energy, or medicine, moves fully into the roots.

Birch

Dwarf birch can be used topically for pain relief. BG

THE ENGLISH POET Samuel Coleridge (1772–1834) called the silver birch the "Lady of the Woods" because of the way it gently sways when the wind blows.

It's a beautiful, majestic tree and it has so many gifts to offer.

In the Yukon, we are fortunate to have a family-run birch-syrup operation that provides us with a nutritious, alternative sweetener.

One of the only hardwoods in the Yukon, birch has an array of uses. The wood can be used in fine woodworking, cabinetry, and other crafts. Its outer bark is famous for making canoes and the beautiful baskets that were traditionally made by northern First Nations to collect berries. Interestingly, it turns out that the bark contains antibacterial properties that help keep berries fresh—remember, there are no accidents!

The whitish-pink bark peels off the tree, but I don't recommend doing this because stripping the tree of its bark will kill it.

Medicinal Actions
Anti-inflammatory, anti-rheumatic, antiseptic, astringent, diaphoretic, diuretic, hemostatic, tonic, urinary antiseptic

Medicinal Preparations
Bath, compress, cream, decoction, essential oil, hydrosol, infusion, liniment, oil, poultice, salve, steam, syrup, tea, wash

Medicinal Uses
Dried birch leaves are good to have on hand for ailments such as urinary problems, kidney stones, chronic cystitis, and edema. Birch leaf can be used in a tea to treat high cholesterol.

Birch leaves are bitter, and are considered both a stimulant and tonic to the digestive system. Dried inner bark is good to have in your medicine-making pantry for use as a decoction for treating diarrhea.

Birch is high in methyl salicylate (salicin), the anti-inflammatory that makes birch such good medicine for pain relief. The leaves and branches are full of a resin-like substance that has a wintergreen aroma; they can be used as a tea or decoction, or topically as a compress, poultice, or in a bath for sore, strained muscles, sprains, headaches, rheumatic pain, and gout.

A steam of leaves can help clear sinus congestion. You can also add a teaspoon of strained birch-water infusion to a neti pot nose-wash solution. (A neti pot is a small teapot-like pot used to cleanse the nasal passages.)

The birch buds can be collected in the spring and infused in oil to make an external salve for inflamed skin. The buds can also be blended with others, such as poplar or spruce tips, to create an aromatic, well-rounded topical remedy.

Part II: *Trees*

Top: The birch sap comes out clear. It is in the boiling process that the sap takes on the dark colour of birch syrup. BL

Above: The hot syrup is put through a felt filter and sealed in buckets. Later Berwyn and his crew bottle the syrup in a commercial kitchen in Dawson city. BL

Yukon Birch Syrup

In the Yukon, we are fortunate to have Uncle Berwyn's Yukon Birch Syrup Company. Every year, Berwyn and his crew collect the crystal-clear sap in buckets in the early spring from their Mayo-area birch stand. The sap is then evaporated using a stainless steel, wood-fired evaporator. Once it's made into the dark, rich syrup, it's poured into big buckets and transported through the bush by sled, put onto a truck, and then transported to a commercial kitchen in Dawson City where it's bottled and made market ready. Some of the syrup is then shipped to Whitehorse where the Yukon Brewing Company makes it into a birch beer, only available in one-litre growler jugs that hold draft beer. It goes fast!

Due to changes in weather conditions, each year the sap run is a little different. Berwyn has had seasons ranging from eleven to twenty-one days, though he talked to an old-timer in the Mayo area who once had a month-long run. While in camp, Berwyn and his crew drink the sap like water because they don't have an accessible water source, and because the sap is ninety-nine parts water to one part sugar.

Once boiled down, the unique-tasting syrup is delicious, but in a different way than maple syrup. Birch syrup is darker and richer tasting than maple syrup. This is because maple contains sucrose, the sugar found in white sugar, while birch syrup is fructose sugar, the same sugar found in fruit.

It also takes only forty litres of maple sap to make one litre of syrup—while Berwyn says it takes eighty litres of birch sap to make one litre of syrup.

"[That's] a lot more work for the same amount of syrup!" says Berwyn. "Comparing the two is like comparing apples and oranges."

Northern chefs covet birch syrup. They use it in many interesting ways, including in marinades, barbeque sauces, vinaigrettes, desserts, and as a glaze for vegetables. In *The Boreal Gourmet: Adventures in Northern Cooking*, chef Michele Genest has some great recipes using birch syrup including a cedar-planked salmon with whiskey birch and maple syrups sauce, and a glaze for cinnamon rolls.

Food Uses

In the early spring, birch trees around the world are tapped to make syrup. The clear, free-flowing sap is also enjoyed by many as a spring tonic.

I make a birch granola that I love! It's great first thing in the morning topped with a big dollop of homemade yogourt.

Fresh birch leaves infused in apple cider vinegar for a month or so make an excellent nutritive tonic that is rich in vitamins and minerals that are both bone-building

Birch

and strengthening to the immune system. The vinegar can be added to salad dressings. A tablespoon in water works as a quick pick-me-up and can help relieve a headache.

Birch trees also contain a sweetener called xylitol. The Finnish were the first to discover this and began processing the inner bark to make toothpastes, mouthwashes, and gums. Xylitol is the sweetener in many natural gums and toothpastes and can even be bought in bulk in health markets.

It's claimed that this natural sweetener kills bacteria and helps reduce cavities. Unfortunately, xylitol can be fatally poisonous to dogs, and there have been a number of cases where dogs have died after eating products containing xylitol.

Nutritional Profile
Syrup: Vitamin C, iron, riboflavin, zinc, manganese, calcium, thiamine, magnesium, and potassium

Cosmetic Uses
A stronger astringent decoction of the leaves can be used topically for acne, or in a compress for boils, wounds, psoriasis, eczema, dandruff, and for hive-like rashes.

Alternatively, you can steep ¼ cup (60 mL) of dried leaves in 4 cups (1 L) of boiled water for four hours. After straining the leaves, use the infusion topically as a wash to soothe and heal the skin.

The branches and leaves can be used in an oily-skin facial steam for clearing congested pores.

Other Uses
Northern birches can also be used to make essential oils. Yukoner and wild-plant enthusiast Birch Kuch and I have distilled dwarf birch together into a hydrosol (water solution) to add to Arnica Ease, an ache-and-pain-relief cream we make at my store, Aroma Borealis. The oil is very refreshing and uplifting to the emotions.

Steve Johnson, founder of the Alaskan Flower Essence Project and author of *The Essence of Healing*, says that paper birch is indicated for "confusion or disorientation about the direction life should take," and when "unable to connect with deeper levels of insight regarding life purpose." He adds that paper birch "encourages a gentle unveiling of the true and essential self that is present within; helps us gain a clearer perspective of our life purpose and how to live it."

The buds of boreal birch's European cousin, *Betula alba*, are steam distillated to make a

---

Chaga: Birch's Medicinal Sidekick

Chaga (*Inonotus obliquus*) is a parasitic birch fungus also known as "clinker polypore" and "birch-canker polypore." Commonly found growing on living birch trees, it's characterized by a black, burnt-looking exterior and rust-red interior (though it can be confused with other fungi). It grows in North America, Poland, Siberia, and Russia. Chaga tea has been traditionally used to treat many types of cancers.

Medicinal Actions
Antifungal, anti-inflammatory, antioxidant, anti-tumour, immune stimulating, liver supporting

Medicinal Preparations
Decoction, tea, tincture

Part II: *Trees*

fresh, minty wintergreen-smelling essential oil. It has long been employed by the cosmetic industry in toothpastes, hair care, and soap making. It's also used to treat chronic skin conditions, particularly psoriasis and eczema.

The Finns are known to use a bunch of birch twigs, known as a *vihta* or *vasta*, to slap the skin and encourage blood circulation while in the sauna. This is considered purifying.

Cautions
None known.

ADDITIONAL RECIPE

Respiratory Steam, page 329

Birch-Syrup Granola

2 cups (500 mL) rolled oats
2 teaspoons (10 mL) wheat germ
¼ teaspoon (1 mL) salt
¼ cup (60 mL) dried nuts or seeds (sunflower, sesame, hemp, almonds, etc.)
¼ cup (60 mL) butter (melted) or sunflower oil
¼ cup (60 mL) birch syrup
1 teaspoon (5 mL) vanilla extract
¼ cup (60 mL) dried cranberries or blueberries (optional)

Combine dry ingredients (but not the berries) in a bowl.

In a separate bowl, blend the oil or melted butter, birch syrup, and vanilla. It should be runny.

Pour the liquid into the dry ingredients and stir thoroughly, making sure to coat the oats.

Spread onto a cookie sheet and toast in the oven at 250°F (120°C). Bake 20 to 40 minutes (depending how dark you like your morning granola), stirring the mixture at least once every 5 minutes to make sure it toasts evenly.

Once cooked, add the dried fruit, allow it to cool, and store in an airtight container.

SUBALPINE FIR

Abies lasiocarpa

Other Names and Etymology
Alpine fir, Canadian balsam. "Fir" is derived from the old English *furh*, the Old Scandinavian *fyri*, and the Danish *fyr*, each meaning "fire" or "firewood." *Abies* is derived from the Latin *abeo* meaning "to rise" or "arising," and is in reference to the height some species reach. *Lasiocarpa* is from the Greek *lasi*, meaning "hairy" or "shaggy," and *carpos*, "fruit," refers to the fir's rough-textured cones. In Ogham, the Celtic tree alphabet, fir is called, *ailm*, meaning "power."

Family
Pinaceae (pine family)

Botanical Description
A small evergreen tree becoming shrub-like or even prostrate at timberline; bark smooth, greyish, and covered with conspicuous resin-filled blisters. Needle leaves linear, 5–20 mm long, flattened in cross-section, leathery, shiny green above, somewhat whitish beneath, blunt-tipped, all tending to turn upward. Unlike other conifers in the North, the cones sit erect on the branch, a slate-grey to purple colour, usually oozing with sap; cones mature

Left: Female subalpine fir cones dwarf the male cones on the same branch. PL

Right: The subalpine fir is the official tree of the Yukon territory. The whole tree is considered powerful medicine among First Nations PL

Wild Food and Medicine Plants of the North | 247

Part II: *Trees*

Fir bark is smooth and grey, with resin blisters. Northern First Nations "milk" the trees for use as medicine. CA

in their first year and then promptly disintegrate, with the spike-like central stalks remaining long after the rounded, fan-shaped scales have fallen.

Habitat and Range
Mountain slopes, usually on acid rock, extending up to treeline. Found throughout British Columbia to just north of latitude 64° north in the Yukon territory; disjunct in south-central Alaska. (Note: Occasionally trees survive in valley bottoms, such as at the Carcross Dunes, near Bennett Lake in the Yukon, where the trees are much taller.)

Plant Parts Used
Bark, needles, resin (pitch)

Harvest Time
Bark and needles: throughout the year. The vitamin C content in the needles is highest in mid-winter. Pitch: spring and throughout the summer

THE SUBALPINE FIR is the official tree of the Yukon territory and the whole tree is considered powerful medicine among First Nations people. I find the energy of fir very clearing and cleansing to my spirit. At least once or twice a year I go to the mountains to sit with the fir trees, meditate, and listen to their wisdom and teachings. Once I have connected to the energy of the tree, I can then carry it with me. Its medicine seems to then run through my veins: I stand taller and feel more space in my mind, body, and spirit.

Medicinal Actions
Antioxidant, antiseptic, diaphoretic, disinfectant, diuretic, emetic, expectorant, laxative, mucolytic, pectoral, rubefacient, stimulant, tonic

Medicinal Preparations
Cream, essential oil, foot soak, hydrosol, infusion, liniment, oil, poultice, salve, steam, syrup, tea, wash

Medicinal Uses
Balsam fir was listed in *The United States Pharmacopoeia* from 1820 to 1916 and the *United States National Formulary* since 1926.

Subalpine Fir

Its bark is smooth and grey, with resin blisters. Northern First Nations "milk" the trees for the aromatic resin (pitch) and use it as medicine in topical preparations to treat skin ailments such as boils, wounds, cuts, bug bites, and acne. It's also used to help fight infections and respiratory congestion, and to expectorate phlegm, heal a sore throat, and get rid of colds and flus.

The pitch can be chewed to clean teeth or made into an infusion and used as a gargle or rinse for mouth or throat infections, and to treat bad breath. The pitch has also been used to cleanse and detoxify the body. The resin contains turpentine and is considered to act as a stimulant, diuretic, diaphoretic, and, topically, as a rubefacient.

A decoction of the bark is used as a tonic in the treatment of colds and flu.

The flat needles can also be made into a tea infusion that can be taken to treat the coughing up of blood—often the first sign of tuberculosis—and as a laxative, for chest congestion caused by viral infections, to induce sweating, and to treat a fever. Sore muscles, aches, and pains can also benefit from a fir liniment, poultice, massage oil, or by adding the needles and bark to a bath along with epsom salts.

Finely ground needles can be sprinkled on open cuts to help heal the wounds; you can also put them in the bottom of your shoes to ward off foot odour, or in water for a foot soak. Your feet will absorb some of the beneficial properties of fir, which will help keep you healthy and alert throughout the day.

The smell of fir is very uplifting and can help people struggling with fatigue. The boughs can be used in a sauna, steam bath, or placed in a pot of water on a wood stove.

BG

"And this, our life, exempt from public haunt, finds tongues in trees, books in the running brooks, sermons in stones, and good in everything."

—William Shakespeare, *As You Like It* (1599)

Food Uses

Fir is high in vitamin C, so it's nice to make an infused tea of the needles. The needles are at their nutritional peak in mid-winter.

It's reported that, traditionally, First Nations dried the inner bark, ground it into a powder, and then combined it with other flours to make bannock or bread. The cones can also be used as food. They can be ground into a fine powder, which in the past was mixed with fat. The result was considered both a delicacy and a digestive aid.

Nutritional Profile

Vitamin C

Part II: *Trees*

The Celts considered the fir tree a symbol of honesty. Evergreens signified hope, promise, and renewal in the midst of the long, dark winter months and were used to mark the winter solstice and/or Christmas celebrations. These celebrations symbolize rebirth, the death of the old year, and the birth of the new year.

Other Uses

Fir needles can be used as incense or a smudge for cleansing the spirit.

Aromatic, sweet-balsamic body oil can be made by soaking the needles in olive oil, or an odourless oil like jojoba.

The essential oil of fir is obtained through steam distillation of the needles, bark, and resin. It's considered to be antiseptic, astringent, and wound healing. It helps regenerate the skin cells and can be used to treat cuts, rashes, burns, and sores. It's also used to relieve heart and chest pains, coughs, and a variety of infections. Its emotional effects are grounding, opening, sedative, and uplifting. Use the essential oil as an inhalant, in a diffuser, or on skin in a carrier.

Aromatherapist Suzanne Catty, author of *Hydrosols: The Next Aromatherapy*, says, "Balsam fir hydrosol is a good general system tonic, it is antiseptic and seems to boost the immune system. It is of great benefit to sufferers of SAD (Seasonal Affective Disorder)." She recommends smelling it, or adding to a bath or bottom of the shower (with the plug in) once or twice a week. She also recommends using it in a compress for sore, aching joints or muscles, or as a stimulating addition to a footbath. Catty says that it's gently stimulating to the circulatory system, while at the same time calming to the mind. She adds that energetically it's very expansive and opening.

Cautions

Fir tea should be used in moderation as it can cause skin reactions in some people. Also, fir's emetic properties can cause vomiting if taken in large doses.

ADDITIONAL RECIPES

Cold-and-Flu Bath, page 326

Three-Tree Healing Salve, page 313

Tree-Medicine Purification Bath Oil, page 325

Sweet-Balsamic Body Oil

This massage and bath oil can act as a base for medicinal salves, ointments, and creams.

1 part fir needles (bark and pitch can also be included)
2 parts sunflower oil

Put needles into the top of the double boiler; pour oil over the needles. Bring the water on the bottom of the double boiler to a low simmer, and let the oil slowly heat for 30 to 60 minutes, stirring constantly to make sure the oil is not overheating.

The lower the heat is, and the longer you can infuse the oil, the better it will be. Let the oil fully cool, strain through a cheesecloth, then bottle it in a glass jar. Once labelled, the oil should be stored in a cool place.

LODGEPOLE PINE

Pinus contorta

Other Names and Etymology
The common name refers to the fact that the tree is used to make teepee poles. *Pinus* is most likely derived from the Latin word *picnus*, meaning "pitch." *Contorta*, also Latin, means "contorted," though the pine trunk is actually a straight and slender tree.

Family
Pinaceae (pine family)

Botanical Description
Evergreen tree reaching 25 m or more in height in the boreal forest; bark thin and scaly, brownish to blackish. Leaves are needles in groups of two, 1.5–7.5 cm long, semi-cylindrical, often twisted, persistent. Cones are 3–5 cm long, lopsided, hugging the stems with scales bearing a sharp prickle; cones persist on the trees for many years, opening in the heat of forest fires to release seeds.

P. contorta

Left: With its tall, thin trunk, it's easy to see why lodgepole pine is favoured for making teepee poles.

Right: Pine cones can cling to their tree for many years.

Wild Food and Medicine Plants of the North

Part II: *Trees*

A decoction of inner pine bark is used as a tonic in the treatment of colds and flu. BG

Your Dentist Would Approve

If you ever forget to pack your toothbrush on a backcountry trip, there's no need to live with plaque buildup. Simply cut a small twig from a pine tree, peel back the outer skin, then gently scrub the twig tip along the tooth surface. You can also use it to gently massage the gums.

Habitat and Range

A tree of western North America, found from Oregon, Nevada, and Colorado, north to the central Yukon territory, eastward to Saskatchewan and the southern Mackenzie Mountains in the District of Mackenzie. Lodgepole pines form dense stands on sandy soil and rocky ridges in central-southeastern Yukon, but are usually scattered in well-drained situations elsewhere.

Jack pine (*Pinus banksiana*) replaces *P. contorta* in the eastern boreal regions. *Pinus contorta* var. *yukonensis* grows in the south-central alpine regions on the Yukon.

Plant Parts Used

Inner bark, needles, pitch, twigs

Harvest Time

Spring, but anytime, as needed. (Note: Use the branches of the pine to harvest the bark—do not take from the trunk as this will kill the tree.)

PINE IS CONSIDERED the tree of peace. First Nations people have long used pine and all the trees in the pine family for food, medicine, and shelter. When early settlers came north, the First Nations shared their knowledge of the properties and usage of the trees, which helped the settlers survive starvation, scurvy, and other fatal illnesses.

I once fell asleep under a pine tree. I was half-awake and half-asleep when I had the sensation that my energy was merging with the tree's. I felt my energy align with the tree, and I held on as long as I could to the feeling, realizing only much later that the tree had imprinted on me. Years later, I had an artist paint this image for me. It's a powerful symbol for me of my commitment to nature, the trees, my own inner growth, and inner and outer peace.

Medicinal Actions

Analgesic, antifungal, antimicrobial, antiseptic, disinfectant

Medicinal Preparations

Cream, essential oil, infusion, liniment, oil, poultice, salve, soak, steam, tea, wash

Lodgepole Pine

Medicinal Uses

Pine is good medicine! The needles and pitch can be of benefit to the skin when used topically in a poultice, oil, salve, or cream, and can help conditions like cuts, abrasions, eczema, boils, and acne, because of the needles' antimicrobial, antifungal, and antiseptic abilities.

The needles and pitch are also excellent when used in a tea or steam for lung congestion: their expectorant and camphoraceous properties help open up the respiratory passageways. Being very antiseptic, the tea can also be used to help with pneumonia, whooping cough, croup, and tuberculosis. As a liniment, salve, or oil, pine works well for joint and muscle pain. The needles can also be used in the bath and as a facial or bronchial steam.

Pine pitch is excellent to extract splinters. Just put a fresh dab over the splinter and literally wait for it to be drawn out.

> **Male and Female Pine Cones**
>
> In her book, *Reading Yukon Forests*, author Lori Schroeder offers this interesting fact about pines and other evergreens. "All conifers bear male (pollen) cones as well as the more recognizable female (seed) cones. One reason the male cones are not often noticed is that they are smaller than the seed cones. The male cones also shrivel and dry up as soon as they release their fertilizing pollen in the spring. Keep an eye open for them shortly after the snow leaves."

Food Uses

An aromatic tea made with young pine needles is high in vitamin C and beta carotene. It's great to drink around the campfire or for use if you find yourself in a survival situation as a way to get a boost of vitamin C and maintain energy.

The inner bark (cambium layer) has long been used as a survival food and can also be eaten in raw slices. I like to use the soft, moist, white, inner bark for making summer pesto. Most pesto recipes call for pine nuts. But one day, when I was making pesto I didn't have any around. Remembering the flavour of pine's inner bark, I thought, Why not? I'll try it. It was wonderful—I haven't used pine nuts since!

The inner bark contains lots of starch and many sugars and can also be boiled or ground and then added to soups and stews.

Nutritional Profile

Vitamin C, beta carotene, starch, sugars

Other Uses

Pine makes an excellent household cleaner. The needles can be infused in vinegar, or you can use essential oil drops in water, or make a strong tea to clean surfaces. Diluted, these can be used in a bucket to wash floors.

Pine's aroma is clean and strong. In aromatherapy, these properties are said to help with nervousness, fatigue, and exhaustion. Pine also helps relieve a tired mind.

Gurudas, author of *The Spiritual Properties of Herbs*, says that "pine increases the ability to activate energies of the third eye."

Daniel Tigner, creator of Canadian Forest Tree Essences, says that "Lodgepole pine is nourishing, expanding and invites us to rediscover our center and to remain firmly rooted in the midst of external dramas." He adds, "Being centered we face life with equanimity and calmness. We are able to stand tall and to speak our truth."

Part II: *Trees*

"Between every two pines is a doorway to a new world."

—John Muir, naturalist (1838–1914)

Dr. Edward Bach, the father of flower essences, said that pine essence is "for those who blame themselves. Even when successful they think they could have done better, and are never satisfied with the decisions that they make."

Cautions
Use tea in moderation. Avoid during pregnancy.

ADDITIONAL RECIPES

Extra-Green Pesto Sauce, page 356

Oh-My-Aching-Bones Bath, page 326

Pine-Forest Smoothie, page 338

Skin-Disinfectant Spray, page 295

Three-Tree Healing Salve, page 313

Tree-Medicine Purification Bath Oil, page 325

Pine Tree Ache-and-Pain Liniment
For soothing tired, sore, and inflamed muscles.

1 cup (250 mL) pine needles
¼ cup (60 mL) of spruce pitch
2 cups (500 mL) witch hazel or rubbing alcohol

Place pine needles and pitch in a jar with witch hazel or rubbing alcohol.

Let mixture sit for 6 weeks, shaking daily.

When fully infused, strain and bottle.

Liniments are for external use only; make sure to label your bottles appropriately to ensure the liniment won't be ingested.

For more information, see Tincture Method, page 293.

Pine Household Cleaner
For sinks and countertops.

2 teaspoons (10 mL) borax
10 drops pine oil
2 cups (500 mL) water

Mix ingredients together. Pour into a spray bottle. Shake before using.

Lodgepole Pine

Pine Bark Bread (Crackers)

The following recipe is adapted from a recipe given to me by Laila Spik, a Sami ambassador from northern Sweden. The original recipe in featured in her book, How to Cook a Reindeer: The Reindeer Recipe Book.

¾ cup + 2 tablespoons (200 mL) pine bark flour (see sidebar)
3¾ cups (900 mL) whole meal flour
1 tsp (5 mL) salt
About 1¾ cups (400 mL) cold water

Pre-heat oven to hottest possible setting.

Mix all ingredients and form a dough. Add a small amount of extra water or flour if necessary to ensure dough is neither too sticky nor too dry.

Roll out into very thin sheets and prick with a fork. Cut into bite-sized pieces. Place on a lightly oiled pan.

Bake in the oven for about 3 minutes. Flip sheets halfway through. The crackers will have a pita-bread-like consistency when they first come out of the oven, but will harden as they dry out.

Pine bread is delicious when served with goat cheese and a dollop of fireweed jelly (page 94) or with juniper butter (page 357).

To obtain bark flour: remove pieces of outer bark from a pine tree. Allow to dry, then break into small pieces. Place pieces in a mixer and grind into a fine powder. Sieve to remove any hard residue.

Top to bottom: Outer pine bark, pine bark flour, pine bark bread.

BALSAM POPLAR

Populus balsamifera

Other Names and Etymology
Cottonwood. *Populus* comes from the Latin name for poplar and means "tree of the people." *Balsamifera* comes from the Hebrew *besem*, meaning "sweet smell." The Celtic tree name for poplar is *edad*, which means "victory, transformation, and vision."

Family
Salicaceae (willow family)

Botanical Description
Tall deciduous trees to 40 m (though usually reaching no more than 10–12 m in the Yukon), often much shorter; bark smooth and greenish when young but becoming dark grey and deeply furrowed with age. Leaves 4–12 cm long, shiny above, with lower surface pale. Female flowers in catkins 6–16 cm long, developing into pear-shaped capsules 4–7 mm long. When mature, the capsules split into two valves to release seeds with long, silky plumes. In winter, the much-tapered, pointed buds are covered in a fragrant, sticky orange sap.

Habitat and Range
Balsam poplar grows in moist soils throughout the boreal regions. In North America they grow from Newfoundland to Alaska, and south to New York, Michigan, Wisconsin, Wyoming, and Oregon; in the Yukon found as far north as the British Mountains.

Plant Parts Used
Buds, catkins, inner bark, leaves

The winter buds, leaves, and bark of the balsam poplar can all be used medicinally.

Balsam Poplar

Harvest Time
Winter buds: early spring. Bark: spring. Gather it from pruned branches

THE WINTER BUDS of the balsam poplar are large, resinous, and very fragrant. I love making poplar-bud infusions and salve, and have been making what I call "Boreal Balm" from these winter buds for more than a decade. The feedback I get from people is phenomenal. I have had people tell me that the balm has healed eczema, psoriasis, weird and itchy skin conditions, cracked feet and hands. Some people have used it topically over the lung area to help clear respiratory congestion.

I first made boreal balm because we had a large poplar on our property and the buds were screaming to be picked. So I gathered as many as I could, infused them in oil, and let them sit for about six months before I got around to straining it. I ended up giving every jar of balm away. People who still have some of this original batch say it hasn't gone off. That's because the bud's organic chemical properties stopped the carrier oil (olive oil) from going bad. You've got to love how nature works!

Medicinal Actions
Alterative, analgesic, anticoagulant, anti-inflammatory, anti-rheumatic, antiseptic, astringent, bitter, cholagogue, diuretic, expectorant, febrifuge, pectoral, (mild) sedative, tonic, vulnerary

Medicinal Preparations
Bath, cream, infusion, oil, liniment, poultice, salve, steam, tea, tincture

Medicinal Uses
The winter buds, leaves, and bark can all be used medicinally. Being in the willow family, the poplar contains compounds that help reduce pain and inflammation.

An inner bark tincture contains salicin and is a natural remedy for fevers, rheumatism, arthritis, and diarrhea. The inner bark is strong medicine and can be very stimulating in high dosages.

The buds are emollient and help soften, soothe, and protect the skin. They are also vulnerary, helping heal cuts and wounds; and demulcent, helping soothe and protect irritated or inflamed tissue.

A salve made with the winter buds is traditionally called "balm of Gilead," but I call the one I make for my store, Aroma Borealis, "Boreal Balm" because we are in the boreal forest, after all! The resinous buds are collected in the early spring and soaked in olive oil to extract the medicinal resin. (Note: When collecting, your fingers will get really sticky and full of resin when picking buds. To remove the resin, apply oil to cut and lift the

Top: Balsam poplar catkins are high in vitamin C and can be eaten raw. RF

Above: My great-great-grandfather made and sold a healing ointment from the buds of the poplar tree, naming it "Mitchell's Genuine Balsam" in 1900. BG

Balsam poplar is known as "bee glue" among apiarists. Bees gather it onto their thighs and then use it to seal up crevices in their hives.

Wild Food and Medicine Plants of the North | 257

Part II: *Trees*

> "Every tree and plant in the meadow seemed to be dancing, those which average eyes would see as fixed and still."
>
> –Rumi, Sufi mystic and poet

resin. I generally use olive oil and then rub the oil on my shoulders.)

Poplar oil or salve can be used to remedy and soothe eczema skin, burns, cuts, rashes, and other skin irritations. The remedy can also be used for colds, coughs, and irritated nostrils.

The buds can also be made into tea to help clear respiratory congestion.

Food Uses

The catkins are high in vitamin C and can be eaten raw. They can also be added to any campfire soup or stew. The cambium, the thin layer between the outer bark and the inner sap-wood, can be a very nourishing survival food.

Nutritional Profile

Catkins: vitamin C. Leaves: high in tannins. Inner bark: contains a small amount of protein, fat and carbohydrates

Other Uses

Steve Johnson, founder of the Alaskan Flower Essence Project and author of *The Essence of Healing*, says that balsam poplar is indicated for "inconsistent emotional and sexual response, often resulting from shock and trauma or a lack of grounding." He adds that poplar is useful for "the release of physical and emotional tension associated with sexual trauma; balances the circulation of life-force energy in the body; helps to ground and synchronize our sexual energy with planetary cycles and rhythms."

Cautions

Poplar buds can cause an allergic reaction in people with tree allergies.

Boreal Healing Oil

ADDITIONAL RECIPES

All-Purpose Botanical Ointment, page 313

Boreal Balm, page 314

1 part poplar buds
2 parts olive oil

Place buds and olive oil in a double boiler. Warm slowly and let simmer for 30–60 minutes, stirring continuously. Strain through cheesecloth, let the mixture cool, and bottle.

SPRUCE

Picea glauca (white spruce)
Picea mariana (black spruce)

Other Names and Etymology
The Van Tat Gwich'in word for both white and black spruce is *ts'iivii*. Their word for the soft spruce gum is *dzìh kò*, spruce pitch is *dzìh tl'ùu*, and the spruce bark is called *ts'iivii nèechùu*.

The scientific name *Picea* is most likely derived from the Latin word *pix*, meaning "pitch."

Family
Pinaceae (pine family)

Left: White spruce needles are stiff, sharply pointed, and green to bluish-green. CA

Right: Black spruce is a slender tree with a top-heavy crown; its blunt-tipped needles are a pale bluish-green. PL

Wild Food and Medicine Plants of the North | 259

Top: Spruce tips prime for the picking. BG

Above: The bark is the bottle that holds the medicine that heals our wounds. BG

Botanical Descriptions

Picea glauca (white spruce): Tree growing up to 30 m tall with a trunk diameter up to 50 cm. Branchlets are glabrous. Needles are stiff, sharply pointed, green to bluish-green. Lines of tiny white dots can be seen on all sides of the needles. When crushed, the needles have a distinctive, pungent smell. Cones are sub-cylindrical with thin, rounded, smooth-margined, deciduous scales. Another variety, *Picea glauca* var. *porsildii* (Alaska white spruce) has bark that is smooth in youth with resin blisters similar to fir bark; cone scales, more rounded. Grows on rich, alluvial soils in lowlands. *Picea mariana* (black spruce): A small, slender, often scrubby tree with a top-heavy crown, up to 25 m; bark rough and scaly; branchlets pubescent (usually a fine fuzz of rust-coloured hairs on twigs). Needles are pale bluish-green; blunt-tipped. Lines of tiny white dots are prominent on the undersides of the needles. Cones more squat and egg-shaped than white spruce, with rounded, irregularly toothed scales; cones persist for several years, often forming tight masses in treetops.

Habitat and Range

Picea glauca (white spruce): Grows in a variety of soils and climatic conditions. White spruce is common throughout much of boreal North America, from Newfoundland and Labrador to British Columbia and Alaska, and in the Yukon territory and the District of Mackenzie, in upland and lowland sites north to 69° North. *Picea mariana* (black spruce): Usually a slow-growing tree in treed wetlands, black spruce grows in lowland muskegs, on poorly drained clays, and glacial tills in alpine valleys and slopes. It also occurs on upland sites where its growth form resembles white spruce. The species is found throughout Canada, and the Great Lakes and northeastern United States. It's found across boreal North America; its distribution is patchy in the Yukon, extending northward to the Porcupine River valley. Both white and black spruce are an indicator of boreal forest.

Spruce

Plant Parts Used
Inner bark, needles, pitch, tips, twigs

Harvest Time
Spring, but anytime it's needed

I FEEL MUCH GRATITUDE toward spruce. Like many northerners, I owe my life to this tree. My house is made of Yukon spruce logs. We burn spruce in our wood stove to keep us warm throughout the long, cold winters. I use spruce medicine to prevent illness and, when I do get sick, it's also there to help me get better.

Spruce is also nature's bandage! Years ago, my old dog Hanna got a cut on her nose. It looked somewhat deep, but as it was a long weekend, I didn't want to make a trip to town to the vet and have to pay the extra on-call fee. I called my neighbour and asked her to come over and give me an opinion. After looking at the cut, she said, "Just use spruce pitch." So I did. The wound healed quickly with no infection or scarring.

Medicinal Actions
Analgesic, antifungal, antimicrobial, antiseptic, disinfectant

Medicinal Preparations
Cream, essential oil, foot soak, hydrosol, infusion, liniment, oil, poultice, salve, steam, syrup, tea, wash

Medicinal Uses
According to *Land of My Ancestors: Trees and Forests* (published by the Council of Yukon First Nations), "The sap of the spruce is a tonic and is used each spring to clean the blood. The inner bark of a spruce was made into a tea and strained and was used for stomach upsets, ulcers, weak blood, mouth sores, and sore throats. Spruce cambium was emergency food and often used by trappers or hunters when they were on the trail. People used the cambium during famines to keep them from getting scurvy. This also saved many of the early explorers and traders."

First Nations people use the spruce gum (pitch) as a lozenge for coughs and sore throats. They mix the pitch with grease to treat cuts and topical infusions. It cleanses the wound and protects it from germs.

Spring spruce tips and pitch can also be of benefit to the skin when used topically in a poultice, oil, salve, or cream. Because of the tree's antimicrobial and antiseptic properties, it can help conditions like cuts, abrasions, eczema, boils, and acne.

Fresh or dried spruce tips are also excellent for lung congestion when used in a tea or

Part II: Trees

steam. Spruce's antiseptic properties help with pneumonia, whooping cough, and croup. As a liniment, salve, or oil, spruce works well for joint and muscle pain. Springtime spruce tips can be dried for use year-round. Gwich'in Elder Ruth Welsh says, "If I had nothing else in my medicine cabinet, pitch could do it all."

Food Uses

Springtime spruce tips are tasty and very high in vitamin C. They can be eaten raw, made into a tea, or added to salads, stews, and soups.

The tips are famous for their use in making spruce-tip jelly. This light green conserve makes a great topping for toast, goes well with meats and poultry, or can be eaten on its own as a little afternoon pick-me-up.

The tips can also be soaked in oil to act as a base for salad dressing or cooking. Syrup can also be made to use over desserts, such as ice cream, sorbet, or cheesecake.

"Sprucesicles" were a favourite with my kids when they were little. We just added fresh spruce tips to the end of fruit popsicles. They loved it and it's a great way to enrich the treat with a little more vitamin C.

Of course, spruce beer is a northern staple. Many folks brew beer from spruce tips, including the Alaskan Brewing Company. Its Alaskan Winter Ale was inspired by a recipe found in Captain Cook's seafaring logbook chronicling his exploration of the Alaska coastline. The vitamin-rich beer goes down smoothly, and was a perfect beverage for the seafarers of Cook's time who needed to ward off scurvy. In her book, *The Boreal Gourmet*, Michele Genest has a recipe for spruce-tip salt that she uses to add an interesting and distinctly northern flavour to many of her dishes.

Nutritional Profile

Beta carotene, starch, sugars, vitamin C

Below: A drop of golden tree medicine. BG

Right: Tender young needles are famous for their use in making spruce-tip jelly, tea, and medicinal skin oil. BG

Spruce

Cosmetic Uses
The fresh spruce tips can be used in the bath and as a facial steam for oily skin.

Other Uses
Spruce-root baskets are not only amazing looking but also very functional. I received one as a gift from my friend and artist Helen O'Connor and used it for years. In the summer I used it to gather herbs and vegetables from the garden; in the winter it ends up holding fruit, keys, letters, etc.

Suzanne Catty, aromatherapist and the author of *Hydrosols: The Next Aromatherapy*, says that black spruce does wonders for the adrenal glands. "Use in combination with the essential oil at the change of each season for a three week protocol." She recommends drinking 30 mL (2 tablespoons) of hydrosol in 1.5 L of water (6 cups) every day, and applying 15 drops of undiluted essential oil to your adrenal and kidney areas before your morning shower, then rinsing off with cool to cold water. She says you will be amazed at how you feel.

Black spruce hydrosol is also excellent for boosting the immune and respiratory systems, and can be used as a compress to ease painful and inflamed joints. Suzanne writes that black spruce makes a stimulating and restorative body spray and aftershave that can connect us with the ancient wisdom of the trees.

Steve Johnson, founder of the Alaskan Flower Essence Project and author of *The Essence of Healing*, says white spruce is good for "information overload; feeling dis-integrated; unable to apply knowledge to life's challenges; difficulty integrating how one feels with how one thinks." He adds that white spruce's healing qualities "ground spiritual wisdom into the body; helps us bring logic, intuition, and emotion together into unified action in the present moment."

Steve says that black spruce is helpful for "contracted view of life; tendency to forget information learned from past experiences; out of touch with the wisdom of the soul family." He goes on to say that it "promotes the integration of information from past lessons and experiences into present-time awareness; helps us access eternal and archetypal wisdom from the collective consciousness of the earth."

According to Herbalist Robert Rogers, "Black spruce essence allows us the clarity to see our own shadow. It also ignites our inner wisdom by letting some of our shadow qualities come to light and be accepted, or dissipate. This is a grounding essence that helps us be more realistic. Visiting this essence is like taking a voyage to a monastery. It gives us strength during times of frustration, and helps bring clarity to those involved with addictions."

Cautions
The tea should be used in moderation. Avoid during pregnancy.

> "Spruce gum was excellent for waterproofing birch baskets and sealing boats. Spruce chewing gum is the dried sap on the bark. Break off a small piece and chew it. At first it is sort of crumbly, then sticky. Add another small piece and it will become more like gum. It takes a few practices to get it just right."
>
> —Council of Yukon First Nations,
>
> *Land of My Ancestors: Trees and Forests* (1993)

Part II: *Trees*

ADDITIONAL RECIPES

All-Purpose Botanical Ointment, page 313

Cold-and-Flu Bath, page 326

Cough-and-Cold Syrup, page 300

Pink Drink (High-C Smoothie), page 340

Spruce Tip Jelly, page 352

Three-Tree Healing Salve, page 313

Tree-Medicine Purification Bath Oil, page 325

Collecting spruce pitch for making salve. BG

Old Man's Beard

Old man's beard (*Usnea* species) is the greenish, hair-like lichen that grows on spruce trees—it's also very valuable medicine.

Usnea contains usnic acid that, in one form or another, has been used as a mild antibiotic for hundreds of years. It's reported to have antiviral, antifungal, and antibacterial actions, and has even been used to kill *Streptococcus* and *Staphylococcus* bacteria.

It can be used topically as an infused oil for skin infections and as an alcohol tincture for colds, lung infections, and sore throats. A spray can be made for a sore throat.

Cautions: Some people are seriously allergic to usnic acid. For those people, even touching old man's beard can lead to a rash and swallowing usnea could be harmful. Lichens can also irritate the kidneys if ingested over a long period of time.

Spruce Winter Salve

½ cup (125 mL) of spruce pitch
1 cup (250 mL) olive oil
30 mL melted beeswax

In a double boiler, heat olive oil and spruce pitch until pitch has melted. Pick out any remaining black gobs. Add beeswax and slowly heat the oil and beeswax mixture until it melts together, stirring occasionally. Turn heat off and remove the top pot from the double boiler and make sure the bottom is wiped dry (to insure that no water falls into the measuring cup that the oil will be poured into). From the measuring cup, pour the hot liquid salve into jars. When the salve is fully cooled the lids can be put on. If salve is not fully cooled condensation will form and can cause the salve to go bad. Label the jars.

TAMARACK (LARCH)

Larix laricina

Other Names and Etymology
Alaskan larch. The Van Tat Gwich'in word for tamarack is *ts'iiheenjoo*. The word "tamarack" may be derived from an Algonquin word meaning "used to make snowshoes."

Family
Pinaceae (pine family)

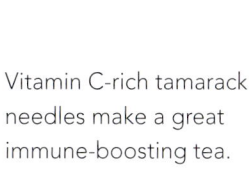

Vitamin C-rich tamarack needles make a great immune-boosting tea.

Left: PL, below: BG

Part II: *Trees*

A tamarack-needle foot soak is refreshing after a long day on your feet. PL

Botanical Description
The only native conifer in the Yukon that loses its needles in the autumn, this tree can grow to 20 m in height (taller in the south). Its bark is thin and scaly. Needle leaves are in small clusters of twelve to twenty, pale green, turning bright yellow in autumn before falling off. Male and female cones globular or egg-shaped, at the ends of very short peg-like shoots; female cones erect, their scales dark red at flowering time, but becoming leathery and brown in age.

Habitat and Range
Tamarack grows in wet places like muskegs, usually preferring lime soils. Boreal North America, from Newfoundland and Labrador to central Alaska; all the Great Lakes states and northeastern United States. In the Yukon territory, found in the southeastern part and in the drainage of the Peel, Liard, and upper Porcupine rivers in the north.

Plant Parts Used
Bark, needles, resin

Harvest Time
Needles: in the spring when new and fresh. Resin: softest in the spring and summer. Inner bark: best gathered before the needle bloom (early spring) or after the needles have dropped (in the autumn)

We PLANTED A TAMARACK after the homebirth of our second child. To nourish its roots, we buried the placenta under the tree. The tree is a reminder of the miracle of birth and how special the family bond is.

It's interesting to see the tree's needles turn orange and fall off every autumn, while all the other conifers still fill the air with their green, intoxicating beauty.

Medicinal Actions
Alterative, anti-inflammatory, astringent, disinfectant, diuretic, expectorant, immune stimulant, laxative, tonic

Medicinal Preparations
Inner bark: capsules, poultice, salve, tea/decoction. Needles: infusion, salve, tea. Resin: decoction, poultice, raw (for chewing), salve

Tamarack

Medicinal Uses

Larch-gum resin can be chewed to treat infected gums, or made into a decoction and used as a mouth gargle for treating infected gums and sore throats. In combination with inner bark and needles it can also make a healing salve.

The resin and the inner bark can also be used topically as a poultice or compress, or as a wash for wounds, infected cuts and abrasions, boils, rashes, frostbite, and skin conditions such as psoriasis and eczema. It can also be drunk as a tea for muscle aches and pains, including arthritis.

A needle tea can be drunk for its vitamin C content to help boost the immune system, and to help with respiratory complaints due to cold and flu.

A foot soak made with the needles is refreshing after a long day on your feet. In fact, Herbalist Robert Rogers writes that decocted tamarack bark is one of the best cures for gout—so a footbath in the decocted bark may help this as well.

A primary active compound in the wood of larch species is arabinogalactans that belongs to a group of carbohydrates called polysaccharides. Studies have shown that larch arabinogalactans help to stimulate the immune system to fight cancer, and act as a prebiotic to help stimulate and promote good bacteria in the digestive tract, and help to produce digestive enzymes.

Tamarack can also be used for treating the common cold, flus, liver disease, inner-ear infection. Some people use it to provide dietary fibre, lower cholesterol, and to boost the immune system.

The starch-like larch arabinogalactan is used in foods as a stabilizer, emulsifier, binder, and sweetener. The fibre ferments in the intestine and increases good probiotic intestinal bacteria, such as *Lactobacillus*. It has other effects that could be beneficial to digestive-tract health such as good bowel movements.

The commercially available larch arabinogalactan is generally produced from western larch (*L. occidentalis*), however, larch arabinogalactan can also be produced by other larch-tree species including *L. laricina*.

> "Except during the nine months before he draws his first breath, no man manages his affairs as well as a tree does."
>
> —George Bernard Shaw, dramatist (1856–1950)

Food Uses

Many First Nations peoples traditionally used larch sap as a natural sweetener, and the tasty, fresh spring shoots as an edible green. The inner bark was also ground into a powder and used as flour.

In *Rogers' Herbal Manual*, Herbalist Robert Rogers gives a recipe for tamarack bread: "Scrape off the soft wood and inner bark of tamarack, mix with water, and ferment into dough to be mixed with rye meal. Bury under the snow for a day. As fermentation begins the dough can then be cooked as camp bread or as dumplings, the sweet wood pulp acts as a sugar for the yeast in the rye."

Part II: *Trees*

Nutritional Profile
Needles: high in vitamin C

Other Uses
Steve Johnson, founder of the Alaskan Flower Essence Project and author of *The Essence of Healing*, says that tamarack (*Larix laricina*) is indicated for use for people with "no confidence in one's unique skills and potentials; weak sense of self-identity; lacks awareness of what one is capable of." He adds that the tree's healing qualities help to "promote self-confidence by helping us reach a deeper understanding of our unique strengths and abilities; encourages the conscious development of individuality."

Dr. Edward Bach, who developed a larch essence for the Bach flower remedies, said that it's "For those who do not consider themselves as good or capable as those around them, who expect failure, who feel that they will never be a success, and so do not venture or make a strong enough attempt to succeed."

Larch wood is a good, strong, and pliable wood and has been used for making snowshoes, toboggans, and canoes. Its roots were split and made into rope and nets for catching fish.

Cautions
Use in moderation.

Tamarack Tea

ADDITIONAL RECIPE

Footbath, page 327

Gather a pinch of fresh spring needles, place them in a teacup, pour steamy hot boiled water over them, cover, and let steep for 15 minutes. Lift the cover, take a deep inhalation of the steam, sip slowly, and enjoy!

WILLOW

Salix species

Other Names and Etymology
The Van Tat Gwich'in name for willow is *k'àii*. *Salix* is derived from the Celtic *sal* which means "near," and *lis* for "water." It may also be derived from *saille*, the Celtic tree alphabet name for willow. "Willow" is derived from an old Anglo Saxon word *wileg*.

Family
Salicaceae (willow family)

Botanical Description
There are over forty-five species of willow in the Yukon, only about ten of them grow big enough to be considered trees. To identify the various species and their local forms requires special training and dedication. Willows vary in growth from dwarf, trailing, Arctic, or alpine shrubs to trees. In winter, willows can be distinguished from other woody plants by their buds: A single scale encloses each bud, making it appear seamless. Leaf and twig characteristics can vary markedly within and among species. Individual willow plants produce either male or female flowers (i.e., they are dioecious plants), making cross-pollination and fertilization necessary. Flowers, which lack showy petals, are clustered in catkins. They either flower before the leaves have expanded in early spring, or later in sync with leaf development, depending on the species. Mature female flowers look like little

Left: Willow bark comes in a multitude of colours. PL

Middle and below: Some catkins flower in early spring, others in time with the development of the leaves. Middle: PL, below: CA

Part II: *Trees*

Herbalists say that Aspirin contains "a willow-like substance"—and not the other way around. The medicine is strongest in the inner bark and milder in the leaves. BG

pears (the pistils) protected by bracts. Mature male flowers have bright yellow anthers that release large quantities of pollen.

Habitat and Range
Grows in a variety of habitats from sedge meadows to dry, sandy tundra and heath sites, to forested areas within the mountains, to lakeshores, riverbanks, wet meadows, thickets on mountain slopes, Arctic and alpine tundra. They form thickets along streams and grow in black spruce and white spruce woodlands. Willows grow throughout boreal North America and many species are circumpolar.

Plant Parts Used
Inner bark, catkins, leaves

Harvest Time
Bark: in spring when the sap is running (which is good for making baskets and medicine). Leaves: throughout the summer. Branches: in the autumn (also good for basket making)

IN HERBAL CIRCLES the saying goes that Aspirin contains "a willow-like substance" and not the other way around.

A few years ago, we were on a day trip on the Yukon River when I started to develop a swollen gum from a lodged popcorn kernel. When we stopped for a break, I picked some willow leaves, chewed them up, and stuffed them in my cheek over my sore tooth and gum. Within twenty minutes the pain was gone, and by the end of the trip the kernel had dislodged!

With so many different willow species, it can be difficult to tell the species apart, but the good news is that all willows are medicinal. Their bark and leaves all contain salicin and make excellent medicine.

Medicinal Actions
Analgesic, anti-inflammatory, antioxidant, antiseptic, anodyne, astringent, bitter, diuretic, tonic, vulnerary

Medicinal Preparations
Willow's inner bark is best known for its analgesic properties to help with pain and inflammation. The salicin found in the inner bark and leaves is responsible for its starring role in herbal pain relief!

Its antiseptic properties make it excellent for topical use for cuts, abrasions, wounds of all sorts, insect bites and stings, and as a tea for inflammations of the urinary tract.

Willow

> "Going to plant a weeping willow,
>
> On the bank's green edge it will grow, grow, grow."
>
> —Words by Robert Hunter performed by the Grateful Dead,
>
> "Brokedown Palace" (1970)

Pussy willows are unopened catkins! Once expanded, the sweet nectar from their flowers attracts bees and flies that in turn attract hungry migrating birds such as orange-crowned warblers. This makes willow an important part of the northern ecosystem.

Using the inner bark in a tea, capsule, or tincture can help relieve headaches, and is reported to not cause gastrointestinal upsets like some pharmaceutical pain relievers.

Taken internally or used in massage oil, willow inner-bark and leaves can help relieve pain and inflammation from lower back pain, sciatic nerve pain, osteoarthritis, tendonitis, bursitis, gout, muscle aches, sprains, and strains.

For menstrual cramping willow inner bark tincture or tea, on its own or in combination with raspberry leaf and crampbark, will help to alleviate pain and allow you to carry on as usual.

Willow can also help to reduce the discomforts of fever from a cold or flu.

The tincture is so versatile. It can be added to the bath or to a topical compress. The leftover inner bark can be air dried or put in a dehydrator on the herb setting to dry. Once dried, it can be ground in a coffee grinder (there will be long fibres in it; just pick them out) and put into capsules for therapeutic use. Put the capsules in a dark jar or bottle and store in the freezer.

The medicine is strongest in the inner bark and milder in the leaves.

Medicinal Uses
Bath, capsule, cream, decoction, infusion, liniment, oil, salve, soak, steam, tea, tincture

Food Uses
The cambium layer of willow bark is very bitter, so it doesn't have many food uses. That said, the spring catkins are high in vitamin C and can be eaten, on their own, fried up with other plants, or in soups. In a survival situation the inner bark could be eaten.

Nutritional Profile
Vitamin C

Part II: *Trees*

The willow moose by Bob Atkinson. It used to stand outside of Aroma Borealis, but people kept stealing its antlers. BG

Bruce Lee, the famous martial artist and actor, once said, "Notice that the stiffest tree is most easily cracked, while the bamboo or willow survives by bending with the wind."

In fact, willow is so flexible it's regularly used for basket weaving and willow-furniture making. When I was pregnant with my second child, Markie, I weaved over thirty baskets throughout the winter and spring to pass the time. We also made a willow chair and a laundry hamper as a gift to our midwife.

In the Yukon we have some pretty amazing willow artists, including "Willow" Bob Atkinson, who once made me a full-sized willow moose for the front of my store Aroma Borealis, and Shiela Alexandrovich, a world-renowned willow weaver, beader, and overall natural-materials artist. Shiela once told me about a "living" willow table she made—the legs rooted and began growing just from being placed on the ground!

Willow is such a diverse and interesting plant species. I don't think it could ever be overharvested by medicine and basket makers because as soon as you harvest a branch a bunch more sprout up in its place! And, as it so happens, these new shoots are excellent for basket making.

Willow shoots and water make an excellent rooting hormone. BG

Other Uses

Steve Johnson, founder of the Alaskan Flower Essence Project and author of *The Essence of Healing*, says that willow is indicated for "resistance to taking responsibility for one's actions or for the life one has created; unaware of how thoughts create reality." He adds that its healing quality "stimulates mental receptivity, flexibility, and resilience; helps us remove our resistance to consciously creating our lives."

Dr. Edward Bach, the creator of the Bach flower remedies, wrote that willow essence is good for "those who have suffered adversity or misfortune and find these difficult to accept, without complaint or resentment, as they judge life much by the success which it brings. They feel that they have not deserved so great a trial that it was unjust, and they become embittered. They often take less interest and are less active in those things of life which they had previously enjoyed."

You can make your own rooting hormone with the spring shoots. Simply cut a handful of new, fresh green shoots, chop them up and put in a canning jar with water. Let it sit for a few days, strain, and then use the water to soak root cuttings, or dip in the roots of transplants to help them establish a solid foundation. You can also water new transplants with the willow water to help growth. I put willow branches in my greenhouse water barrel.

Willow

Cautions

Willow can irritate the stomach if taken over a long period of time. Salicin does not cause bleeding in the stomach or thin the blood like Aspirin.

Willow Inner-Bark Tincture

1 part fresh willow inner bark, peeled and cut
1 part vodka (enough to cover the willow)

In a jar, add the vodka to the willow bark. Shake every day for 4 to 6 weeks, strain, bottle, and label. (See Tincture Method, p. 293.)

Note: I find it best to make a willow tincture in the early spring, before the leaves bud. Simply cut some healthy stems, cut and peel off the outer bark, and tincture the cream-coloured inner layer.

ADDITIONAL RECIPES

Good-for-Gout Infusion, page 284

Juniper-Willow Liniment, page 317

Muscle-and-Pain Relief Oil, page 307

Muscle-Ease Cream, page 317

Oh-My-Aching-Bones Bath, page 326

Pain-Aid Capsule, page 301

Pain-Aid Tincture, page 294

Rejuvenating Facial Steam, page 329

Skin-Disinfectant Spray, page 295

Urinary-Tract Tincture, page 294

Inner willow bark. BG

PART 3
PLANT PREPARATIONS and RECIPES

MEDICINAL PREPARATIONS AND RECIPES

HERBAL TEAS 279
   Basic Herbal-Tea Recipe 280
Herbal Tea or Tisane 280
   Spirit-of-the-Boreal-Forest Tea . 281
   Afternoon Floral Tisane 281
Herbal Infusions 281
   Nourishing Infusion 281
   High-Calcium Infusion 282
   Good-for-Fever Infusion 282
   Good-Sleep Infusion 283
   Get-Up-and-Go Infusion 283
   Lung-Tonic Infusion 283
   Detoxifying Infusion 283
   Good-for-Gout Infusion 284
   Moon-Time Infusion 284
   Tea-for-Two Pregnancy
     Infusion 284
   Good-for-Nausea Infusion 284
   Beautiful Hair, Skin, and Nails
     Infusion 285
   Dandelion-Root Coffee 285
Decoctions 286
   Spring-Cleanse
     Dandelion Decoction 286
   Rosehip Decoction 286
Solar and Lunar Infusions 287
Iced Teas 289
   Red-Clover Tea Aid 290
TINCTURES, VINEGARS, ELIXIRS,
   AND BITTERS 291
   Tincture Method 293
   Pain-Aid Tincture 294
   Athlete's-Aid Tincture 294
   Cold-and-Flu Tincture 294
   Urinary-Tract Tincture 294
   Prostate-Support Tincture 295
   Skin-Disinfectant Spray 295
   Sore-Throat Spray 295

Vinegars 296
   Nourishing Vinegar Tonic 296
Elixirs and Bitters 297
   Preparing Elixirs
     and Bitters 297
   Boreal Bitters 297
   After-Dinner Elixir 297
SYRUPS 298
   Basic-Syrup Recipe 298
   Rosehip Syrup 299
   High-Iron Syrup 299
   Cough-and-Cold
     Medicinal Syrup 300
CAPSULES 301
   Preparing Capsules 301
   Allergy-Aid Capsule 301
   Sleep-Aid Capsule 301
   Pain-Aid Capsule 301
PLANT ESSENCES 302
   Preparing Plant Essences 303
TOPICAL TREATMENTS 304
Medicinal Oils 304
   Basic Oil-Infusion Recipe 305
   Traditional Method 305
   Sun-Infused Oil Method 306
   Double-Boiler Method 306
   Muscle-and-Pain Relief Oil 307
   Boreal Skin Oil 307
   Dandy-Lioness Oil 308
Salves, Ointments,
   and Balms 309
   Salve Double-Boiler Method 312
   Basic-Salve Recipe 312
   All-Purpose Botanical
     Ointment 313
   Three-Tree Healing Salve 313
   Hot-Water Bath or Bain-Marie
     Method 313

   Aromatic Herbal Balm 314
   Boreal Balm 314
   Moisturizing Lip Balm 314
Healing Herbal Creams 315
   Basic-Cream Recipe 315
   Making Cream 316
   Muscle-Ease Cream 317
   Boreal Botanical
     Skin Cream 317
Liniments 317
   Juniper-Willow Liniment 317
Poultices 318
   Plantain Poultice 319
   Spit Poultices 319
Fomentations 319
   Wild-Onion Fomentation 319
Cold Compresses 320
Styptic Powder 320
   Stop-the-Bleeding
     Styptic Powder 320
ESSENTIAL OILS AND
   HYDROSOLS 321
   Preparing Hydrosols 322
BOREAL BATHS AND STEAMS .. 324
   Tree-Medicine Purification
     Bath Oil 325
   Boreal Healing Bath 325
   Flower-Power-Peace Bath 325
   Cold-and-Flu Bath 326
   Oh-My-Aching-Bones Bath 326
Footbaths 327
Sitz Bath 328
Steam 328
   Respiratory Steam 329
   Rejuvenating Facial Steam 329

Pages 274–275: Arnica flower essence. BG
Previous pages: Boreal bounty. BG

HERBAL TEAS

SITTING DOWN WITH A CUP of steamy hot herbal tea invokes a wonderful feeling of well-being and a connection to a timeless global ritual.

Drinking herbal teas is one of the simplest ways to receive the health benefits of wild plants. Herbal teas are caffeine free, nourishing, relaxing to drink, and most smell good. Ingesting herbal teas on a regular basis can help to tone, soothe, nourish, and balance the body.

Tea has been used in rituals since the beginning of time. Herbal teas are an infusion made from plants other than *Camellia sinensis*, which originated in China and the Middle East and is made into black and green teas.

When preparing herbal teas, water is used to remove the chemical constituents of the plant. Water removes vitamins, minerals, starches, gums, sugars, tannins, acids, as well as some glucosides and volatile oils.

Preparing Teas

Herbal teas should be made daily because water has no preservative qualities unless refrigerated or frozen (think iced tea and herbal ice cubes—delicious and convenient).

Herbal tea can be made with fresh or dried flowers, leaves, seeds, fruits, or shredded

Receiving tea during a Sami ceremony in Maze, Norway. BG

Part III: *Medicinal Preparations and Recipes*

Conifer tea is high in vitamin C. CA

roots. The different parts of the plant can be prepared in a variety of ways to get the optimum qualities you are looking for, including flavour.

Whether using one plant at a time or creating a tea blend, consider the end result you would like to achieve, including taste. I am a great believer in making medicine that tastes good!

Below are explanations and examples of the three types of tea preparations referred to throughout this book: tisanes, infusions, and decoctions. Most of the recipes use the simpler's method of measurement (see page 37) so that you can make as much, or as little tea, as you'd like.

Basic Herbal-Tea Recipe

Dried Herb: 1 teaspoon (5 mL) dried herbs to 1 cup (250 mL) boiling water.

Fresh Herb: 1 tablespoon (15 mL) fresh herbs to 1 cup (250 mL) boiling water.

HERBAL TEA OR TISANE

A herbal tea or tisane is a light infusion of the delicate parts of the plant in hot water. The word "tisane" is derived from *ptisánë*, a Greek word used to describe pearl-barley water.

Most people today drink herbal teas by steeping a prefabricated tea bag in hot water for a few minutes. In general, I prefer using loose plant matter because the plants are fresher and less processed, and create a more full-bodied, aromatic, flavourful, and healthful tea. Why is this? Because herbs that are in tea bags have been ground up and lost many of their volatile oils, which in turn leaves the herbs "inactive." Plus, in many cases the tea bag is a bleached pouch fastened with glue and attached to a bleached cotton string.

The good news is that the quality of some tea bags is getting better. There are companies that have made great strides in creating bags that use unbleached paper or fabric and are sewn together. Unfortunately, these teas generally cost a lot more. So it's more economical and environmentally friendly to go out and gather, dry, and use your own wild mint than to purchase boxes of mint tea.

To prepare a tisane, place a generous pinch of your wild herbs in a teacup or teapot, pour boiling water over them, let steep for a few minutes, strain, and sweeten with honey, if desired.

Herbal Infusions

Spirit-of-the-Boreal-Forest Tea

1 part sweetgrass
1 part horsetail
2 parts Labrador tea
2 parts bedstraw leaf
2 parts mint leaf
2 parts crushed rosehips

Labrador tea infusion. CA

Afternoon Floral Tisane

1 part wild chamomile
1 part fireweed flowers
1 part rose petals
1 part raspberry leaf

HERBAL INFUSIONS

Herbal infusions differ from herbal teas. The primary difference is a longer steeping time that makes the infusion generally darker in colour and fuller in taste, and allows more of the plant's water-soluble constituents—minerals, vitamins, and various other medicinal properties—to be released. A herb's nutritional properties are better absorbed and its therapeutic purposes are more effective when taken as an infusion.

Infusions are made from the leaves, flowers, seeds, and sometimes the roots of a plant.

To prepare an infusion, place dried or fresh plant parts into a teapot, large Mason jar, or any other container that has a tight lid, and fill with boiling water. Cover. Let steep for 5 to 45 minutes or overnight. After herbs are strained, drink either cold, hot, or at room temperature.

Use 1 teaspoon (5 mL) of the dried herb to 1 cup (250 mL) of water.

Use 4–6 tablespoons (60 to 90 mL) per 4 cups (1 L) of water or 6-cup (1.5-litre) teapot.

Nourishing Infusion

I like letting stinging-nettle (*Urtica dioica*) tea steep overnight, as there is nothing quite like a nice cup of room temperature, dark green nettle-infusion first thing in the morning! Stinging-nettle infusion contains an abundance of vitamins and minerals, including calcium, in each cup. Nettle helps to strengthen adrenal functioning, promote restful sleep, increase energy, prevent allergic reactions, strengthen blood vessels, and can prevent hair loss.

Part III: *Medicinal Preparations and Recipes*

It is so rewarding to gather and preserve herbs for food and medicine. CA

High-Calcium Infusion

1 part stinging-nettle leaf
1 part raspberry leaf
1 part horsetail leaf
3 parts rosehips (crushed)
2 parts mint leaf

Can also be made into a nourishing vinegar, or a glycerin tincture.

Good-for-Fever Infusion

1 part raspberry leaf
1 part plantain leaf
2 parts mint leaf
2 parts elder flowers
2 parts yarrow flower/leaf

Can also be made into an alcohol, glycerin, or vinegar tincture.

Herbal Infusions

Good-Sleep Infusion

2 parts wild chamomile
2 parts raspberry leaf
2 parts red clover leaf/flower
1 part valerian root

Can also be made into an alcohol, glycerin, or vinegar tincture.

Get-Up-and-Go Infusion
Great added to your morning smoothie!

1 part Labrador tea
1 part mint leaf
2 parts nettle leaf
2 parts rhodiola root

Can also be made into an alcohol, glycerin, or vinegar tincture.

Lung-Tonic Infusion

1 part coltsfoot leaf
1 part lungwort leaf
1 part yarrow flower
2 parts plantain leaf

Can also be made into an alcohol, glycerin, or vinegar tincture.

Detoxifying Infusion

1 part plantain
1 part mint
1 part crampbark
1 part stinging nettle
1 part bedstraw
2 parts dandelion root
2 parts red clover

Can also be made into capsules, or an alcohol, glycerin, or vinegar tincture.

Wild Food and Medicine Plants of the North

Part III: *Medicinal Preparations and Recipes*

Gathering wild mint. BG

Good-for-Gout Infusion

1 part willow bark
1 part raspberry leaf
1 part wild chamomile
2 parts celery seeds (optional)
2 parts birch leaves
3 parts stinging-nettle leaf
3 parts dandelion root and leaf

Can also be made into an alcohol, glycerin, or vinegar tincture.

Moon-Time Infusion

1 part mint leaf
1 part stinging-nettle leaf
1 part red clover
2 parts raspberry leaf
2 parts crampbark

Can also be made into an alcohol, glycerin, or vinegar tincture.

Tea-for-Two Pregnancy Infusion

1 part mint leaf
2 parts stinging-nettle leaf
4 parts raspberry leaf

Good-for-Nausea Infusion

1 part mint
2 parts wild chamomile

Can also be made into an alcohol or glycerin tincture.

Herbal Infusions

Beautiful Hair, Skin, and Nails Infusion

1 part mint leaf
1 part chickweed leaf
2 parts horsetail leaf
2 parts stinging-nettle leaf

Can also be made into a glycerin or vinegar tincture.

Dandelion-Root Coffee

When properly brewed, roasted dandelion-root coffee closely resembles coffee in flavour and body.

6, or more, large and fresh dandelion roots

Preheat oven to 250°F (120°C).

Wash dirt off the roots, and finely dice them.

Spread the chopped roots, about ½ in. (1.25 cm) deep, on cookie sheets and put in the oven. This dries and roasts the roots at the same time.

After the roots dry they will begin to roast, turning from a blonde colour to a dark coffee colour. Stir roots often to assure even drying and roasting. Be careful not to burn them! The roasting process takes about two hours.

Freshly dug, autumn, dandelion root—chopped and ready to roast or decoct. BG

Let the roots cool, then grind them in a coffee grinder. Store in a glass jar.

Use 2 teaspoons of roasted root grind for each cup of water.

You can make the coffee in a coffee press, with a tea strainer, or, like instant coffee, just add the grind to your cup.

Serve hot. You can add rice milk, almond milk, soy milk, cow milk, and to sweeten add honey or birch syrup. Try experimenting with the dandelion-root coffee: use it to make chai, lattes, and mochas.

Part III: *Medicinal Preparations and Recipes*

DECOCTIONS

Decoctions are made from the tougher parts of the plant, such as the roots and bark, that tend to be harder to extract constituents from.

Decoctions are made by slowly simmering or infusing the roots or bark of a plant overnight.

To prepare a decoction, place the herbs in a small stainless-steel pot and cover with cold water. Heat slowly and simmer with the lid on for 5 to 30 minutes. The longer you simmer the herbs, the stronger your brew will be. Ideally, without boiling the herbs, you want to warm the liquid enough that it will be reduced by half. Sometimes when I'm in a hurry, I boil the water in the pot, turn the heat down, put herbs in the pot and simmer for a few minutes, turn off the heat and let it steep for many hours. I often do this before I go to bed so that when I wake up I have a strong brew ready to go.

Use 1 teaspoon (5 mL) of the dried herb to 1 cup (250 mL) of water. Use 4–6 teaspoons (20–30 mL) per litre of water.

Spring-Cleanse Dandelion Decoction

In the North one of the first spring medicines is dandelion root. The optimum time for digging the roots is just as the green shoots are peering through the soil. At this time all of the plant's energy is in its roots, ready to feed and nurture the growth of the dandelion's contrasting green leaves and radiating yellow flowers. By ingesting the roots at this time, we absorb this fertile life-force energy (prana) and cleanse our entire system.

1 part dandelion root
2 parts water

Rosehip Decoction

Over the years many of my clients have asked me how to make the optimal rosehip tea—this is the answer!

1 part dried rosehips
2 parts water

Bring water to a boil, add rosehips, simmer on low for 20 minutes with the lid on. Turn heat off and let steep for one hour. Reheat if you want a hot cup of tea.

Some people like to add a dollop of honey. I like to let the tea steep overnight and the next day make a delicious rosehip iced tea with a squeeze of lemon.

Herbal Infusions

Beautiful Hair, Skin, and Nails Infusion

1 part mint leaf
1 part chickweed leaf
2 parts horsetail leaf
2 parts stinging-nettle leaf

Can also be made into a glycerin or vinegar tincture.

Dandelion-Root Coffee

When properly brewed, roasted dandelion-root coffee closely resembles coffee in flavour and body.

6, or more, large and fresh dandelion roots

Freshly dug, autumn, dandelion root—chopped and ready to roast or decoct. BG

Preheat oven to 250°F (120°C).

Wash dirt off the roots, and finely dice them.

Spread the chopped roots, about ½ in. (1.25 cm) deep, on cookie sheets and put in the oven. This dries and roasts the roots at the same time.

After the roots dry they will begin to roast, turning from a blonde colour to a dark coffee colour. Stir roots often to assure even drying and roasting. Be careful not to burn them! The roasting process takes about two hours.

Let the roots cool, then grind them in a coffee grinder. Store in a glass jar.

Use 2 teaspoons of roasted root grind for each cup of water.

You can make the coffee in a coffee press, with a tea strainer, or, like instant coffee, just add the grind to your cup.

Serve hot. You can add rice milk, almond milk, soy milk, cow milk, and to sweeten add honey or birch syrup. Try experimenting with the dandelion-root coffee: use it to make chai, lattes, and mochas.

Part III: *Medicinal Preparations and Recipes*

DECOCTIONS

Decoctions are made from the tougher parts of the plant, such as the roots and bark, that tend to be harder to extract constituents from.

Decoctions are made by slowly simmering or infusing the roots or bark of a plant overnight.

To prepare a decoction, place the herbs in a small stainless-steel pot and cover with cold water. Heat slowly and simmer with the lid on for 5 to 30 minutes. The longer you simmer the herbs, the stronger your brew will be. Ideally, without boiling the herbs, you want to warm the liquid enough that it will be reduced by half. Sometimes when I'm in a hurry, I boil the water in the pot, turn the heat down, put herbs in the pot and simmer for a few minutes, turn off the heat and let it steep for many hours. I often do this before I go to bed so that when I wake up I have a strong brew ready to go.

Use 1 teaspoon (5 mL) of the dried herb to 1 cup (250 mL) of water. Use 4–6 teaspoons (20–30 mL) per litre of water.

Spring-Cleanse Dandelion Decoction

In the North one of the first spring medicines is dandelion root. The optimum time for digging the roots is just as the green shoots are peering through the soil. At this time all of the plant's energy is in its roots, ready to feed and nurture the growth of the dandelion's contrasting green leaves and radiating yellow flowers. By ingesting the roots at this time, we absorb this fertile life-force energy (prana) and cleanse our entire system.

1 part dandelion root
2 parts water

Rosehip Decoction

Over the years many of my clients have asked me how to make the optimal rosehip tea—this is the answer!

1 part dried rosehips
2 parts water

Bring water to a boil, add rosehips, simmer on low for 20 minutes with the lid on. Turn heat off and let steep for one hour. Reheat if you want a hot cup of tea.

Some people like to add a dollop of honey. I like to let the tea steep overnight and the next day make a delicious rosehip iced tea with a squeeze of lemon.

SOLAR AND LUNAR INFUSIONS

An ancient way of preparing your herbal infusion is to use the sun's fire energy or the moon's water energy to help extract healing constituents from the plants.

Solar Infusions

Plants and their parts that can be used for solar infusions include rose petals, fireweed flowers, red clover, spruce tips, wild chamomile, dandelion flowers, and goldenrod flowers—to name just a few.

To prepare a solar infusion, place fresh or dried herbs and water in a clear glass jar. Place the jar in direct sunlight for several hours, allowing the warm, radiating energy of the sun to gently heat the water, releasing the plant's fragrance and other offerings into the water.

It's uplifting just to look at a solar infusion steeping in the sun. BG

Part III: *Medicinal Preparations and Recipes*

For the sun tea ceremony I use a special bone china wild-rose teacup that belonged to my grandmother. BG

Boreal Summer-Solstice Tea Ceremony

Various cultures have tea ceremonies that involve connecting to the sacred. Creating your own tea ceremony can bring the reverence back to the preparation of tisanes. As all northerners know, the summer solstice in the North is a magical time, when the sun never sleeps and neither do we. The wild roses are usually in full bloom during the solstice and the bright pink petals make a wonderful sun tea—and an easy start to creating your own rose-tea ritual.

To begin, sit quietly among the wild roses, close your eyes and tune into yourself, your breath, and the intoxicating scents being released from the wild-rose petals.

When you are feeling calm and centred, start to sense the energy of the plants, tune into the essence, and wait for insight. When you are ready, open your eyes and start to study the physical aspects of the plant. Memorize the lines and scent of it, touch the leaves, flowers and stem, feel the texture, and the taste. Look at all of the plant's subtleties and open your mind and heart to what it's telling you.

Check your feelings and emotions and what thoughts come into your head. At this point set your intention for your Summer-Solstice Tea Ceremony.

Start to feel the energy between you and the roses. Make an offering with gratitude and start to slowly gather petals for your tea—leave behind one petal on each flower to attract the bees. When you have enough petals, place them in your clear glass container filled with water and set in a sunny spot outdoors or on a sunny windowsill to let the sun's energy infuse the flavour and essence of the plant into the water.

When you're ready to drink the sun tea, pour it into your cup. I don't strain out the petals, I like to eat them as I sit and sip my tea, reflecting on the magic of the day.

To finish the ceremony, say thank you in your own way. It's nice to take time to write down insights that have come to you during your ceremony. The ceremony can be done at anytime, either by yourself or with others. It's fun to do with children. I like to start the ceremony at noon and drink the tea at midnight.

Iced Teas

When making the solar infusion, the jar or container should be open. If flies or other insects are a concern, place a piece of muslin cloth over the opening.

I usually just intuitively sprinkle the chosen plants into the jar as this feels in tune with the essence of these infusions, but for those who need a guide, try 1 tablespoon (15 mL) of the dried herb to 1 cup (250 mL) of water. Use 4–6 tablespoons (60 to 90 mL) per 4 cups (1 L) of water or a 6-cup (1.5-litre) teapot. For fresh herbs use double the amount of plant matter.

> "The monks gathered before the image of Bodhi Dharma and drank tea out of a single bowl with the profound formality of a holy sacrament."
> —Okakura Kakuzo, *The Book of Tea*

Lunar Infusions

A lunar infusion captures the essence of the reflective light source and also the energetic influences that the moon has on us, helping bring our attention to our creative forces and our emotions.

Plants that can be used for lunar infusions include wild sage, wild chamomile, yarrow flowers and leaves, Labrador tea, and any other plants that remind you of the moon's energies.

To prepare a lunar infusion, place fresh or dried herbs and water in a clear glass jar and leave outside overnight. The jar or container opening can be covered to keep out night bugs and nocturnal animals like mice and cats! If these are not problems, leave the jar open.

ICED TEAS

Iced tea is made with a cooled infusion or decoction. It tastes great, is beautiful to look at, and is a refreshing way to get all the healthful benefits from plants.

Plants that make great iced tea include cranberry, rosehip, blueberry, red clover, Labrador tea, and nettle. You can also trying combining berries and fruits with herbs such as mint, dandelion root, rose petals, or red clover to create a cooling, nourishing, and healing beverage.

Homemade syrups (see page 298) like rosehip, cranberry, blueberry, and high-iron syrup, can be added to your iced tea for extra flavour or medicine.

To prepare an iced tea, pour cooled tea into a jug, straining the herbs. Stir in a squeeze of lemon and, if desired, a sweetener such as honey, maple syrup, agave nectar, birch syrup, or a home-prepared syrup such as rosehip or mint. Add a handful of ice, stir, and let cool. Serve with a sprig of mint or with wild cranberries floating on top.

There's nothing like a glass of thirst-quenching iced tea on a hot summer's day. BG

Part III: *Medicinal Preparations and Recipes*

ADDITIONAL HERBAL-TEA RECIPES:

Bearberry Tea, page 54

Mountain Sorrel Iced Tea, page 159

Rhodiola Decoction, page 140

Tamarack Tea, page 268

Wild Chamomile Tea, page 62

Red-Clover Tea Aid

This herbal lemonade is a refreshing summer drink, and a wonderful way to add some nutritious herbs to your family's diet.

2 cups (500 mL) fresh red-clover blossoms
4 cups (1 L) water
½ cup (125 mL) honey or birch syrup
1 cup (250 mL) lemon juice

Gently simmer clover blossoms in a covered pot for 10 minutes. Add honey, stirring until it dissolves. Cover and let steep and cool for several hours or overnight. This makes a strong, potent tea, and maximizes the release of calcium and other nutrients from the clover.

Strain out blossoms. Add lemon juice and chill in the fridge.

TINCTURES, VINEGARS, ELIXIRS, AND BITTERS

MAKING HERBAL PREPARATIONS in the summer, when a plant is fresh, is a good way to capture and preserve the plant's constituents and essences. Not to mention that after you have finished making your preparations, you will have a convenient and abundant supply of energy- and nutrient-rich herbal medicines that you can use anytime you need to.

It's worth noting that your complete attention is required when making medicines. By being observant while medicine making, you will be fully in the moment.

Using the simpler's measurement method (see page 37) you're permitting your intuition to play a vital part in the creation of medicines.

While the maturation time needed to make herbal oils, tinctures, etc., can vary, I like to make them during moon cycles: from new moon to full moon; from full moon to full moon; or, new moon to new moon.

TINCTURES

Tinctures are very concentrated and can last a long time, so you won't have to make them every year. I find that some tinctures I use rarely, such as shepherd's purse, and so it lasts for years. However, I make nettle tincture almost every summer because I use so much of it.

Tinctures don't take up much room, so are convenient to take travelling, and good to include in a herbal first-aid kit.

Tinctures are different from teas in that water is not used to extract the properties of the plant. Alcohol, glycerine, or vinegar is used as a "menstruum" (solvent) in tinctures to remove a plant's medicinal properties, or "marc." This is because the marc of some plants is not water soluble. The process of extracting the constituents from the plant is called "maceration."

Tinctures are applied internally under the tongue, for quick absorption. However, some people find the taste too strong and prefer to add tinctures to juice, water, or tea. I also like adding tinctures to food preparations such as soups, sauces, dips, and to my morning smoothie. People who prefer not to ingest alcohol can add the tincture to a steaming hot cup of water or herbal tea—the hot liquid will cause most of the alcohol to evaporate.

Above: Antique medicine bottles at Matthew Watson General Store, Carcross, Yukon. BG

Below: Tincturing is a good way to preserve the healing properties of herbs for a long time. BG

Part III: *Medicinal Preparations and Recipes*

> When using tinctures, the standard therapeutic dosage for an adult is 10–30 drops 3 times a day. (For more on determining dosage, see page 38.)

Menstruums

Alcohol, glycerine, and vinegar can all be used as menstruums to make tinctures. Each of these has its own unique characteristics.

Alcohol (Ethyl Alcohol)

Alcohol dissolves resins, tannins, organic acids, alkaloids, balsams, and glucosides, but usually not gums and starches. The alcohols that are most readily available for medicine making are vodka, brandy, rum, scotch, and other grain alcohols. Never use rubbing alcohol.

Glycerine

Glycerine is a constituent of all oils derived from animals or vegetables. Glycerine is a sweet tasting, nutritive solvent that has good preservation qualities, but it does not extract as many chemical constituents from plants as alcohol or water. Glycerine does dissolve fixed alkalis, minerals, vitamins, gums, starches, acids, oils, and tannins. Glycerine is good to use in preparations for children, alcoholics, and people who are sensitive or allergic to alcohol. Glycerine tinctures can be taken in the same way as alcohol tinctures.

Vinegar

Vinegar extracts the same constituents as alcohol and also minerals such as calcium, magnesium, potassium, and vitamins. This makes the resulting infused vinegar an excellent base for a nutritive tonic to add to food preparations. Apple cider vinegar is your best choice for making medicinal and nutritive vinegars; it has its own nutrient value and acts as an antiseptic. It has a shelf life of about one year.

Preparing Tinctures

Tinctures can be made from dried herbs, but are best made from fresh herbs.

When using fresh herbs: Break up the botanicals (this helps activate the extraction process) and completely fill a jar with them. Pour the menstruum over the botanicals until they are covered, making sure no plant matter is sticking out of the liquid.

When using dried herbs: Break up botanicals and fill half of a jar with them. Fill the jar to the top with the menstruum.

See the opposite page for method. To help preserve the remedy, store in a dark glass bottle out of direct sunlight.

Tincture Recipes

Single-plant herbal tinctures can be made from boreal plants such as rhodiola root, usnea lichen, bearberry leaves, chickweed leaves, dandelion root, valerian root, mint leaves, plantain leaves, and juniper berries.

The following multi-plant recipes can be made using alcohol, glycerine, vinegar, or made into herbal teas. Each plant can also be tinctured separately and then blended into the specific formulas. Most of the time this will make sense because many of the plants in these recipes are ready for harvesting at different times. Tincturing each plant separately also gives you a broader spectrum of possibilities for use in other formulas.

Tinctures

Tincture Method

1. Break up or garble herbs into a jar.

2. Pour menstruum over the herbs until they are fully covered with liquid.

3. Cover the jar with a lid. When using vinegar, use a lid made of plastic, as metal will rust when it comes into contact with acid. If you don't have a plastic lid, place a plastic or wax paper barrier between the lid and the jar.

4. Label your preparation. (For details on labelling, see page 35.)

5. Place out of direct sunlight and let steep for several weeks. Generally, you will want to let the tinctures infuse for at least two weeks. I recommend letting them steep from the new moon to a full moon, or for one full-moon cycle.

6. When your preparation is ready, strain through a piece of cheesecloth, wringing out the cloth to get every last precious drop. Compost the spent botanicals.

CA

Part III: *Medicinal Preparations and Recipes*

Note: If preparing tinctures for commercial usage, there are resources available that give detailed information on marc and menstruum ratios for each plant. When preparing tinctures for yourself, the simpler's method of measurement will give you the freedom to make any amount you want. (For more on the simpler's method of measurement see page 37.)

Before using the tinctures, please be sure to cross-reference the plants in *Part 2: Plant Profiles* of this book, to make sure they do not contraindicate with any conditions you have or medications you are taking.

Pain-Aid Tincture

Antispasmodic, analgesic, anti-inflammatory, nervine, and sedative. Good for pain relief, headaches, sleep and nervous disorders, and menstrual cramps.

1 part crampbark
1 part valerian root
1 part willow bark
1 part devil's club inner bark

Athlete's-Aid Tincture

Rhodiola root helps increase energy. Great for active people!

1 part rhodiola root
1 part mint leaf
1 part stinging-nettle leaf

Cold-and-Flu Tincture

Great as a preventative, or if you're in the thick of a cold or flu!

1 part mossberry bark
1 part yarrow flowers and leaves
1 part plantain leaf
1 part rosehips
1 part bearberry leaf
1 part raspberry leaf
2 parts usnea lichen

Urinary-Tract Tincture

1 part bearberry leaf
1 part bedstraw leaf
1 part willow bark
1 part fireweed flowers
1 part horsetail leaf
1 part yarrow flowers
1 part cranberries
1 part goldenrod

Usnea tincture. BG

Tinctures

Prostate-Support Tincture

1 part fireweed root
1 part fireweed flowers
1 part fireweed leaves
1 part mossberries

Note: Fireweed root is best gathered in the early spring; the leaves as they unfold; the flowers as they bloom; and the mossberries as they ripen. Make separate tinctures and then blend together after they've been strained and decanted.

Sprays

When combined with water, the following tinctures can be used as sprays. You can use them on the skin for topical infections, or in the mouth for treating sore throats. These spray-tincture recipes can also be stored in a regular bottle and used topically in a compress or as a gargle for treating sore throats.

Skin-Disinfectant Spray

Use topically for cuts, abrasions, and infected skin. You can use this for sore throats as well.

1 part usnea tincture
1 part chickweed tincture
1 part mossberry bark tincture
1 part fireweed-flower tincture
1 part willow inner-bark tincture
1 part pine- or fir-needle tincture
1 part wild-sage-leaf tincture

Mix tinctures in a spray bottle, top up with freshly boiled water (optional), and leave to cool. When fully cooled, install the spray pump and use spray as necessary.

Sore-Throat Spray

1 part usnea tincture
1 part chickweed tincture

ADDITIONAL TINCTURE RECIPES:

Clear-Skin Tincture, page 72

Gentian Tincture, page 97

Nettle Tincture, page 128

Plantain Quit-Smoking Spray, 135

Shepherd's Purse Tincture, page 156

Valerian-Root Tincture, page 177

Part III: *Medicinal Preparations and Recipes*

VINEGARS

Nourishing Vinegar Tonic

A daily spoonful of this tonic helps keep the body healthy. It can also be used as a salad dressing, or as a base for making mayonnaise and creamy dressings. Nourishing Vinegar Tonic is high in minerals such as calcium, magnesium, potassium, and vitamins. A tablespoon can also be added to water and drunk in the morning.

1 part stinging nettle
1 part raspberry leaves
1 part horsetail
1 part pineapple weed
1 part rose petals
1 part yellow-dock root
2 parts peppermint
2 parts strawberry leaves
2 parts chickweed
3 parts rosehips
Apple cider vinegar

Pour the apple cider vinegar over the botanicals until they are covered, making sure no plant matter is sticking out of the liquid. Shake daily.

ADDITIONAL VINEGAR RECIPE:

Moon-Cycle Nourishing Vinegar, page 203

Nourishing Vinegar Tonic is nutritious and delicious. BG

Elixirs and Bitters

ELIXIRS AND BITTERS

Elixirs and bitters are prepared in the same way as tinctures, but brandy or wine is used to extract the medicinal properties from the herbs. Elixirs are sweetened tinctures and are often used with very bitter herbs to improve the palatability of the remedy.

Bitters are, like their name suggests, a very bitter tincture with no sweetener added. Bitters are generally taken before a meal to stimulate the secretion of the digestive juices and bile. This increases the appetite and helps activate digestion and bowel movements. Good bitter herbs are gentian root, dandelion root, yarrow leaves and flowers, and yellow-dock root.

Preparing Elixirs and Bitters

Single herbs or a combination of herbs can be used. Place herbs in a jar and fill with brandy, or a burgundy or currant wine.

Place in a cool, dark place for four to six weeks, shake daily.

When ready, strain. To each cup (250 mL) of liquid add ½ cup (125 mL) of black-cherry concentrate (available in health-food and herb stores), rosehip syrup (see page 299), honey or maple syrup. Mix well and bottle. Shake before using.

A bitter made from gentian root is one of the best-known digestive remedies. PL

Boreal Bitters

A daily digestive remedy that can be taken before meals.

1 part gentian root
1 part plantain
1 part mint
1 part crampbark
2 parts dandelion root
2 parts red-clover flowers
2 parts yellow-dock root
20 parts brandy

After-Dinner Elixir

1 part mint
1 part pineapple weed
1 part yarrow
2 parts dandelion root
8 parts brandy
2 parts black-cherry concentrate or rosehip syrup (sweetener is optional)

SYRUPS

Syrups are a way to make sweet-tasting herbal preparations. They are simple to make, easy to store, usually last from up to six months to one year, and are an easy way to give herbs to children (though it is recommended you not give unpasteurized honey to children under a year old). I like to use honey in my syrups, but you can use other sweeteners such as cane sugar, vegetable glycerine, maple syrup, or birch syrup. When making syrups, always mix in a small amount of brandy at the end to help preserve the formula. The rule of thumb is 1 tablespoon (15 mL) of brandy for 1 cup (250 mL) of liquid. Your homemade syrups will last for up to one year if preserved with brandy and stored in the refrigerator.

> It's a good practice to use sterilized, distilled, spring, or reverse osmosis water when making your herbal remedies. That said, when I'm making medicines for my family, I use the water from our tap, which comes from our well that I know is a healthful source of clean water. Tap water can be sterilized by bringing it to a rolling boil for five minutes to kill bacteria and funguses. It can be stored in a sterilized bottle in the fridge for a couple of days.

Basic-Syrup Recipe

1 part herbs or berries (If you do not have fresh herbs on hand, it's okay to use dried herbs.)
1 part honey or other sweetener
4 parts water
Brandy

Add botanicals and water to a stainless-steel or glass cooking pot. Simmer over low heat, concentrating the liquid down to half while pulping the fruits with a potato masher to help draw out the juices. Remove from heat and let cool.

Strain botanicals through a sieve lined with cheesecloth into a measuring cup so that you can see how much liquid infusion you have. (Do not allow any plant matter to remain in the liquid, otherwise the liquid may start to ferment and spoil your preparation.)

Pour the liquid back into the pot. Simmer over low heat for one minute or so. Add sweetener. Heat only enough to allow the liquid and sweetener to blend well. For 1 cup (250 mL) of infusion, add 1 cup (250 mL) of honey, and 1 tablespoon (15 mL) of brandy.

Syrups

Remove from heat, cool, and add brandy. Pour into a sterilized bottle. It's best to have a large storage bottle or jar, and to then decant into smaller bottles for frequent use. This will help preserve the syrup longer.

Label and date your syrup, then refrigerate.

Rosehip Syrup

Rosehip syrup can be used alone or blended with tinctures or other herbal syrups, as a base for tonics, as a topping for desserts or pancakes, as part of a fruit-leather mixture, or as a concentrate for making iced or hot herbal teas.

1 part rosehips
6 parts water
Brandy

Follow the Basic-Syrup Recipe instructions.

Note: Rosehips and some berries absorb a greater volume of water than leaves do. So by adding a couple more parts of water during the cooking process you'll end up with more volume.

High-Iron Syrup

High in iron and other minerals, this tonic is easily absorbed and tastes great! This formula can also be made into a tea or vinegar infusion.

1 part yellow-dock root
2 parts rosehips
2 parts stinging-nettle leaves
2 parts strawberry leaves
2 parts horsetail

Follow the Basic-Syrup Recipe instructions.

Top: Autumn rosehips. RF

Above: Strawberry leaves. BG

Part III: *Medicinal Preparations and Recipes*

Add a tablespoon of your medicinal syrup to a glass of cold water, throw in an ice cube and a splash of lemon or lime, and—voilà—iced tea! Or, add a tablespoon of syrup to a glass of sparkling water to make your own carbonated herbal drink.

Cough-and-Cold Medicinal Syrup

This syrup is antispasmodic, antimicrobial, expectorant, and astringent, and is good for coughs, colds, and respiratory congestion.

1 part yarrow
1 part crampbark
1 part horsetail
1 part mint
1 part spruce tips
1 part plantain
1 part fireweed leaves and flowers
1 part goldenrod herb
4 parts coltsfoot
Water
Honey
Brandy

Simmer the herbs in water until half of the water remains. Cover the pot so that all the volatile oils of the herbs don't escape, but remember to stir the mixture occasionally. Strain herbs, put the liquid back into the pot at a low heat, and add the honey, stirring until it blends together. Cool and add brandy. Bottle and keep refrigerated or in cool place.

Take 1 teaspoon (5 mL) as needed for a cough, cold, or sore throat. A variation on this recipe is to use usnea tincture with, or instead of, brandy to help preserve the syrup. This formula can also be made into alcohol tincture, or the herbs dried to make into a tea. If you decide to use the tincture, a few drops can be added to a spoonful of honey or to a cup of hot tea sweetened with honey.

Fireweed leaves and flowers for fighting respiratory ailments. BG

CAPSULES

CAPSULES ARE A VERY CONVENIENT WAY to take dried botanicals, especially those that taste strong and bitter. Vegetarian capsules or the standard gelatin capsules can be filled with your herbs of choice for therapeutic use.

Preparing Capsules

Powder botanicals in a blender, or a clean coffee or spice grinder. If using more than one herb, you can blend them either separately or together. Place powder in a bowl. If using more than one herb, make sure they are all blended together at this stage.

Separate a capsule and fill each side with powder. Push capsule ends together. An easier and more convenient method of encapsulation is to use a small, manual, capsule-filling machine called Cap-M-Quik, which can make up to fifty capsules at a time. It is inexpensive and available online at *www.Cap-M-Quik.com*.

Capsule sizes vary: The "00" capsule is a standard size that holds approximately 950 mg of ground herb. The "0" capsules hold approximately 680 mg.

Allergy-Aid Capsule

1 part dried mossberries
1 part plantain leaf
2 parts stinging-nettle leaf

Sleep-Aid Capsule

1 part pineapple weed
1 part valerian root

Pain-Aid Capsule

1 part valerian root
1 part crampbark
2 parts willow inner bark

PLANT ESSENCES

PLANT ESSENCES ARE a wonderful way to help people shift gears emotionally, mentally, physically, and spiritually.

The concept of plant essences was developed in the 1930s by the English physician and homeopath Dr. Edward Bach. Bach created these homeopathic-like remedies with the intended purpose of using them primarily for emotional and spiritual conditions, including, but not limited to, depression, anxiety, insomnia, and stress.

Several of the plants Bach mentions in his teachings grow in the boreal forest. For example, Dr. Bach used the aspen essence for "vague unknown fears, for which there can be given no explanation, no reason."

Plant essences are an effective addition to your herbal repertoire. They are easy to make and are an excellent way to connect to a plant's spirit (for more on this, see Sacred-Spirit Plant Healing, page 28). In the summer, boreal forest dwellers have a unique opportunity to make the essences, as the sun is in the sky for so long and the boreal plant life is abundant and mostly untouched.

On the summer solstice I like to combine arnica and dandelion flowers in a quartz bowl to make Aroma Borealis's Solar Plexus Vibrational Essence. BG

302 | The Boreal Herbal

Plant Essences

Preparing Plant Essences

Fill a clean glass or crystal bowl with pure spring water.

Choose an area where the plant you wish to connect to is growing abundantly. Create a sacred space, and tune into the nature and the spirit of the plant you want to create an essence from. Make an offering of gratitude to the plant for sharing its gifts with you. (For more on this, see page 29.)

When you are feeling attuned to the spirit and vibration of the plant, carefully snip the blossoms, (or other desired part of the plant) with scissors, letting them fall onto the surface of the water. Make sure your hands don't touch the petals or the water.

Set the bowl down in front of plants you have picked from. Allow the sun to penetrate the water and plant matter for about three hours, or until blossoms or plant matter begin to look faded or spent, and the water has a light and bubbly luminescence to it.

Left: Aroma Borealis Root Chakra Vibrational Essence. BG

Below: Dandelion flowers being made into an essence. BG

"Health depends on being in harmony with our souls."

—Dr. Edward Bach

Gently remove the plant material with tongs. Measure plant water in a glass measuring cup, add equal amounts of brandy to help preserve and carry the essence. You have just created the "mother essence." Pour it into a clean jar or bottle.

The mother essence is used to prepare smaller stock bottles for use. To make a stock bottle, add 2–10 drops of mother essence to a 30-mL dropper bottle filled with brandy or water. To ingest, add 2–4 drops to a cup of water or tea, then drink. Essences can also be used directly under the tongue, in the bath, in a water fountain, or anywhere you want the essence to permeate.

All boreal forest plants can be made into a plant essence. Wild roses are a wonderful choice for a heart chakra flower essence. Essences can be made with other parts of a plant as well, including catkins, buds, leaves, and berries.

ADDITIONAL PLANT ESSENCE RECIPE:

Twinflower Vibrational Essence, page 173

TOPICAL TREATMENTS

USING WILD PLANTS topically is a wonderful way to become aware of and to connect with their healing properties. From the simplicity of a spit poultice to the complexities of making herbal creams, in this section I hope to introduce you to the knowledge and skills needed to make your own topical remedies.

If you have never made a herbal oil infusion, a salve, or a herbal cream before, I hope that you will find the instructions easy to navigate. I organized these recipes so that each type of remedy would build upon the next. This section begins with making infused herbal oils and ends with making a cream.

What follows are just guidelines for making topical treatments. As you become familiar with each process and recipe, you can start to create your own recipes. These remedies all use the simpler's method of measurement (see page 37) as this will allow you to make the remedies in any quantity you want.

MEDICINAL OILS

Making infused oils with wild plants is easy. These are great as massage and bath oils, and act as a base for medicinal salves, ointments, and creams.

The oil acts as the solvent, or "menstruum," for removing the oil-soluble medicinal properties from the plants, which include essential oils, resins, balsams, waxes, and vitamins.

Herbal oils will last up to one year depending on the type of carrier oil that you use. Olive oil is a popular choice for medicinal recipes because it has a long shelf life, is nutrient rich, and is easily absorbed through the skin. Lighter oils, such as sunflower, almond, or grapeseed, can also be used. Adding natural vitamin-E (tocopheral) oil squeezed from a vitamin-E capsule will extend the shelf life of herbal oils. Essential oils of benzoin or tea tree will also help to preserve infused oils. Medicinal oils can also be made with hard vegetable oils, such as coconut, or animal fat. Many Yukon First Nations use bear grease or other animal fats as a base for medicinal preparations.

Fresh or Dried Botanicals?

Making oil with fresh plants is wonderful, but it's very important to first let the fresh plant matter wilt for a few hours, or overnight, so that excess water evaporates. By doing this, you will stop your preparation from growing mould and spoiling. When I make infused oils from fresh botanicals, I don't put a tight fitting lid on immediately. Instead, I cover the oil with a cheesecloth to allow any excess moisture to evaporate.

Poplar buds infusing using the basic infusion method. CA

Medicinal Oils

You can still make beautiful infused oils with fully dried plants. The rule of thumb for infusing dried herbs is use half the amount of plant matter you would use for fresh herbs.

Each plant can also be infused in oil separately and then blended into the specific formulas. Most of the time this will make sense because many of the plants in these recipes are ready for harvesting at different times of the year. Infusing each plant separately also gives you a broader spectrum of the possibilities for use in other formulas. If you do this then you do not need to add the carrier oils that are listed in the recipes because they are already infused.

> I generally like to let oils infuse with the moon cycles. From a new moon to a full moon is optimal, or a full-moon cycle, from new moon to new moon, or full moon to full moon.

Basic Oil-Infusion Recipe

1 cup (250 mL) fresh or ½ cup (125 mL) dried garbled herbs
2 cups carrier oil
¼ teaspoon (1 mL) vitamin-E oil (approx. 3 capsules) as a preservative
Optional: Essential oils (in addition to or in place of vitamin E)

Use one of the following infusion methods:

Traditional Method

To begin, make sure the jars, utensils, and equipment you will be using are clean and dry.

Begin by breaking up or garbling herbs into a jar by hand. This bruises the herbs and releases the volatile oils.

Pour oil over herbs and cover jar with a tight-fitting lid. If you are using fresh herbs, cover with cheesecloth to allow excess moisture from the plants to evaporate.

Label and store out of direct sunlight (unless following the Sun-Infusion Method, described on the following page).

Shake or stir daily. When using fresh herbs it's important to watch for any changes to the oil—like mould or cloudiness caused by excess plant moisture. If this does start to happen, save your oil by immediately straining the plant matter.

After two to four weeks, strain the oil through a strainer lined with cheesecloth into a bowl or large measuring cup with a spout. When straining herbs from oil, fold the cheesecloth into two layers so that none of the plant material will get through. Unbleached coffee filters or muslin cloth can also be used.

Add vitamin-E and/or essential oils.

Pour oil into a clean, and preferable dark glass bottle. Label and store in a cool place out of direct sunlight.

> To get the maximum yield of oil from your infusion, wring out excess oil from the cheesecloth. The rung-out cheesecloth can be put in a plastic bag or container in the freezer and used later as a topical treatment for injuries or wounds. CA

Part III: *Medicinal Preparations and Recipes*

Right: Arnica infusing in the sun. BG

Below right: Double-boiler method. CA

Below: Fir needles, fireweed, and yarrow flowers infusing in a double boiler. CA

Bottom: Whichever infusion method is used, a dark bottle will help preserve the oil. CA

Sun-Infused Oil Method

The Sun-Infused Method follows the same steps as the traditional method, described above, with a couple of differences.

If you are using fresh herbs it is very important to make sure you have wilted the plants long enough so that the excess moisture has dissipated. Cover the opening with cheesecloth instead of a lid—if not, you will start to see condensation forming around the top of the jar.

Place the jar on a windowsill in direct sunlight, allowing the heat of the sun to infuse the herbs into the oil for at least two weeks, before straining, bottling, and adding vitamin-E or essential oils to the oil to preserve it.

Double-Boiler Method

The slow heat double-boiler method is good if you want to process herbs immediately. For this method you will need a double boiler (make sure to put enough water in the lower saucepan, as overheating the herbs and oils will compromise the quality of the oil). For large batches, a Crock-Pot can also be used, though first make sure it has a low-heat setting, otherwise, it can get too hot and burn the oil.

Break up herbs by hand into the top of the double boiler.

Pour oil over herbs and bring water (in bottom pot) to a low simmer.

Heat slowly for 30 to 60 minutes, stirring constantly to make sure oil is not overheating. (The lower the heat, the longer you can infuse the oil, and the better the quality of the oil will be.)

Let oil fully cool, then strain through a cheesecloth, add vitamin-E or essential oils to preserve the oil, bottle, and store in a dark place.

Medicinal Oils

Muscle-and-Pain Relief Oil

Topical oil for bruises, muscular aches, pains, and spasms. This recipe can be made with all the herbs listed below, or with just a few of them as they all have anti-inflammatory qualities!

1 part willow inner bark and leaves
1 part juniper berries, crushed
1 part devil's club root
2 parts arnica flowers
5 parts crampbark
½ part Labrador tea leaves
8 parts olive oil
2 parts St. John's wort oil (available at herb shops)
Vitamin-E oil
4 drops mint essential oil (optional)
6 drops fir essential oil (optional)
6 drops juniper essential oil (optional)

Follow Basic Oil-Infusion Recipe to yield two cups of infused oil, for larger batches, double or triple the recipe. Choose method of preparation and follow instructions. This oil can be used on its own or as a base for herbal salves or creams.

Boreal Skin Oil

Helps alleviate itching, chapping, and drying. Also makes a good infusion base for a diaper-rash ointment.

1 part plantain leaf
1 part yarrow flowers
1 part fireweed flowers
2 parts chickweed herb
3 parts rose petals
15 parts grapeseed oil
1 part rosehip seed oil
Vitamin-E oil

Choose method of preparation and follow instructions. This oil can be used on its own, or as a base for a herbal salve or cream.

Part III: *Medicinal Preparations and Recipes*

ADDITIONAL MEDICINAL-OIL RECIPES:

Ancient Aromatic Panax Oil, page 86

Bedstraw Ear Oil, page 58

Boreal Healing Oil, page 258

Golden Flower Body-and-Massage Oil, page 101

Plantain Herbal Healing Oil, page 136

Dandy-Lioness Oil

Dandelion blossoms and roots make an excellent infused oil that is great for massage, and releases tension and toxins from the body. It's even reported to reduce breast cysts.*

1 part dandelion blossoms and roots
2 parts olive oil
Vitamin-E oil

Choose method of preparation and follow instructions. This oil can be used on its own, or as a base for a herbal salve or cream.

Dandelion blossoms, roots, and olive oil make an excellent medicinal oil. BG

---

* Sat Dharam Kaur, N.D., author of *The Complete Natural Medicine Guide to Breast Cancer: A Practical Manual for Understanding, Prevention, and Care*, writes that [dandelion] oil works, "well for relaxing tense back and neck muscles and is great for use in deep tissue work, as it helps release stored tension and toxins. Dandelion oil can also be used to reduce breast cysts." Kaur also writes that using red-clover blossom oil will "help to remove breast lumps, discourages breast cancer, and improves lymphatic circulation. It can be used daily for breast self-massage."

Salves, Ointments, and Balms

SALVES, OINTMENTS, AND BALMS

Herbal oils provide you with the medicinal base for making a salve, ointment, or balm. You can also combine your oils, depending on what your preparation is intended for. For example, you could make a salve for treating chapped skin, relieving sore muscles, and for overall moisturizing.

Salves, ointments, and balms are semi-solid, fatty herbal mixtures prepared for convenient topical use. Ointments generally have more oil content in the preparation than a salve, and balms are generally a lot more aromatic. All three are created using the same method with slight variations.

When applied, the emollient properties of these topical treatments are absorbed into the skin, while the beeswax—used in the preparation—forms a breathable, protective coating on the skin's surface.

Balms, cooling. BG

Tips for Preparing Salves, Ointments, and Balms

In general, use 2 tablespoons (30 mL) of beeswax to one cup (250 mL) of herbal oil.

When using vitamin-E oil as a preservative, add 1 teaspoon (5 mL) to every 4 cups (1 L) of oil.

Adding a few drops of essential oil will extend the shelf life of your preparation, and will infuse the mixture with the essential oil's healing properties and pleasant aromas.

Before pouring your salve, remove a spoonful and let cool to see if you like the consistency. If you find that your salve, ointment, or balm is too soft, add more beeswax; if it's too hard, add more oil.

Salves and ointments can also be made with a hard vegetable oil, such as coconut oil, or an animal oil such as bear grease, and can be used topically for burns, cuts, abrasions, and rashes.

Once poured into jars, always let the preparations cool fully before putting the lids on. Otherwise condensation will form, which can lead to mould-growth and cause the salve to go rancid.

Part III: *Medicinal Preparations and Recipes*

CA

Carrier Oils

Carrier oils, also known as base oils, come from vegetables, nuts, and grains. Infused carrier oils are your best choice for spreading botanical extracts over a large surface of the body, and they are highly absorbable.

Vegetable oils have nourishing and therapeutic properties of their own, and often contain vitamins and minerals. Using herb-infused oils allows you to receive the therapeutic values of both the botanical and the carrier oils, without irritation to the skin.

There are a variety of carrier oils available, all with their own specific properties and uses.

Almond oil *(Prunus dulcis)* is extracted from the almond kernel. It's odourless, lubricating and has great penetrating qualities. It's rich in vitamin D and helps rejuvenate skin. It's suitable for all skin types, but particularly good for dry, sensitive, and mature skin.

Avocado oil *(Persea americana)* is high in essential fatty acids, lecithin, protein, pantothenic acid, and vitamins A, B1, B2, D, and E. It's good for all skin types and improves elastin. It's particularly good for dry, dehydrated, and eczema-prone skin. Blend in a 1:10 ratio with another carrier oil (1 part avocado oil to 10 parts other oil), or 10% of your oil blend.

Coconut oil *(Cocos nucifera)* is high in fatty acids and good for dry, itching, and sensitive skin. It's very good for use in making suppositories and boluses. A bolus is generally made by mixing dried powdered herbs with coconut oil and formed into a bullet-like shape to be inserted into the vagina and used for yeast infection and other reproductive complaints. Powdered herbs like yellow dock and chickweed can be used.

Castor oil *(Ricinus communis)* is high in linoleic acids. A thick, amber-coloured oil, it can dissolve cysts, remove warts, soften corns, help prevent scars, and soothe chapped skin. Use in a warm compress pack for cysts. In a blend with other carrier oils use 10%, or a 1:10 ratio of castor oil (1 part castor oil to 10 parts other oil).

Grapeseed oil *(Vitis vinifera)* is extracted from grape seeds and is odourless, lubricating and non-allergenic. It contains the antioxidant pycnogenol that has an anti-aging effect.

Salves, Ointments, and Balms

Hemp oil (*Cannabis sativia*) is high in essential fatty acids, like linolenic acid. It is very moisturizing and soothing for the skin and muscles.

Jojoba oil (*Simmondsia chinensis*) is a fine and stable oil with a long shelf life. It is anti-inflammatory and nourishing, and its chemical composition is close to the skin's own oil (sebum), making it good for all skin types. It encourages new hair growth, has natural emulsifying properties, is rich in vitamin E, and can also be used in other carrier oils as a preservative. Jojoba is great for treating skin disorders, infections, eczema, and psoriasis. It's an excellent oil for scalp and hair products, and it's useful for all skin types: helping balance dry or oily skin.

Olive oil (*Olea europaea*) is extracted from olives and is rich in vitamins and minerals. It has exceptional disinfecting and wound-healing properties. Taken orally, olive oil helps prevent heart disease, high cholesterol, constipation, fatigue, hypertension, and rheumatism. It can be used topically to soothe dry skin and ease rheumatic pain.

Rosehip seed oil (*Rosa mosqueta*) is high in minerals and many other nutrients. It is extracted from rosehip seeds and is excellent for dry, scaly, fissured skin; dull, acne-prone skin; eczema; psoriasis; over-pigmented skin; scars; and ulcerated veins. Use neat or in a 10% dilution with other carrier oils.

Sesame oil (*Sesamum indicum*) is high in vitamin E, minerals, proteins, lecithin, and amino acids. This oil is good for treating psoriasis, eczema, rheumatism, and arthritis. It's nourishing for all skin types.

St. John's wort (*Hypericum perforatum*) is made from flowers infused in oil (generally olive oil). It's used topically for treating neuralgia, muscular pain, sciatica pain, rheumatism, gout, and arthritis. It's an excellent treatment for all types of sores, wounds, and burns.

Sunflower oil (*Helianthus annuus*) is made from sunflower seeds and is a light oil that is rich in essential fatty acids. It's used as an emollient to help the skin retain moisture, preventing excessive dryness. It's easily absorbed and is high in vitamins A, B, D, and E.

Part III: *Medicinal Preparations and Recipes*

Basic-Salve Recipe

1 cup (250 mL) infused oil
2 tablespoons (30 mL) beeswax
¼ teaspoon (1 mL) vitamin-E oil (2 capsules)
4 drops benzoin oil (a fixative made from benzoin tree sap)
Essential oils

To make a balm, use more aromatic herbs and essential oils. To make an ointment, add a little more carrier or infused oil. To prepare salve, follow either double-boiler method below, or hot water bath (bain-marie) method opposite.

Salve Double-Boiler Method

1. Measure out herb-infused oil.

2. Melt beeswax in the top of the double boiler, then add the infused oil and stir with a stainless steel spoon.

3. When the oil and beeswax have melted together add vitamin-E oil, benzoin, and/or essential oils. Pour the mixture into a measuring cup. (Remember to dry the bottom of the pot so water from the bottom of the pan does not drip into the oil mixture.)

4. Carefully pour into clean, sterilized jars.

5. Wait until the salve has fully cooled before putting on the lids.

6. Cap and label your salve.

Salves, Ointments, and Balms

All-Purpose Botanical Ointment

An antimicrobial and wound-healing ointment for cuts, rashes, scrapes, stings, slivers, blisters, sore muscles, and chest congestion.

Prepare a herbal oil infusion (see page 305) using:

2 parts plantain leaf
2 parts yarrow leaves and flowers
2 parts mossberry branches, broken up
1 part poplar buds
1 part pineapple weed
1 part usnea lichen
1 part wild-sage leaf
1 part bearberry leaf
20 parts olive oil, or any other carrier oil
2 parts St. John's wort oil (optional)

After the infused oil is made, follow the Basic-Salve Recipe instructions, using your preferred heating method.

Each plant can also be separately infused in oil and then blended into the specific formulas. Most of the time this will make sense because many of the plants in these recipes are ready for harvesting at different times of the season. Infusing each plant separately also gives you a broader spectrum of the possibilities for use in other formulas. If you do this then you do not need to add the carrier oils that are listed in the recipes because they are already infused.

Three-Tree Healing Salve

Smells like a walk in the boreal forest! Soothing, healing, and very aromatic!

Prepare herbal oil infusion (see page 305) using:

1 part pine needles
1 part fir needles
1 part spruce tips
6 parts sunflower oil

After the infused oil is made, follow the Basic-Salve Recipe instructions, using your preferred heating method.

Hot-Water Bath or Bain-Marie Method

1. Choose a pot and a heat-resistant glass measuring cup that will fit comfortably inside it.

2. Fill half the pot with water and place on the stove over medium heat.

3. Place measuring cup in the pot. Add beeswax to the measuring cup and allow it to melt. Add the oil.

4. Stir occasionally with a stainless steel spoon.

5. When oil and wax are melted together, add vitamin-E oil, benzoin, and/or essential oils.

ADDITIONAL OINTMENT, SALVE, AND BALM RECIPES:

Arnica Ointment, page 50

Wild-Rose Petal Healing Ointment, page 148

Part III: *Medicinal Preparations and Recipes*

Aromatic Herbal Balm
A beautiful wild and soothing aromatic balm.

Prepare a herbal oil infusion (see page 305) using:

- 1 part rose petals
- 1 part pineapple weed
- 1 part wild sage
- 1 part sweetgrass
- 1 part Labrador tea flowers
- ½ part valerian root
- 7 parts jojoba oil

After the infused oil is made, follow the Basic-Salve Recipe instructions, using your preferred heating method. In this recipe, jojoba oil is used because it's odourless and has a very long shelf life. **Note:** Less carrier oil is used for the aromatic balm in order to enhance the aroma.

Below: Boreal Balm—a long-time favourite of Aroma Borealis customers, this is one of the first healing balms I made. Everyone I gave it to loved it and I kept getting requests for refills, so from it a business was born! BG

Boreal Balm

- 1 cup (250 mL) infused poplar-bud oil
- 2 tablespoons (30 mL) beeswax
- ¼ teaspoon (1 mL) vitamin-E oil (2 capsules)

After the infused oil is made, follow the Basic-Salve Recipe instructions, using your preferred heating method.

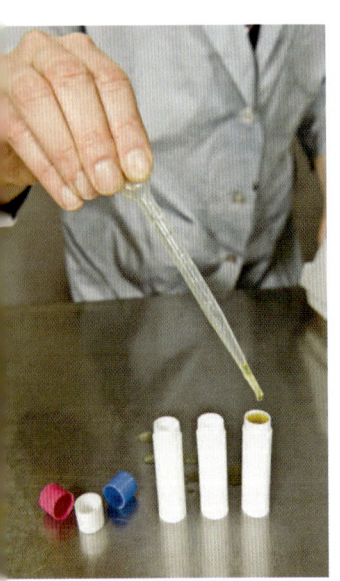

Bottom: Making lip balm at Aroma Borealis. CA

Moisturizing Lip Balm
For smooth, supple, and kissable lips!

Prepare a herbal-oil infusion (see page 305) using:

- 2 parts rose petals
- 1 part fireweed flowers
- 1 part plantain leaf
- 1 part mint leaf
- 1 part chickweed leaf
- 12 parts sunflower oil

When the oil is infused, prepare the lip balm following the Basic-Salve Recipe and using your preferred heating method. Pour into a jar or lip balm tubes.

Healing Herbal Creams

HEALING HERBAL CREAMS

Go to any cosmetics counter and you will find an abundance of creams promising to make you look younger, remove fine lines, and essentially give you a whole new face. I'm always skeptical of these products, because they are made up of ingredients that can contribute to serious illness, such as petroleum by-products, alcohol, and funky chemicals with long, hard-to-pronounce names. Once you have made your own creams, you will never want to go back to expensive, chemical-laden commercial creams.

Creams differ from salves, ointments, and balms in that they contain water or hydrosols, and a blender is used to facilitate the emulsification process. They are also a little trickier to make, as both timing and temperature are factors in making a successful cream.

Once you have your basic cream-making skills down, you can start to experiment with different measurements of water, herbs, oils, and other ingredients. If you're just beginning though, start out by following the Basic-Cream Recipe.

Basic-Cream Recipe

1 cup (250 mL) infused oil
2 tablespoons (30 mL) beeswax
1 teaspoon (5 mL) vitamin-E oil
Essential oils (optional)
¼ cup (60 mL) distilled water or hydrosol (see page 322)

Follow Basic-Cream method on following page.

> Natural creams don't have many preservatives. To keep your cream fresh longer, avoid sticking dirty or wet fingers in the jar. You can avoid contaminating the cream by using a toothpick, cotton swab, or clean Popsicle stick as a scoop.

Part III: *Medicinal Preparations and Recipes*

Making Cream

1. In a double boiler or bain-marie, heat beeswax until melted. Add infused herbal oil. Slowly heat oil and beeswax mixture until it melts together, stirring occasionally.

2. Turn off heat. Add vitamin-E and/or essential oils and pour warm mixture into a blender.

3. Blend on low.

4. Slowly pour room-temperature distilled water or hydrosol into the mixture. Increase the speed of the blender. This is where the magic happens—within minutes the mixture will start to thicken and look like cream.

5. Turn off blender. Use a spatula to scrape excess oil mixture off the insides of the blender and drop it back into the developing cream. Blend again until the mixture is fully combined—it should have a smooth and runny texture.

6. When cream has reached the desired consistency, pour into dry, sterilized jars.

7. When the cream has fully cooled, secure lids and label jars.

Creams and Liniments

Muscle-Ease Cream

1 cup (250 mL) Muscle-and-Pain Relief Oil (page 307)
2 tablespoons (30 mL) beeswax
¼ cup (60 mL) distilled water or hydrosol (see page 322)
1 teaspoon (5 mL) vitamin-E oil

Follow Basic-Cream method on previous page.

Boreal Botanical Skin Cream

1 cup (250 mL) Boreal Skin Oil (page 307)
2 tablespoons (30 mL) beeswax
¼ cup (60 mL) distilled water or hydrosol (see page 322)
1 teaspoon (5 mL) vitamin-E oil

Follow Basic-Cream method on previous page.

Creams cooling. BG

ADDITIONAL CREAM RECIPE:

Sacred-Spirit Sweetgrass Cream, page 170

LINIMENTS

Liniments are used as topical disinfectants and are a good choice to rub into tired, sore muscles. They are made the same way as tinctures (see Tincture Method, page 293) only the solvent used is rubbing alcohol or witch hazel.

Note: Liniments are for external use only. It's important to label bottles appropriately to make sure your liniments are not ingested.

Juniper-Willow Liniment
Great for disinfectant purposes, or to soothe sore, inflamed muscles.

1 part juniper berries, crushed
1 part inner willow bark
2 parts rubbing alcohol or witch hazel

Infuse botanicals with witch hazel or rubbing alcohol. Let mixture sit for up to 6 weeks, shaking daily. When fully infused, strain, bottle, and label.

ADDITIONAL LINIMENT RECIPE:

Pine Tree Ache-and-Pain Liniment, page 254

Part III: *Medicinal Preparations and Recipes*

POULTICES

Poultices are made with fresh or dried botanicals, and are used to regenerate, soothe and heal tissue, stimulate circulation, and to relax and warm muscles. Poultices are excellent for drawing out toxins and foreign objects from the skin.

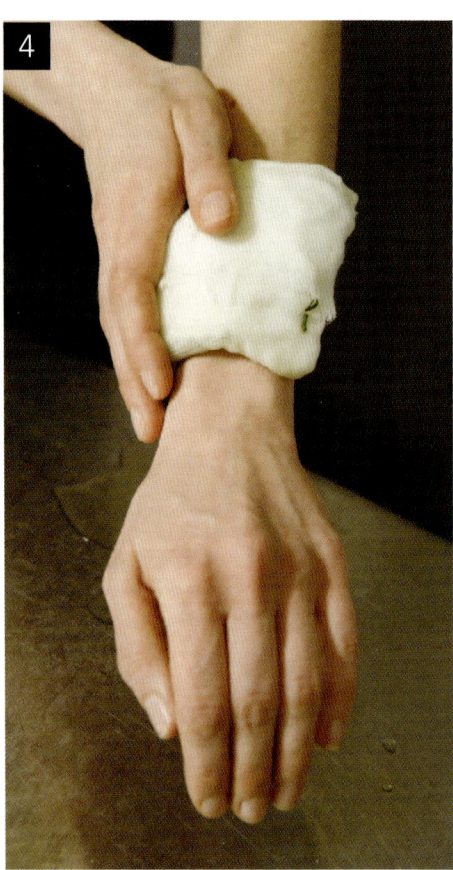

Preparing Poultices

1. Barely cover mixed fresh or dried herbs with boiling water in a heat-resistant measuring cup or saucepan. Let herbs steep until preparation drops to room temperature (17°C) and the herbs have absorbed the water.

2. Pour or scoop herbs onto a cheesecloth

3. Wrap the herbs in a couple of layers of cheesecloth.

4. Place the poultice over the affected area and cover with a dry cloth to retain heat and absorb excess poultice juice.

CA

318 | The Boreal Herbal

Poultices and Fomentations

Plantain Poultice
Excellent for drawing out foreign objects, and to help heal and soothe cuts, scrapes, rashes, and burns.

1 part dried plantain leaf, or ½ part fresh leaf
Water to cover herbs

Follow method on previous page or simply steep the leaf in hot water until soft and place over the affected area.

Spit Poultices

My favourite and most frequently used remedy in the summer! Spit poultices are effective and easy to make. They are great for relieving the itching and swelling of bug bites, to stop bleeding, and to promote healing. Yarrow, chickweed, plantain, and lamb's quarters all make great spit poultices.

To prepare a spit poultice chew, or macerate, herb with your teeth. Place over affected area.

ADDITIONAL POULTICE RECIPES:

Alder Poultice, page 237

Arnica Poultice, page 49

CA

FOMENTATIONS

Fomentations are prepared by using the liquid from a water infusion or decoction (see herbal infusions and decoctions section). A fomentation is used for stimulating circulation, soothing tissue, or to help break up chest congestion (when placed over the lung area).

Preparing Fomentations
Prepare hot infusion or decoction with desired herbs. Dip a soft cotton cloth in the preparation. Ring out the cloth and let it cool (so that it doesn't burn the skin). Place over affected area and cover with a couple of towels to keep the heat in. After fomentation has cooled, remove, re-soak, and re-apply.

Wild-Onion Fomentation
Excellent for clearing up lung congestion.

Wild-onion chives, chopped
Water

Put water in a saucepan and add chives. Bring to a boil and let simmer for 10 minutes, with lid covering pot. Let cool until fomentation will not burn the skin. Dip in cloth and place over the chest. Cover cloth with dry towels to keep heat in.

ADDITIONAL FOMENTATION RECIPE:

Lungwort Fomentation, page 117

Part III: *Medicinal Preparations and Recipes*

COLD COMPRESSES

Cold compresses are used for removing excess heat from the body. Because they stimulate the production of red and white blood cells, they are very useful for inflammation, swelling, and fevers. Yarrow is an excellent herb to use in a cold compress to help break up fevers, while a chickweed compress is a great choice for healing eye irritations.

Preparing Cold Compresses

A cold compress is made in the same way as a fomentation. After the preparation has cooled, place it in the fridge or freezer so it becomes really cold.

Dip a soft cotton cloth in preparation and ring out.

Place over affected area and cover with a towel.

After the compress has warmed up, remove, re-soak, and re-apply.

After using the compress, dry the skin thoroughly.

STYPTIC POWDER

A herbal remedy must-have! Styptic powder helps stop bleeding from cuts and wounds—so it's good to have on hand in your herbal apothecary collection, and/or first-aid kit. You can apply the powder directly on a wound, put it in a poultice or wash, or make it into an ointment.

Stop-the-Bleeding Styptic Powder
The herbs can be used individually or in any combination, in equal parts.

- Plantain
- Yarrow
- Horsetail
- Nettle leaf
- Shepherd's purse
- Goldenrod

Powder dried botanicals in a grinder, either together or separately. Strain powder through a sieve to remove any lumps. Pour into a dark glass jar with a tight-fitting lid. Label and store in a cool area. Make up smaller jars to use at your convenience.

ESSENTIAL OILS AND HYDROSOLS

MAKING ESSENTIAL OILS and hydrosols is a sensuous process that is guaranteed to bring out your inner mad scientist or alchemist. Using the distillation process, hot steam is forced through copious amounts of plant matter, then travels through cooling tubes where it's transformed back into a liquid before dripping into a beaker. The distillation process is beautiful to watch and absolutely heavenly to smell!

Two products result from steam-distilling plants: essential oil and hydrosol (water solution). The essential oil floats on top of the hydrosol. It's then separated and skimmed off until you end up with a vile of pure, unadulterated essential oil and a large bucket that is full of beautiful hydrosol.

In the North there is a handful of people who make essential oils and hydrosols on a very small scale.

Essential oils are concentrated and powerful plant extracts that are 40% to 100% stronger than herbs. For instance, distillers of commercial rose oil report it takes approximately 70 large damask roses to produce one drop of oil. Wild boreal roses would yield very little essential oil but they produce a beautiful, elegant hydrosol.

Essential oils have complex organic chemical makeups that include chemicals such as alcohols, esters, ketones, phenols, and terpenes. The chemical arrangement of each essential oil is determined by the oil's individual chemistry, which in turn determines the oil's properties and influences the human body and its systems. Essential oils are pure and natural, and work in harmony with the body. Beware of some "imposter" oils such as "fragrance oils" or "nature identical oils" that contain questionable ingredients and leave a residue on your skin.

Essential oils from related plants have similar chemistries, which means they share certain properties and therapeutic uses. For example, the mint family (*Lamiaceae* or *Labiatae*) produces the largest variety of essential oils. Plants in this family include not only mint, but also other familiar herbs such as basil, clary sage, lavender, marjoram, oregano, and patchouli. All of these plants contain terpene alcohols that are considered to be therapeutic tonics of life. Plants in this family are all known to be antiviral and antibactericidal, and they all affect the central nervous system.

Dwarf birch being distilled into a hydrosol. BG

Part III: *Medicinal Preparations and Recipes*

Collection of boreal hydrosols made by Yukoner Birch Kuch. BG

Hydrosols contain water-soluble essential-oil molecules that flowed through the plant cells when the plant was collected, as well as constituents that are not present in the essential oil. Although hydrosols are often called "floral water" or "flower water," they can be derived from all plant parts including flowers, leaves, stems, seeds, barks and roots.

Hydrosols are potent, yet subtler than essential oils. Hydrosols may be applied to the skin as sprays, compresses, and soaks, either straight or diluted with purified water. They are used in creams, lotions, facial masks, and in cooking. Under the guidance of an aromatherapist or naturopath, hydrosols can also be used internally in very small dosages.

In her book *Hydrosols: the Next Aromatherapy*, Aromatherapist Suzanne Catty talks about the internal usage of hydrosols and provides recommended usages and dosages for each plant. I highly recommend adding this book to your herbal resource library because it's the most comprehensive book on the use of hydrosols.

There are a number of methods used to extract essential oils from plants. The method depends primarily on the characteristics of the plant itself. To make essential oils and hydrosols for commercial use, it's best to invest in a good still. Hydrosols, however, can be made at home with a crude setup that includes basic kitchen tools.

Preparing Hydrosols

Beautiful hydrosols can be made from a range of boreal-plant parts, including spruce tips, dandelion flowers, fireweed flowers, rose petals, mint leaves, wild-sage leaves and flowers, wild chamomile, fir needles, sweet-scented bedstraw, sweetgrass, juniper berries and needles, Labrador tea, and valerian root—to name just a few.

Botanicals
Water
Ice cubes

Tools
Large stainless steel pot with lid
Heat-resistant bowl
Clean brick or flat rock

Place the brick or rock in the bottom of the pot. Add flowers, leaves, or needles around brick or rock, making sure that the plant matter doesn't rise above the brick or rock. Cover the mixture with water. Place the bowl on top of brick or rock.

Place pot on stove and turn element to medium-high. Be careful not to burn plant matter.

Put the lid upside-down on pot and fill with ice cubes. As the water starts to boil, herbal-filled steam will hit the cold lid, and the droplets will fall into the bowl.

Essential Oils and Hydrosols

Monitor the pot closely and pay attention to your sense of smell to ensure the water doesn't run out and the botanicals don't burn.

After the ice cubes have melted, take a peak in the pot to make sure that the water has not run dry and to see how much hydrosol has been collected. If the bowl is full you're done.

Turn off the heat and let sit for 5 to 10 minutes, to cool down. This also allows the rest of the condensation from the steam to settle into the bowl. Carefully remove the lid and take a big whiff of the beautiful hydrosol you have created.

Remove the bowl and let hydrosol fully cool.

Bottle in a dark glass bottle, label, and store in a cool place out of direct sunlight. (I like to keep mine in the fridge. If you make an abundance of hydrosol you can also freeze it for future use.)

You can use the hydrosol instead of water in cream making, to make iced tea, or in a spritzer bottle for use on your face and body.

MAKING A HYDROSOL

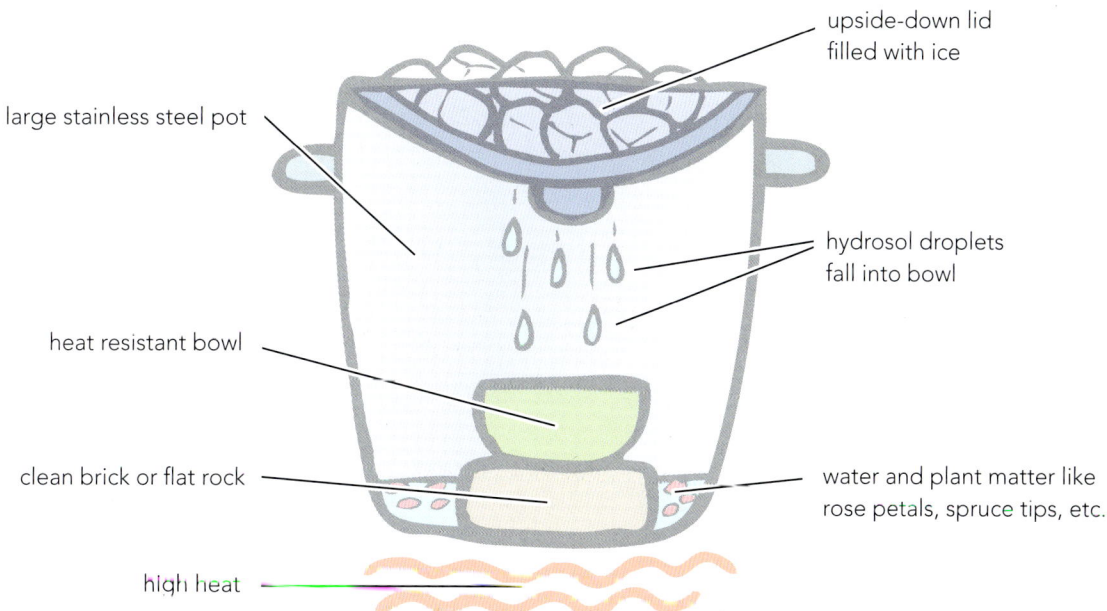

Illustration by Tanya Handley adapted from *Discovering Wild Plants* by Janice Schofield.

BOREAL BATHS AND STEAMS

USING WILD BOREAL BOTANICALS for a home spa treatment is a great way to benefit from the plants' aromatic, emollient, and antimicrobial healing properties. Adding these to baths and steams is an easy and luxurious way to experience the relaxing and cleansing effects the botanicals offer.

BATHS

Taking a bath filled with wild, aromatic botanicals is a therapeutic and beneficial way to integrate the healing power of plants into your life. Add candles and soft music, and you've created a relaxing retreat. A luxurious bath for one can be just the self-care treat you need to nurture and pamper yourself. A luxurious bath for two can be just what you need to nurture your relationship!

Administering herbs in the bath is a useful and enjoyable way to use them because your body will absorb some of the herbal constituents through your skin. You can use herbs in a full bath, footbath, hand bath, or sitz bath.

Herbal Bath Basics

Herbal baths are easy to prepare. Simply place botanicals in a cotton bag or tie them up in a face cloth. Place in bath as hot water is running, get in, and soak. You can squeeze the bag intermittently to release the maximum amount of herbal properties. The herbal bag can also be used as a washcloth. My favourite way to use herbs in the bath is to sprinkle them in loose—the only downside to doing this is they may clog your pipes, but this can be remedied by placing a strainer in the drain.

Another way to prepare herbs for bathing is to take 4 cups (1 L) of strained infusion or decoction (see page 286), or 1 cup (250 mL) of a herb-infused vinegar (see page 297) and pour into the bath.

All of the following bath recipes can be prepared using the methods described above.

Self care! Ahh… CA

Baths

Tree-Medicine Purification Bath Oil

The needles of conifer trees are full of nutrients, especially vitamin C, and their camphoraceous natures help us to breathe deeply and bring more oxygen into our bodies.

1 part pine needles
1 part spruce tips
1 part fir needles
6 parts carrier oil
Sea salt, Epsom salts, or Dead Sea salts (optional)

Infuse the oil (see page 281). I like using sunflower oil in this recipe because of its nourishing nature, and because it's a product of the Canadian Prairies.

After the infusion is prepared and ready to be used, combine ¼ cup (60 mL) of oil to ½ cup (125 mL) of salt and add to your bath.

Boreal Healing Bath

Great before bedtime to help you relax and get a great night's sleep.

1 part yarrow
1 part wild chamomile
1 part chickweed
1 part rose petals
1 part horsetail
4 parts sea salt, Epsom salts, or Dead Sea salts
½ part carrier oil
Essential oil of chamomile or lavender (optional)

A bath full of boreal botanicals. BG

Powder herbs, stir into salts, then stir in carrier oil. You can make enough for one bath or make a big batch to enjoy for many baths. Store in an airtight container.

Flower-Power-Peace Bath

Bring summer's energy into your self-care practice with this beautiful and soothing bath.

1 part fireweed flowers
1 part rose petals
1 part pineapple weed
(continued next page)

Part III: *Medicinal Preparations and Recipes*

1 part dandelion flowers
1 part horsetail
1 part sweetgrass
4 parts powdered milk, full fat if possible
1 part sea salt, Epsom salts, or Dead Sea salts (alone or in combination)
1 part carrier oil

Powder herbs, stir into salt, then stir in carrier oil. Once combined, add powdered milk, and stir until completely mixed (I use my hands). You can make enough for one bath or make a big batch to enjoy for many baths. Store in an airtight container.

Cold-and-Flu Bath

A healing and nurturing bath for when you feel a cold or flu coming on, or for use as a preventative. I like to make this up ahead of time so that it's easily available when needed.

1 part yarrow
1 part fir or pine needles, or spruce tips (one or any combination of these)
1 part Labrador tea leaves
1 part juniper berries
1 part wild sage
4 parts Epsom salts

Powder herbs, stir into salts, and pour into the bath. While bathing, drink a cup of Good-for-Fever Infusion (see page 282) or use 10 drops of Cold-and-Flu Tincture (see page 294) to 1 cup (250 mL) of hot water. You can also add 30 drops of the tincture to your bath water.

Oh-My-Aching-Bones Bath

A wonderful combination of botanicals to help relieve muscle and joint pain.

2 parts juniper berries
2 parts willow inner bark
1 part pine needles
2 parts Epsom salts

ADDITIONAL BATH RECIPE:

Juniper Sea-Salt Scrub, page 213

Powder herbs, stir in salts, and pour into the bath. While bathing, drink 20 drops of Pain-Aid Tincture (see page 294) in 1 cup (250 mL) of hot water.

FOOTBATHS

The world may be in your hands, but you walk through it with your feet, and they need to be taken care of. Our feet carry us many thousands of miles in our lifetimes, so when they are tired, sore, ulcerated, infected, cracked, dry, or itchy, this can stress our entire body.

I am always amazed by how relaxing and effective foot soaks are to help heal the feet. All the boreal trees and herbs can be used in a footbath: fir, spruce, willow inner bark, plantain, pineapple weed, rose, and fireweed, just to name a few.

If you have athlete's foot, make a mossberry-twig vinegar infusion (see vinegar-making, page 296) to help treat your feet.

Preparing a Footbath

Soaking your feet in plain hot water is the simplest footbath you can take. It doesn't have to be complicated: you just put on the kettle and snip a few yarrow flowers or wild-sage leaves from your yard and add them to a bowl of hot water. This said, adding wild botanicals, salts, and essential oils can make the experience truly luxurious and healing.

A footbath is a simple way to pamper yourself. CA

To begin, pour boiling water into a container large enough to cover your feet. Then add:

- 1 cup (250 mL) of herbal infusion, decoction, or dried or fresh herbs. If you are using loose herbs, I find adding them to the bowl first and then pouring the boiling water over them makes for a better infusion.
- ½ cup (125 mL) Epsom salts
- ¼ cup (60 mL) apple cider vinegar or a couple of drops of your favourite essential oil.

Let the water cool to toe temperature, immerse your feet, and soak for 15 to 20 minutes.

Dry your feet off immediately and slip into your favourite pair of wool socks or slippers.

If a cold or flu is trying to take hold of you, wrap yourself in a blanket, add Cold-and-Flu Bath blend (see previous page) to hot water and soak your feet in it for 15 to 20 minutes.

Drink 1 cup (250 mL) of hot yarrow infusion while soaking to help you sweat.

Dry feet, put on warm socks, and go to bed to sweat it out. This recipe can also be used as a hand bath.

ADDITIONAL FOOTBATH RECIPE:

Arnica Footbath, page 49

Part III: *Medicinal Preparations and Recipes*

SITZ BATH

A sitz bath is specifically used to reduce the swelling of hemorrhoids, or help heal an episiotomy or tear of the perineum after childbirth. Plantain leaf, pineapple weed, peppermint, and fireweed are good healing herbs for treating these kinds of conditions.

Preparing a Sitz Bath

Make a strong infusion or decoction (see page 286). Strain herbs.

Fill a large basin or metal bowl halfway with warm water and add approximately 4 cups (1 L) of the infusion or decoction. (Some pharmacies sell special sitz-bath basins.)

It's best to place the basin in a bathtub so that the overflow goes down the drain and not all over the floor. Sit in basin until the water cools.

STEAMS

A steam is an effective way to help clear up respiratory congestion and clean facial skin. You can use herbs such as peppermint, plantain, coltsfoot, or lungwort leaves for a decongestant steam bath. Pine, fir, or spruce needles are also good choices. For a facial steam, use herbs such as chamomile, peppermint, fireweed flowers, yarrow, and rose petals. Essential oils can also be used in steams—add only a couple of drops per bath.

Steaming is an ancient and effective way of relieving sinus and skin congestion. CA

Steams

Preparing a Steam Bath

Place a handful of herbs in a cooking pot and cover with water. Place the lid on the pot, and simmer for 10 to 15 minutes. Pour preparation into a large bowl.

Place bowl on a table and sit comfortably in front of it. Put your head above the bowl, then cover your head and the bowl with a large towel (making a tent). Steam for as long as you can handle it.

Another way to prepare a steam is to add the herbs to the bottom of the bowl and pour boiling water over top. Cover for 5 minutes and then make your tent and allow the volatile plant oils and steam to help heal you.

Respiratory Steam

- 1 part fir needles
- 1 part devil's-club root, shredded
- 1 part juniper berries, crushed
- 1 part mint leaves
- 1 part coltsfoot leaf
- 1 part birch leaf

Birch leaves are a good addition to a respiratory steam. CA

Rejuvenating Facial Steam

- 1 part fireweed flowers
- 1 part rose petals
- 1 part plantain
- 1 part willow inner bark
- 1 part horsetail
- 1 part bearberry leaf

Sauna Steam

If you, like many northerners do, have an outdoor sauna with a woodstove, you can place boreal herbs in a pot of water directly on top of the stove. Labrador tea adds a great aroma to any sauna, as does wild sage, wild chamomile, birch leaves, evergreen needles, and juniper. We have only wood heat at home and I often have herbs strewing in a pot on top of the stove to vaporize into the air.

GETTING WILD IN THE KITCHEN

FOOD AND BEVERAGE RECIPES

Cooking with Wild Foods 333
Raw Juices and Smoothies 334
   Basic Berry Juice Recipe 334
   Cool Cranberry Juice 335
   Strawberry Fields Forever 335
   Revolutionary Raspberry
      Rosehip 335
   Blueberry–Mossberry Juice .. 336
Green Smoothies 336
   Spring-in-the-Boreal
      Smoothie 336
   Wise-Weed Smoothie 337
   Sunshine Daydream
      Smoothie 337
   Toon-Town Smoothie 337
   Pine-Forest Smoothie 338
   Hot Tomato 338
   Kalerific! 338
   Chickweed Shooter 339
Fruit-and-Yogourt Smoothies 339
   Athletes' Blend Smoothie 340
   Herb-and-Berry Antioxidant
      Smoothie 340
   Pink Drink: High-C
      Smoothie 340
   Wild Rose and Mint Lassi 341
Muffins, Biscuits,
   and Pancakes 342
   Blueberry Muffins 342
   Cranberry-Mint Muffins 343
   Red-Clover Tea Biscuits 344
   Dandelion-Petal Pancakes 345
   Wild-Berry Buckwheat
      Pancakes 346
   Dandelion-and-Birch
      Cornbread 347

Jams, Jellies, Chutney,
   and Topping Syrups 348
   Wild-Blueberry Jam 349
   Dandelion Jelly 350
   Rose-Petal Jelly 351
   Spruce Tip Jelly 352
   Wild-Blueberry Chutney 353
Topping Syrups 354
   Basic Syrup Recipe 354
   Herb-and-Berry Sweet
      Summer Days Syrup 355
   Dandelion-Petal Syrup 355
Sauces and Dressings 356
Sauces 356
   Extra-Green Pesto Sauce 356
   Juniper Butter 357
   Nettle Pasta Sauce 358
   Highbush Cranberry
      Applesauce 358
   Dandelion-Petal Mustard 359
   Green Sauce 360
Dressings and Vinaigrettes 361
   Wild-Weed Dressing 361
   Wild-Rose Petal
      Vinaigrette 361
   Raspberry-and-Rosehip
      Vinaigrette 362
   Dandelion Dressing 362
Soups and Salad 363
   Green Bouillon 363
   Cream of Wild-Weed Soup .. 364
   Wild-Weed Salad 365

Main Courses and Sides 366
   Piquant Plantain 366
   Wild-Weed Spanakopita 367
   Healthful Weed Pie 368
   Wild-Greens Stir-Fry 370
Desserts and Sweet Treats 371
   Dandelion-Petal Cake 371
   Wild-Blueberry Fruitsicles 373
   Creamy Cranberry-
      Vanilla Ice 373
   Roasted Dandelion-Root
      Ice Cream 374
   Flower-Delight Tempura 375
Fruit Leather 376
   Raw Uncooked Fruit
      Leather Process 376
   Living Fruit Leather 377
   Cooked Process 377

Previous pages: Dandelion leaves. BG

COOKING WITH WILD FOODS

BRINGING WILD, LOCAL INGREDIENTS into our kitchens and incorporating them into our diets make for more flavourful, healthful, and interesting meals. Many of the people I know want to eat more locally produced food and are interested in exploring the concept of a "100-Mile Diet" (where everything eaten is harvested from within a one-hundred-mile radius), but to many modern-day northerners, eating this way seems impossible. Yet this was a way of life for many First Nations and early settlers of the boreal forest, so we know it can be done.

Even if we can't commit to being 100 percent pure "locavores" we can help contribute to our local economies while being stewards of the Earth by going out and gathering wild plants as ingredients for our foods, medicines, and body-care products, and by supporting local producers who do this for us.

When we gather plants, we get in touch with the environment and with our ancestral roots: our primal selves. We become a more integrated part of the ecosystem end of the food chain—even when we are just gathering dandelion greens from our lawns—there is a part of us that instinctively knows and connects to this process.

For me, cooking is an intuitive process: most often I glance at a recipe and then follow my instincts. I encourage you to also be intuitive in the kitchen, be creative with the wild plants, and let your imagination go wild. The culinary possibilities are truly endless!

A few notes about the recipes in this book: Whenever possible, I use organic ingredients, because organic farming is better for animals, for the soil, for human health, is more nutritious, and is usually more sustainable. This said, I realize it's not always available and often more expensive than conventionally grown or raised food, so it might not always be practical for people to use organic ingredients in these recipes.

For health reasons, I don't include much sugar in my diet, but when needed for cooking, I use organic cane sugar whenever possible. Sugar cane contains vitamins and minerals, and is not as highly processed as commercial white sugar (which is often made from sugar beets). If I can use honey, birch syrup, or maple syrup in a recipe then I will. So far, though, I haven't figured out how to successfully make preserves with any of these.

I also use spelt flour in most of my baked goods because some members of my family are allergic to whole-wheat flour. Spelt is an ancient grain in the wheat family and is high in protein, has a nutty flavour, and contains gluten. In these recipes, wheat or spelt flour can be used interchangeably.

RAW JUICES AND SMOOTHIES

THE FLAVOUR SENSATIONS, aromas, and nutritional value of wild plants make them the perfect ingredients for healthy raw juices and smoothies. Wild-plant juices and smoothies are a great way to start the day and are excellent for your health, and all you need is a blender to get started.

Note: The recipes in this chapter make two to four servings.

Above: Blueberry harvest. MC

Below: Cranberries. FM

RAW JUICES

The simplest way to make a nutritious raw juice is to macerate fresh or dried herbs or fresh or frozen berries in a blender with cold water or a cooled herbal tea, and sweetener, if desired. Adding lemon or lime juice will enhance the taste and help preserve the juice a little longer. Raw juices are the most healthful of beverages, but they do not last that long—when made in a blender the juice will last up to three days in the fridge (if made in a juicer, the juice will start to oxidize immediately). If you're not going to drink the juice that quickly, freeze it in portion-sized containers, or make home-made fruitsicles with it!

Basic Berry Juice Recipe

Any of the boreal wild berries can be used; mix and match for new and interesting juices! Imagine mossberry-raspberry juice! Yum!

2 cups (500 mL) berries (frozen works best)
4 cups (1 L) cold water or cooled herbal tea
2 tablespoons (30 mL) sweetener (honey, birch syrup, agave nectar, cane sugar, or deseeded dates all work great)
Juice of one lemon

Place all the ingredients in a blender, blend on high speed until smooth (this usually takes about two minutes).

Pour, drink, and, if desired, strain. Enjoy the antioxidant goodness in each sip.

Raw Juices and Smoothies

Cool Cranberry Juice
A tart and colourful juice.

2 cups (500 mL) wild cranberries
1 cup (250 mL) fresh mint leaves
4 cups (1 L) cold water*
2 tablespoons (30 mL) sweetener, or to taste
Juice of 1 lemon (60 mL)

Place all the ingredients in a blender, blend on high speed until smooth (this usually takes about two minutes).

*Notes: If you replace the water portion of the recipe with bearberry tea, you will have an excellent remedy to prevent or help heal a urinary-tract or bladder infection. Add a pure cranberry-juice ice cube to carbonated water for a tasty and natural fizzy pop.

Strawberry Fields Forever
Refreshing and light, bold and bright!

2 cups (500 mL) wild strawberries
1 kiwi fruit, peeled and diced
4 cups (1 L) of cold water
2 tablespoons (30 mL) sweetener, or to taste (optional)
Juice of 1 lemon

Place all the ingredients in a blender, blend on high speed until smooth (this usually takes about two minutes).

Revolutionary Raspberry Rosehip
A delicious, vitamin C-and-bioflavonoid-rich, juice revolution!

1 cup (250 mL) wild raspberries
1 cup (250 mL) wild rosehips (deseeded)
4 cups (1 L) cold water
2 tablespoons (30 mL) sweetener, or to taste (optional)
Juice of 1 orange

Place all the ingredients in a blender, blend on high speed until smooth (this usually takes about two minutes).

Cool cranberry juice. BG

Part III: *Getting Wild in the Kitchen*

Blueberry-Mossberry Juice
Antioxidant power!

1 cup (250 mL) wild blueberries
1 cup (250 mL) wild mossberries
4 cups (1 L) cold water
Sweetener to taste (optional)
Juice of 1 orange

Place all the ingredients in a blender, blend on high speed until smooth (this usually takes about two minutes).

GREEN SMOOTHIES

I love green smoothies! They're packed full of vitamins, minerals, proteins, and antioxidants! Green plants and leaves have been consumed since the beginning of time and are the only living energy on our planet that can transform energy from the sun into food using photosynthesis. Green leaves produce chlorophyll, which is in essence green sunshine! A diet rich in non-starchy, chlorophyll-rich greens helps to support a healthy heart, to cleanse the liver of heavy metals and toxins, and to improve the health of the intestines and lungs. A bonus is that it also helps to improve body odour, is a natural breath freshener, and helps to regulate the acid-alkaline balance in the body.

To make smoothies, all it takes is a blender, creativity, and, in the beginning, a little forethought. After many years of making this daily elixir of life, I can whip up a blender full of smoothie goodness just minutes after waking up in the morning while still half asleep. The possible combinations are endless—so experiment and have fun with it! Although smoothies, as their names suggests, are smooth, you should chew your smoothies, as this will help activate your digestive juices.

Note: If the consistency is too thick, add more liquid; if it's too thin, add more greens.

Spring-in-the-Boreal Smoothie
A sweet-and-creamy green taste sensation!

1 cup (250 mL) dandelion greens
1 cup (250 mL) fireweed greens
1 cup (250 mL) wheat-grass greens or a rhubarb stock
1 banana
2 ripe pears
1 cup (250 mL) cold water

Place all the ingredients in a blender, blend on high speed until smooth (this usually takes about two minutes).

Raw Juices and Smoothies

Wise-Weed Smoothie
A vitamin-and-mineral-rich smoothie that is great for an energetic midday pick-me-up!

2 cups (500 mL) chickweed
1 cup (250 mL) lamb's quarters
1 banana
1 mango
2 cups (500 mL) apple juice

Place all the ingredients in a blender, blend on high speed until smooth (this usually takes about two minutes).

Sunshine Daydream Smoothie
Stimulate and warm your digestive fires!

1 cup (250 mL) fireweed shoots
1 cup (250 mL) dandelion greens
1 cup (250 mL) chickweed
1 teaspoon (5 mL) sliced or shredded ginger root
2 mangoes
¼ fresh pineapple, chopped
1 cup (250 mL) cold water

Place all the ingredients in a blender, blend on high speed until smooth (this usually takes about two minutes).

Green fuel. BG

Toon-Town Smoothie
Purple-and-green goodness all in one glass!

1 cup (250 mL) parsley
1 cup (250 mL) dandelion greens (soak the leaves if the taste is too bitter)
2 cups (500 mL) Saskatoon berries
1 banana
1 cup (250 mL) water

Place all the ingredients in a blender, blend on high speed until smooth (this usually takes about two minutes).

Part III: *Getting Wild in the Kitchen*

Pine-Forest Smoothie

Pine needles add a woodsy flavour to the sweet berries and bitter herbs!

1 cup (250 mL) dandelion greens
1 cup (250 mL) spinach
1 tablespoon (15 mL) crushed pine needles
2 cups (500 mL) strawberries
2 mangoes
2 cups (500 mL) water

Place all the ingredients in a blender, blend on high speed until smooth (this usually takes about two minutes).

Hot Tomato

A smooth and healthy alternative to canned tomato juice—serve with a lemon wedge! High in the antioxidant lycopene, vitamins C and A!

2 cups (500 mL) crushed tomatoes
1 cup (250 mL) plantain leaf
½ cup (125 mL) watercress
1 cup (250 mL) Swiss chard
2 celery stalks
1 apple
1 ripe pear
1 cup (250 mL) water

Place all the ingredients in a blender, blend on high speed until smooth (this usually takes about two minutes).

Kalerific!

Kale and chickweed are excellent sources of vitamins A and C, and manganese. Kale is also a very good source of dietary fibre, copper, calcium, vitamins B6 and K, and potassium.

2 cups (500 mL) kale leaves
1 cup (250 mL) chickweed
1 banana
2 ripe pears
1 cup (250 mL) cold water

Place all the ingredients in a blender, blend on high speed until smooth (this usually takes about two minutes).

Raw Juices and Smoothies

Chickweed Shooter

A concentrated, dark green, chlorophyll-rich drink that is great tasting and nourishing. Chickweed helps support weight loss.

2 cups (500 mL) chickweed
½ cup (125 mL) water

Fill a blender with fresh chickweed, add water, and blend until smooth.

Can be enjoyed as is, or strained through cheesecloth.

Note: Drink a 2 teaspoon (10 mL) shot three times a day. It's best to drink this fresh. The remainder of the drink can be frozen in ice cube trays to enjoy throughout the winter months in soups, stews, smoothies, and "green" drinks.

Chickweed juice can be frozen and used throughout the winter in soups, stews, smoothies, and green drinks. CA

Below: Berries are the jewels of the boreal forest. FM

FRUIT-AND-YOGOURT SMOOTHIES

A fruit-and-yogourt smoothie is a wonderful thing! Packed full of goodness, flavour, and nutrients, the health benefits of smoothies are as diverse as the many ingredient combinations used to create them. Kids love to drink smoothies too!

Frozen wild berries are a great foundation for a smoothie. Tinctures, jams, jellies, syrups, tea infusions (fresh or frozen), and ground herbs can also be added to create a drink that supports your specific health needs. I give a few examples here, but I encourage you to experiment to find your favourite flavour combinations!

Note: If the consistency is too thick, add more liquid; if it's too thin, add more frozen fruit.

Part III: *Getting Wild in the Kitchen*

Athletes' Blend Smoothie

Rhodiola root, mint leaf, and nettle come together to create a stimulating and nourishing support for athletes.

1 cup (250 mL) wild blueberries, fresh or frozen
4 juniper berries
1 banana
1 cup (250 mL) yogourt (dairy or soy)
1 tablespoon (15 mL) hemp powder
Vanilla soy milk (enough to cover other ingredients)
30–60 drops Athletes' Aid Tincture (see page 294) or Get-Up-and-Go Infusion (see page 283)

Place all the ingredients in a blender, blend on high speed until smooth (this usually takes about two minutes).

Herb-and-Berry Antioxidant Smoothie

High in vitamins E and C, and beta carotene. This smoothie protects body cells from the damaging effects of oxidation.

1 cup (250 mL) raspberries, frozen
½ cup (125 mL) mossberries, frozen
¼ cup (60 mL) red-clover tea infusion (see Herbal Infusions, page 281) or 10 drops of red-clover tincture (see Tinctures, page 293)
¼ cup (60 mL) chickweed juice, tea infusion, or frozen chickweed-juice ice cubes
1 cup (250 mL) yogourt
1 cup (250 mL) almond milk, or enough to cover the mixture
2 tablespoons (30 mL) hemp, protein, or goat whey powder

Place all the ingredients in a blender, blend on high speed until smooth (this usually takes about two minutes).

Pink Drink: High-C Smoothie

The Pink Drink tastes great all year round, but it's especially beneficial during cold-and-flu season.

1 cup (250 mL) cranberries, frozen
1 cup (250 mL) rosehip decoction *(continued next page)*
¼ cup (60 mL) lamb's quarters seeds, dried
¼ cup (60 mL) spruce tips, fresh or dried

Raw Juices and Smoothies

1 cup (250 mL) yogourt
1 cup (250 mL) plain or vanilla rice/soy/almond milk
2 tablespoons (30 mL) hemp powder

Place all the ingredients in a blender, blend on high speed until smooth (this usually takes about two minutes).

ADDITIONAL SMOOTHIE RECIPE

The Purple Drink, page 190

Wild Rose and Mint Lassi

Lassi is the smoothie of India. This one has a northern twist.

2 tablespoons (30 mL) plain yogourt
½ cup (125 mL) unsweetened soy milk, or other non-fat milk
2 cups (500 mL) ice, crushed
¼ cup (60 mL) fresh mint leaves, chopped or torn into small pieces
¼ cup (60 mL) rose petals, dried or fresh
1 teaspoon (5 mL) lime juice
Pinch of sea salt, optional

Blend to desired consistency.

Rose petals are a delight to see, smell, and taste. BG

MUFFINS, BISCUITS, AND PANCAKES

I LOVE THE AROMA of baking wafting throughout the house—it's true aromatherapy! Incorporating wild berries, herbs, and flower petals from botanicals is a nice way to add a wild twist to baking and reap the benefits from the nutritional boost they offer.

Note: The following recipes were developed using my old propane stove and I find that its heat distribution differs from an electric stove. So keep a close eye on your baking because there may be variations in temperature and cooking times.

Blueberry Muffins

Yummy healthy muffins! They can be eaten for breakfast or as a snack with your tea. They're also great in the school lunchbox!

2 cups (500 mL) spelt or all-purpose flour
1½ teaspoons (7 mL) baking soda
½ cup (125 mL) cane sugar, date sugar, or honey
1 teaspoon (5 mL) cinnamon
1 teaspoon (5 mL) grated lemon, or lime zest
1 cup (250 mL) yogourt
2 eggs, beaten
¼ cup (60 mL) soft butter, or sunflower oil
1 cup (250 mL) blueberries, fresh or frozen

Preheat oven to 375°F (190°C).

In a bowl, mix dry ingredients thoroughly. (If using a dry sweetener, like cane or date sugar, add it, too.)

In another bowl, mix wet ingredients thoroughly until blended. (If using a wet sweetener, like honey, add it, too.)

Muffins, Biscuits, and Pancakes

Add wet ingredients to dry, mixing until combined. Do not over mix. For fluffy muffins, you want the batter to have a somewhat lumpy texture.

Gently fold in blueberries. Spoon mixture into lightly greased muffin tins, or muffin cups. Bake for 20 minutes, or until tops are golden brown. Insert a toothpick into a muffin: if it comes out clean, they are done! Let cool for at least 5 minutes before eating.

Makes 12 regular-sized muffins.

Cranberry-Mint Muffins
A minty twist on the traditional cranberry muffin!

Cranberry-mint muffins. BG

1½ cups (375 mL) spelt, or all-purpose flour
½ cup (125 mL) cane sugar
1 tablespoon (15 mL) baking powder
Pinch of salt
1 egg, beaten
1 cup (250 mL) milk
⅓ cup (80 mL) butter, melted
1 cup (250 mL) cranberries, fresh or frozen
¼ cup (60 mL) fresh wild mint, chopped, or 2 tablespoons (30 mL) dried
½ cup (125 mL) chopped nuts or hemp seeds (optional)

Preheat oven to 375°F (190°C).

In a bowl, combine flour, sugar, baking powder, and salt.

In a separate bowl, whisk together the egg, milk, and butter.

Add wet ingredients to dry, mixing until combined. Do not over mix. The batter should remain somewhat lumpy; this will give you a fluffy muffin.

Gently fold in cranberries, mint, and nuts or seeds.

Spoon batter into lightly greased muffin tins or cups.

Bake for 35 minutes, or until tops are golden brown. Insert a toothpick into a muffin: if it comes out clean, they are done! Let cool for at least 5 minutes before eating.

Makes 12 regular-sized muffins.

Part III: *Getting Wild in the Kitchen*

Red-Clover Tea Biscuits*

If you feel like experimenting, try substituting red-clover petals with other boreal flowers like rose, fireweed, or dandelion.

2 cups (500 mL) whole wheat or spelt flour
½ cup (125 mL) ground almonds
1 tablespoon (15 mL) baking powder
¼ cup (60 mL) organic butter, room temperature
2 eggs, beaten
½ cup (125 mL) buttermilk or yogourt
¼ teaspoon (1 mL) vanilla extract
1 cup (250 mL) powdered red-clover petals

Preheat oven to 450°F (230°C).

Grind almonds in a blender or food processor.

In a bowl, combine ground almonds, flour, and baking powder.

Add butter and knead to make a crumbly mass.

In a separate bowl mix eggs, buttermilk or yogourt, vanilla, and red-clover petals together. Gradually add to the crumbly mass until it forms dough.

Roll out dough and cut into shapes. I like to use the mouth of a glass or a round cookie cutter—you can also just use a knife and cut into squares or triangles.

Bake on an ungreased cookie sheet for 15 minutes, or till golden brown.

Serve warm with butter and jam, or jelly.

Makes 12 regular-sized biscuits.

BG

*Recipe inspired by the late Rose Barlow.

Dandelion-Petal Pancakes*
A hardy pancake full of dandelion goodness!

A stack of dandelion petal pancakes brings sunshine to the morning. BG

1 cup (250 mL) all-purpose or spelt flour
1 cup (250 mL) cornmeal
1 teaspoon (5 mL) sea salt
2 teaspoons (10 mL) baking powder
1 cup (250 mL) dandelion petals
2 organic eggs, beaten
¼ cup (60 mL) sunflower oil
¼ cup (60 mL) Dandelion-Petal Syrup (see page 355), honey, birch syrup, or maple syrup
1 cup (250 mL) milk

In a bowl, combine flour, cornmeal, salt, baking powder, and dandelion petals.

In a separate bowl, blend eggs, oil, honey or syrup, and milk.

Stir wet mixture into the dry (the batter should be thin enough to pour).

Cook on medium heat in lightly oiled frying pan (I use a cast-iron pan).

Top with butter, jam, birch syrup, or Dandelion-Petal Syrup.

Makes 7–10 medium sized pancakes.

* Recipe inspired by the late Rose Barlow.

Part III: *Getting Wild in the Kitchen*

Wild-Berry Buckwheat Pancakes

Cranberries, blueberries, currants, Saskatoon berries, raspberries, strawberries, and mossberries on their own, or in combination, taste terrific in these pancakes.

ADDITIONAL RECIPE

Cloudberry-Buckwheat Pancakes, page 194

¼ cup (60 mL) buckwheat flour
½ cup (125 mL) spelt flour
1 teaspoon (5 mL) baking powder
¼ teaspoon (1 mL) sea salt
1 egg, beaten
¾ cup (175 mL) milk
1 tablespoon (15 mL) butter, or oil
1 tablespoon (15 mL) honey
1 tablespoon (15 mL) birch syrup, or molasses
½ cup (125 mL) wild berries, fresh or frozen

In a bowl, combine flours, baking powder, and salt.

In a separate bowl, blend egg, milk, butter or oil, honey, and molasses.

Stir wet mixture into dry.

Gently fold berries into batter.

Cook in a lightly oiled frying pan.

Top with butter, jam, birch syrup, or herbal syrup (see page 354 for recipes).

Makes 6 medium-sized pancakes.

Who wouldn't want to wake up to the smell of wild-berry buckwheat pancakes? BG

Muffins, Biscuits, and Pancakes

Dandelion-and-Birch Cornbread*

A yummy snack on its own, or dipped in your favourite soup, stew, or chili recipe.

1 cup (250 mL) cornmeal
1 cup (250 mL) white or spelt flour
2 teaspoons (10 mL) baking powder
¾ teaspoon (3 mL) baking soda
1 teaspoon (5 mL) sea salt
1 cup (250 mL) dandelion petals
2 large eggs, beaten
½ cup (125 mL) birch syrup
¼ cup (60 mL) butter or oil
1 cup (250 mL) milk

Dandelion petals. BG

Preheat oven to 375°F (190°C).

In a bowl, mix together cornmeal, flour, baking powder, baking soda, salt, and dandelion petals.

In a separate bowl, whisk eggs, birch syrup, butter or oil, and milk.

Stir wet mixture into dry, mixing until smooth.

Pour batter into a cake or loaf pan, muffin tin, or cast-iron frying pan.

Bake for 35 minutes.

Serve warm with butter.

Makes one medium-sized loaf.

---

*Recipe inspired by the late Rose Barlow.

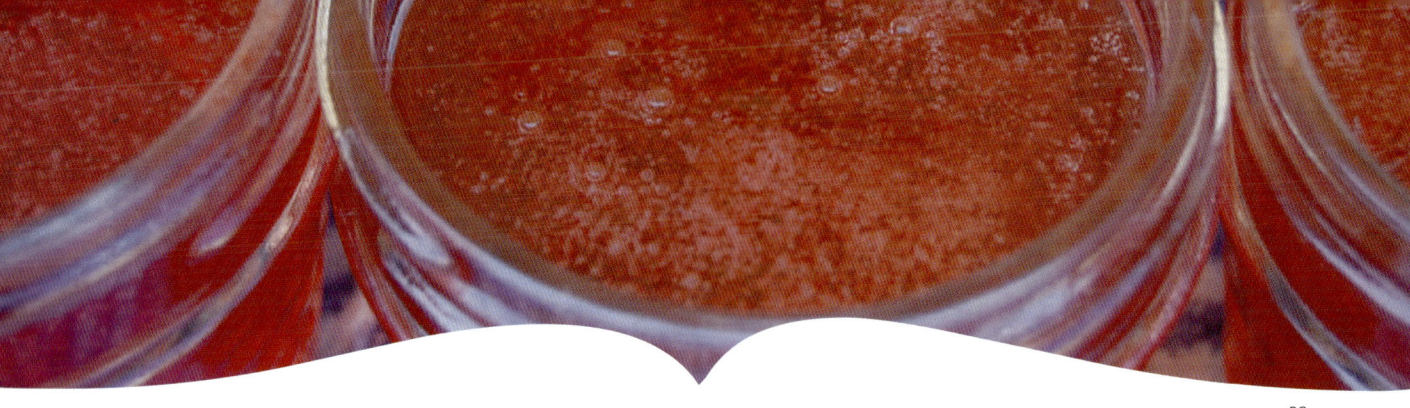

JAM, JELLIES, CHUTNEY, AND TOPPING SYRUPS

THERE IS SOMETHING really satisfying about making your own preserves—seeing the finished jars and all the interesting colours, flavours, textures, and smells. Early in a boreal summer, herbs such as dandelion, rose, and wild mint start to bloom and can be made into lovely jellies. Even the tender spruce tips can be turned into a tart vitamin C-rich jelly to use with wild game, poultry, or in dressing, and in desserts! While I'm making my early summer jellies I like to rout through the freezer to see what leftover wild berries I have from the year before and use them up before the next berry harvest.

Getting out and traversing the wild lands for fresh fruit—often on your hands and knees and in some pretty beautiful places—can be a gathering adventure that will keep you fed and nourished throughout the year.

A pantry full of homemade jams and jellies. BG

Jam, Jellies, Chutney, and Topping Syrups

Wild-Blueberry Jam

A rich, dark blue jam that tastes great on toast or pancakes, or in yogourt or smoothies.

- 4 cups (1 L) crushed wild blueberries
- 2 tablespoons (30 mL) lemon juice
- 1 tablespoon (15 mL) grated organic lemon rind (optional)
- 1 package (57 g) powdered pectin
- 1 cup (250 mL) cane sugar

In a large pot, crush blueberries with a potato masher. (If the berries are frozen wait until they thaw to make jam.)

Add lemon juice, rind, and pectin.

Bring to a rolling boil for a minute over a high heat, stirring constantly.

Add sugar and bring to rolling boil again, stirring constantly. Keep at a hard boil for 1 minute, still stirring.

Remove from heat and skim foam off top.

Pour or spoon into clean jars, seal, and heat process the jars.

Makes 3 cups (750 mL) of jam.

Tips for Successful Canning and Heat Processing

Sterilize all jars and lids that you will be using for your jams, jellies, and chutneys.

When making cooked preserves in jars that will be stored at room temperature, they must be canned in boiling water for up to 20 minutes—the boiling time is dependent on the altitude. The greater the altitude, the longer processing time you'll need for jams and jellies. For example, I live at approximately 640 m (2,100 ft.) above sea level, so I heat process for 10 minutes. If you live at sea level, your heat-processing time will be approximately 5 minutes but check on the directions for the pectin you're using as most include this information. I use Bernardin pectin and it comes with complete jam- and jelly-canning instructions in the box.

When canning make sure the jars are covered by at least 1 in. (2.5 cm) of water.

When you have completed the canning process wait at least 5 minutes then remove the jars without tilting the contents. Cool the jars undisturbed for 12–24 hours, then check that the lids are sealed (they should be curved downward). If a lid doesn't flex up and down, it's sealed. If there is movement when you touch the lid, it's not sealed: refrigerate and eat the contents within three weeks.

Label jars with the type of jelly and date.

The website www.homecanning.ca is an excellent resource for all your canning needs and questions.

Part III: Getting Wild in the Kitchen

Dandelion Jelly

Dandelion jelly has a light, honey-like taste and is great on its own, or added to other sauces and vinaigrettes.

4 cups (1 L) dandelion petals
2½ cups (625 mL) cane sugar
2 tablespoons (30 mL) lemon juice
1 package (57 g) powdered pectin
4 cups (1 L) water

You must pick 8 to 10 cups of dandelion blossoms to have 4 cups (1 L) of dandelion petals. After you remove the green base of each dandelion flower you will be left with approximately 4 cups (1 L) of yellow petals.

Add the petals to the water and simmer for 5 minutes, set aside to steep until infusion reaches room temperature.

Dandelion jelly has a beautiful amber colour. BG

Strain to remove the petals. This should provide you with 3 cups (750 mL) of dandelion-petal infusion.

In a large saucepan, combine dandelion-petal infusion, lemon juice and pectin, and stir until pectin is dissolved. On high heat bring to a full rolling boil for 1 minute.

Add the sugar, stirring until it dissolves, bring the mixture back to a full rolling boil that can't be stirred down. Boil hard for at least 1 minute.

The jelly is ready when it coats the back of a spoon and has a syrup-like consistency. You can check by placing a teaspoon of the jelly on a plate and letting it cool: the surface should wrinkle when pushed with your finger. If still runny, put the mixture back on heat and continue boiling and testing, until the jelly sets.

Skim off any foam on top of the jelly.

Pour into jars, leaving ¼ in. (.5 cm) headspace. Secure the lid and seal.

Makes approximately 5 cups (1¼ L).

Jam, Jellies, Chutney, and Topping Syrups

Rose-Petal Jelly

Lovely sums up the aroma and flavour of this delicate and visually pleasing jelly. Many years ago, I used to make this recipe and sell it in a fancy jar at the annual Christmas Sprucebog craft fair in Whitehorse. The best part of this jelly-making process is going out and gathering the petals among the bees—the colour variation of the petals, from a dark pink to a light pink, is visual medicine, and it feels exhilarating to fill your gathering basket with an abundance of flowers from this aromatic plant.

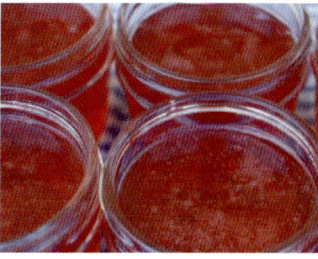

Rose-petal jelly tastes and smells wonderful. BG

- 2½ cups (625 mL) wild rose petals, fresh or dried
- 2 cups (500 mL) water
- 2 cups (500 mL) cane sugar
- ½ cup (125 mL) white grape juice
- ½ cup (125 mL) red grape juice
- 1 package (57 g) powdered pectin
- 2 tablespoons (30 mL) rosewater

Place petals, water, and ⅔ cup (150 mL) of the sugar in a saucepan, bring to a boil.

Reduce heat and simmer 5 minutes.

Remove from heat and let stand overnight so the petals can release their fragrance into the sugar water.

Strain flowers from syrup, and pour the rose syrup in a large pot.

Add the grape juices and pectin, and bring to a boil. Boil hard for 1½ minutes. Add the rest of the sugar and stir; bring the liquid back up to a boil. Boil the mixture hard for 1 minute or more. Remove from heat.

The jelly is ready when it coats the back of a spoon and has a syrup consistency. You can check by placing a teaspoon of jelly on a plate and let cool: the surface should wrinkle when pushed with your finger. If it's still runny, continue boiling and testing, until the jelly sets.

When jelly is ready, stir in the rosewater.

Skim off any foam that has formed on top of the jelly.

Pour into jars, leaving ¼ in. (.5 cm) headspace. Secure the lid and heat seal.

Makes 3 cups (750 mL) of jelly.

Part III: Getting Wild in the Kitchen

Spruce Tip Jelly

A classic northern jelly that's great served on crackers with goat cheese—an interesting conversation starter at potlucks.

4 cups (1 L) spruce tips
4 cups (1 L) water
2 tablespoons (30 mL) lemon juice
1 package (57 g) powdered pectin
1½ cups (375 mL) cane sugar

Add spruce tips to the water and simmer for 5 minutes and set aside to steep until infusion reaches room temperature.

Strain to remove the spent tips.

The spruce tips infusion should measure 3 cups (750 mL).

An abundance of spruce tips! BG

In a large saucepan, combine spruce tip infusion, lemon juice and pectin, and stir until pectin is dissolved. On high heat bring to a full rolling boil for 1 minute.

Add the sugar, stirring the mixture to dissolve the sugar, bringing the mixture back to a full rolling boil that can't be stirred down. Boil hard for at least 1 minute.

The jelly is ready when it coats the back of a spoon and has a syrup-like consistency. You can check by placing a teaspoon of the jelly on a plate and letting it cool; the surface should wrinkle when pushed with your finger. If still runny, put the mixture back on heat and continue boiling and testing, until jelly sets.

Skim off any foam that has formed on top of the jelly.

Pour into jars, leaving ¼ in. (.5 cm) headspace. Secure the lid and heat seal.

Makes approximately 4 cups (1 L) of jelly.

Jam, Jellies, Chutney, and Topping Syrups

Wild-Blueberry Chutney

Perfect served with pork, poultry, red and wild meats, on a turkey or chicken sandwich, or as a topping for brie or cream cheese.

½ cup (125 mL) raspberry vinegar, or any berry vinegar
½ cup (125 mL) cane sugar
1 medium onion, minced
¼ teaspoon (1 mL) minced ginger
1 pinch ground cinnamon
1 teaspoon (5 mL) lemon rind, shredded
1 pinch cayenne pepper
1 pinch sea salt
3 cups (750 mL) wild blueberries
¼ cup (60 mL) wild cranberries

Combine vinegar, sugar, onion, ginger, cinnamon, lemon rind, pepper, and salt in a large cooking pot. On a high heat, bring to a boil and turn down to medium heat, simmer for 15 minutes.

Add cranberries and 1 cup (250 mL) blueberries, and simmer for 20 minutes, stirring frequently.

Add remaining blueberries and simmer for 10 minutes.

Makes about 2 cups (500 mL).

ADDITIONAL RECIPES

Cranberry Chutney, page 198

Fireweed Jelly, page 94

Wildly Mint Jelly, page 122

Raspberry Jam, page 220

Thoughts of wild-blueberry chutney might be just the incentive for coming home from a morning of blueberry picking with a full bucket. CA

Part III: *Getting Wild in the Kitchen*

TOPPING SYRUPS

Wild syrups are so versatile! They can be used as dessert and breakfast toppings, in iced tea or smoothies, in salad dressing and baking. Syrups are good to have on hand in your kitchen or to give as gifts. You could even start your own little bush business making and selling syrups.

Wild boreal syrups, like fruit leathers, can be made from herbs or from boreal berries such as currants, highbush cranberries, cloudberries, juniper berries, Saskatoon berries, raspberries, strawberries, and mossberries. These berries can be used on their own, blended with other berries or wild herbs. Just follow the basic syrup recipe and let your taste buds be your guide.

In the medicine-making section of this book, I have given instruction on how to make medicinal syrup, such as Rosehip Syrup (page 299), that can also used as a topping and in various recipes.

Basic Syrup Recipe
Great on pancakes and ice cream, this recipe can be doubled or even quadrupled, whatever you fancy!

- 4 cups (1 L) berries and/or herbs
- 2 cups (500 mL) cane sugar
- 1 cup (250 mL) water

Blend berries, herbs, sugar, and water in a blender, pour into a cooking pot. If you don't have a blender, mash the ingredients together in a pot.

If you're only using herbs to make the syrup (meaning, you're not adding fruit to your syrup), it's best to add the sugar to the strained juice before heating it.

To make the infusion, bring everything to a boil and then turn down to medium-to-low heat, slowly simmer for 15 minutes. Let it cool (steeping overnight makes for a more flavourful juice).

Strain through a sieve lined with cheesecloth into a bowl. Let the juice drip through, then squeeze the cheesecloth to extract any remaining juice. (This step is optional, when making pure berry syrup I like to have berries in the syrup.)

Pour juice (or the pre-cooked berry purée) into a clean cooking pot. Bring to a boil, then lower to a medium heat. Simmer until syrupy. General simmering time is 20 minutes.

Cool, bottle, and label. Keep refrigerated and it will last for many months.

Jam, Jellies, Chutney, and Topping Syrups

Herb-and-Berry Sweet Summer Days Syrup

The title of this recipe sums up my summers: sweet and filled with an abundance of herbs and berries!

3 cups (750 mL) blueberries
½ cup (125 mL) mossberries
½ cup (125 mL) nettle leaves, dried
2 cups (500 mL) cane sugar
2 cups (500 mL) water

To prepare, follow Basic Syrup Recipe. I like this recipe with the berries left in, so I don't strain them out!

Dandelion-Petal Syrup*

A versatile syrup that is an ingredient in many recipes in this book, including Dandelion-Petal Mustard (page 359) and Dandelion-Petal Cake (page 371).

4 cups (1 L) dandelion flowers
4 cups (1 L) water
3 cups (750 mL) cane sugar
½ cup (125 mL) lemon juice
Zest of one lemon

Follow the Basic Syrup Recipe, making sure to add the sugar and lemon to the strained juice, then heat. Makes about 3 cups (750 mL) of syrup.

ADDITIONAL RECIPE

Mossberry Syrup, page 216

Syrups make great gifts. BG

*Recipe inspired by the late Rose Barlow.

SAUCES AND DRESSINGS

WILD GREENS CAN replace store-bought greens in sauces, adding a wild taste to pasta dishes, rice, grains, and salads. In the summer, I love dunking my favourite fresh vegetables in homemade green dips. My absolute favourite is Chickweed-Garlic Green Dip (see page 66) that I've brought to many potlucks; people are always impressed by the surprisingly delicious taste of chickweed. It's true: you can eat common weeds as food.

SAUCES

Extra-Green Pesto Sauce*

This pesto is great on its own, or served over pasta or rice.

- 1 cup (250 mL) firmly packed chickweed greens
- 1 cup (250 mL) firmly packed basil leaves
- 2–4 cloves garlic, peeled and crushed
- ½ cup (125 mL) olive oil or a vegetable oil
- 2 tablespoons (30 mL) roasted sunflower seeds, pine nuts, or inner pine bark (see page 255)
- 1 cup (250 mL) freshly grated Parmesan cheese

Put chickweed, basil, and garlic in a blender or food processor. Add oil, blend till smooth. Blend in nuts or seeds. Stir in cheese.

If sauce is too thick, add more oil, or even yogourt or cream cheese.

The sauce will keep for 2 weeks in the fridge.

Serves 4 people.

---

*This recipe was created and given to me by Marie-France Campagna, a Yukoner and a passionate organic gardener.

Sauces and Dressings

Laila Spik, Sami Elder

Laila was raised in the traditional Sami ways by her parents. The sharing of traditional knowledge is a big part of Laila's wise and generous spirit. People from all over the world come to apprentice with her at her home in northern Sweden, where she teaches all aspects of traditional Sami life including how to identify, gather, and prepare wild plants and trees of the northern boreal forest for food and medicine.

Laila is also the author of *How to Cook a Reindeer: The Reindeer Recipe Book*. In the cookbook, she shares some of the traditional Sami reindeer recipes that are central to her culture—recipes that she has had the honour of serving to the king of Sweden. Laila has also produced a video on Sami culture's natural remedies and foods called *How Do We Make Use of Nature's Gifts?* and has a website, http://lailaspik.vingar.se.

I had the honour and pleasure of spending time with Laila in Finmark, Norway, during the 7th annual Circumpolar Agriculture Conference in 2010. It was at this conference that she provided the pine bark bread and juniper butter recipes featured in this book.

Juniper Butter

1 cup (200 g) butter
½ a yellow onion
2–4 cloves garlic, crushed
8–10 juniper berries, crushed
2 tablespoons flaxseed oil

Place ingredients in a blender and blend until smooth. Keep refrigerated.

Part III: *Getting Wild in the Kitchen*

Nettle Pasta Sauce*

Serve over your favourite pasta or with spaghetti squash.

4 cups (1 L) spring nettles, fresh
½ cup (125 mL) olive oil
1 onion, diced
2–4 cloves garlic, diced
¼ cup (60 mL) almonds, chopped
¼ cup (60 mL) hemp or sesame seeds
¼ cup (60 mL) Parmesan, cheddar or Swiss cheese (optional)
A pinch of salt and pepper to taste

Steam nettles in a large cooking pot for 5–7 minutes. (For a large amount of greens I pour 1 cup of water in the bottom of a large pot, put my stainless steel pasta strainer on top, cover the pot with a lid, and steam the greens until done.)

Heat 2 tablespoons (30 mL) olive oil in a saucepan. Add onion and garlic and cook on medium heat for 5 minutes, or until the onions are transparent.

Add the rest of the oil, as well as the nuts, seeds, cheese, salt, and pepper.

Stir in nettles and continue to cook on low heat for 5 minutes, or until the ingredients have the consistency of a sauce.

Serves 4 people.

Highbush Cranberry Applesauce**

Delicious on its own or as a topping for desserts and pancakes, or in yogourt. It's also an excellent cold medicine!

4 cups (1 L) highbush cranberries
4 cups (1 L) apple, finely chopped
1–2 teaspoons (5–10 mL) ginger, chopped or grated
1 cup (250 mL) honey or sugar

Squeeze mashed cranberries through a cheesecloth (the large seeds are not edible).

---

*This recipe was created and given to me by Stephen Badhwar, a woodsman and storyteller who lives in Atlin, B.C.
** This recipe was created and given to me by Elissa Miskey, an organic gardener and wildcrafter in Atlin, B.C.

Sauces and Dressings

Combine the cranberry juice, apple, ginger, and sweetener in a pot on medium heat, stir, and bring to a boil.

Reduce heat and simmer on low heat until the apple is soft, stirring occasionally. Mash with a potato masher and let sit for 1 hour to harmonize the flavours.

Keep refrigerated in a sealed container and eat within 2 weeks, or freeze for later.

Makes about 3 cups (750 mL).

Highbush cranberries. RF

Dandelion-Petal Mustard*

As either a condiment or dip, this mustard packs a full flavour and a spicy punch!

1 cup (250 mL) yellow mustard seeds, whole
1¼ cup (300 mL) apple cider vinegar
½ cup (125 mL) Dandelion-Petal Syrup (see page 355) or birch syrup
1 cup (250 mL) fresh dandelion greens, puréed (if you don't have enough dandelion greens, you can add chickweed or lamb's quarters)
½ cup (125 mL) dandelion petals, green part removed
4 cloves garlic, finely chopped
Pinch of sea salt

Soak mustard seeds in apple cider vinegar for several hours

Put in a blender, add remaining ingredients, blend until smooth.

Pour into a jar, seal, and store in the fridge. To help keep it fresh and moist in the fridge, add a teaspoon of olive oil on top of the mustard.

Makes about 2 cups (500 mL).

ADDITIONAL RECIPES

Chickweed-Garlic Green Dip, page 66

Gratitude Gathering Cranberry Sauce, page 199

Wild-Onion Spread, page 131

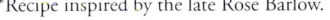

*Recipe inspired by the late Rose Barlow.

Plantain is just one of the wild herbs you can use in green sauce. BG

Green Sauce

This sauce is excellent served with potatoes, as a sandwich spread, or as a vegetable dip. It's a northern variation of an old Irish recipe that was used to top potatoes ... of course!

2 tablespoons (30 mL) apple cider or herbal infused vinegar (see page 296)
2 tablespoons (30 mL) fresh lemon juice
1 large egg
Pinch of sea salt
½ cup (125 mL) wild greens (sorrel leaves, plantain, wild onion chives, dandelion leaves, chickweed, lamb's quarters, etc.)
1¼ cup (300 mL) vegetable oil, or combine olive and hemp oils

Combine vinegar, lemon juice, egg, and salt in a blender, blend until smooth.

Add greens and continue blending until the mixture is uniform. While blending, slowly add oil, in a thin stream, until thoroughly combined.

Kept refrigerated, it will last for up to 4 weeks.

Makes about 2 cups (500 mL).

Sauces and Dressings

DRESSINGS AND VINAIGRETTES

Homemade salad dressings taste great, are easy and inexpensive to make—and they contain no weird artificial ingredients!

Wild-Weed Dressing

This is one of my all-time favourite dressing recipes. It's a wild version of a recipe my friend Nicole Edwards gave me for Christmas one year.

2 cups (500 mL) wild greens (try chickweed, dandelion, lamb's quarters, wild onion)
½ cup (125 mL) apple cider vinegar
1 tablespoon nutritional yeast
3 cloves garlic
1 cup (250 mL) olive oil or oil of your choice
2 tablespoons (30 mL) tamari
2 tablespoons (30 mL) tahini (sesame-seed butter)
2 tablespoons (30 mL) hemp seeds

Combine ingredients in a blender and blend till smooth.

Drizzle over your summer green salads.

Store in a bottle or jar, and refrigerate. It will last for a few weeks in the fridge.

Makes 2 cups (500 mL).

Homemade dressings taste so much better than store-bought varieties. BG

Wild-Rose Petal Vinaigrette

A delicate dressing for a wild-weed salad.

1 cup (250 mL) rose petals
1 cup (250 mL) white-wine or apple-cider vinegar
1½ cups (375 mL) flax seed oil (or your favourite oil)

Infuse rose petals in vinegar, and let sit for 1–2 weeks, shaking every day. (For more on how to prepare herbal vinegars, see page 296.)

Strain and mix with oil.

Bottle and refrigerate. Shake well before using.

Makes 2½ cups (625 mL).

Part III: *Getting Wild in the Kitchen*

Raspberry-and-Rosehip Vinaigrette

A variation on this recipe is to process all the ingredients together in a blender and use immediately as a dressing on fresh green salads.

½ cup (125 mL) raspberries
1 cup (250 mL) apple cider vinegar
2 cups (500 mL) olive oil
2 tablespoons (30 mL) Rosehip Syrup (see page 299)

Infuse raspberries in apple cider vinegar for 2–3 days, shaking daily.

Strain and mix with oil and syrup.

Bottle and refrigerate. Shake well before using. It will last for up to 6 months.

Makes approximately 3 cups (750 mL).

Dandelion Dressing*

Tart and tantalizing!

1 cup (250 mL) olive oil
½ cup (125 mL) hemp oil
¾ cup (175 mL) apple cider vinegar
3 cloves garlic
½ teaspoon (2 mL) sea salt
2 tablespoons (30 mL) Dandelion-Petal Mustard (see page 359)
3 tablespoons (45 mL) Dandelion-Petal Syrup (see page 355) or honey
2 cups (500 mL) dandelion greens, fresh, chopped

Put ingredients into a blender and blend till smooth. Bottle and refrigerate. It will keep for up to 6 months.

Makes approximately 2 cups (500 mL).

ADDITIONAL RECIPE

Borealis Green-Goddess Dressing, page 163

---

* Recipe inspired by the late Rose Barlow.

Sauces and Dressings

DRESSINGS AND VINAIGRETTES

Homemade salad dressings taste great, are easy and inexpensive to make—and they contain no weird artificial ingredients!

Wild-Weed Dressing

This is one of my all-time favourite dressing recipes. It's a wild version of a recipe my friend Nicole Edwards gave me for Christmas one year.

- 2 cups (500 mL) wild greens (try chickweed, dandelion, lamb's quarters, wild onion)
- ½ cup (125 mL) apple cider vinegar
- 1 tablespoon nutritional yeast
- 3 cloves garlic
- 1 cup (250 mL) olive oil or oil of your choice
- 2 tablespoons (30 mL) tamari
- 2 tablespoons (30 mL) tahini (sesame-seed butter)
- 2 tablespoons (30 mL) hemp seeds

Combine ingredients in a blender and blend till smooth.

Drizzle over your summer green salads.

Store in a bottle or jar, and refrigerate. It will last for a few weeks in the fridge.

Makes 2 cups (500 mL).

Homemade dressings taste so much better than store-bought varieties. BG

Wild-Rose Petal Vinaigrette

A delicate dressing for a wild-weed salad.

- 1 cup (250 mL) rose petals
- 1 cup (250 mL) white-wine or apple-cider vinegar
- 1½ cups (375 mL) flax seed oil (or your favourite oil)

Infuse rose petals in vinegar, and let sit for 1–2 weeks, shaking every day. (For more on how to prepare herbal vinegars, see page 296.)

Strain and mix with oil.

Bottle and refrigerate. Shake well before using.

Makes 2½ cups (625 mL).

Part III: *Getting Wild in the Kitchen*

Raspberry-and-Rosehip Vinaigrette

A variation on this recipe is to process all the ingredients together in a blender and use immediately as a dressing on fresh green salads.

½ cup (125 mL) raspberries
1 cup (250 mL) apple cider vinegar
2 cups (500 mL) olive oil
2 tablespoons (30 mL) Rosehip Syrup (see page 299)

Infuse raspberries in apple cider vinegar for 2–3 days, shaking daily.

Strain and mix with oil and syrup.

Bottle and refrigerate. Shake well before using. It will last for up to 6 months.

Makes approximately 3 cups (750 mL).

Dandelion Dressing*

Tart and tantalizing!

1 cup (250 mL) olive oil
½ cup (125 mL) hemp oil
¾ cup (175 mL) apple cider vinegar
3 cloves garlic
½ teaspoon (2 mL) sea salt
2 tablespoons (30 mL) Dandelion-Petal Mustard (see page 359)
3 tablespoons (45 mL) Dandelion-Petal Syrup (see page 355) or honey
2 cups (500 mL) dandelion greens, fresh, chopped

Put ingredients into a blender and blend till smooth. Bottle and refrigerate. It will keep for up to 6 months.

Makes approximately 2 cups (500 mL).

ADDITIONAL RECIPE

Borealis Green-Goddess Dressing, page 163

---

* Recipe inspired by the late Rose Barlow.

Above: Shepherd's purse leaves. BG

SOUPS AND SALAD

ONE OF MY favourite meals is soup because it can be made with just about any combination of ingredients. Many people already have favourite soup recipes that they make almost without thinking. If you're one of these people, consider changing up your recipe by adding wild ingredients like nettle, lamb's quarters, fireweed, and dandelion.

Green Bouillon*
A light, nourishing soup that's great on its own, or as a starter for any meal.

3 cups (750 mL) broth of your choice (vegetable, chicken, wild game, or miso)
1½ cups (375 mL) young nettle shoots, washed
Salt and pepper (optional)

If you are adding miso do not bring to a boil, just heat it and add the other ingredients, keeping it at low temperature to let the ingredients warm and meld. Otherwise, in a saucepan, bring broth to a boil, reduce to a simmer and add nettle shoots.

Simmer with lid on for about 2 minutes or until nettle is tender.

Add salt and pepper if desired.

If the flavour is too strong, add water or more broth. Serve with croutons and chopped chives.

Makes 4 bowls of soup.

---

*This recipe was created and given to me by Mary-France Campagna, a Yukoner and a passionate organic gardener.

Part III: *Getting Wild in the Kitchen*

Cream of Wild-Weed Soup always gets rave reviews. BG

Cream of Wild-Weed Soup

I always make this for participants in my Herb Walk, Talk and Medicine-Making Workshops. It always gets rave reviews so I thought I would include it here for everyone to enjoy!

4 cups (1 L) water
Pinch of sea salt
1 carrot, chopped
1 potato, chopped
4 cloves garlic, chopped
1 onion, chopped
2 cups (500 mL) mixed wild greens (lamb's quarters, nettle, chickweed, red clover, and dandelion)
Herbs and spices, such as pepper, basil, thyme, and nutmeg (optional)

Roux:
⅓ cup (80 mL) flour (I use spelt flour)
⅓ cup (80 mL) butter, melted
2 cups (500 mL) milk

In a large pot, bring water, salt, carrot, and potato to a boil. Cook until soft.

In a pan, fry garlic and onion until onions are transparent, add to the pot.

Mash vegetables in water until semi-puréed.

In another pot, steam greens until tender. Reserve the liquid.

Purée greens in blender, adding the reserved liquid as needed.

Make the roux by whisking flour into melted butter, then whisking in milk, and cooking the roux over very low heat, stirring until the sauce has thickened.

Add wild greens to roux, and then thoroughly blend the roux into the soup.

Cook over low heat, simmer for 10 minutes.

Add herbs and spices, if desired. Serve with a sprig of chickweed, a sprinkle of dandelion petals, and a dollop of plain yogourt.

Makes a nice-sized pot of soup to feed your hungry family! I like to serve with Red-Clover Tea Biscuits (page 344) and a Wild-Weed Salad.

Soups and Salad

Wild-Weed Salad

This recipe is delicious when served with a drizzle of Raspberry-and-Rosehip Vinaigrette (see page 362).

A salad bowl full of wild greens, depending on what's in season: my favourites include dandelion greens*, fireweed greens, lamb's quarters, chickweed, and red clover.
Sprinkling of goat feta to taste
Sprinkling of oven-roasted pecans, warm
Sprinkling of rose petals
Sprinkling of fireweed flowers
Sprinkling of dandelion flowers

Wash and dry greens.

Tear greens into a salad bowl, toss.

Sprinkle feta, pecans, rose petals, and fireweed and dandelion flowers on top.

ADDITIONAL RECIPES

Wild Dock Sorrel-Potato Soup, page 90

River-Beauty Summer Salad, page 143

Feta adds a nice touch to a colourful wild-weed salad. BG

---

* If dandelion greens are bitter, soak them in water for a couple of hours before making the salad. If you don't have enough wild greens, add some garden lettuce, spinach, or kale.

Wild Food and Medicine Plants of the North | 365

MAIN COURSES AND SIDES

Wild greens can be used in a variety of ways in the kitchen. Let your imagination and your taste palate run wild! Once you start getting familiar with the flavours and textures of wild greens you can creatively incorporate them into your favourite dishes—instead of spinach in your lasagna or quiche you could use plantain, nettle, dandelion greens, or lamb's quarters.

Piquant Plantain*
A tart, sweet-and-spicy side dish.

- 4 cups (1 L) plantain leaves, fresh
- 1 tablespoon (15 mL) olive oil
- 1–2 garlic cloves, crushed
- ¼ teaspoon (1 mL) red chili flakes
- Pinch of salt
- 1 tablespoon (15 mL) raspberry or balsamic vinegar, or Raspberry-and-Rosehip Vinaigrette (see page 362)

Rinse or soak plantain leaves clean.

In a large frying pan or wok, heat olive oil over medium-high heat. Add garlic and fry until golden brown.

Add plantain, chili flakes, and salt, stirring to coat leaves with olive oil and garlic. Cover with lid and steam. The plantain will shrink considerably in volume.

When leaves are just tender, add vinegar and stir to coat leaves. Allow to cook for 1 minute more. Serve drizzled with leftover juice from the pan.

Serves 2 people.

---

* This recipe was created and given to me by Nadine Pedersen.

Main Courses and Sides

Wild-Weed Spanakopita*

2 cups (500 mL) feta cheese, crumbled
5 eggs
2 tablespoons (30 mL) flour
2 cups (500 mL) cottage cheese
Pinch of sea salt, pepper, oregano and basil
4 cups (1 L) fresh weeds (lamb's quarters, chickweed, stinging nettle)
1 cup (250 mL) fresh spinach
1 onion, chopped
6 garlic cloves, chopped
3 tablespoons (45 mL) butter
1 box phyllo pastry, thawed
1 cup (250 mL) butter, melted

Preheat oven to 375°F (190°C).

To prepare the filling
In a bowl mix together crumbled feta cheese, eggs, flour, cottage cheese, herbs, and spices.

Clean, stem, and chop wild greens and spinach. Place greens in a frying pan (preferably cast iron) on low heat for up to 5 minutes. Do not add water. When greens have wilted add to the feta mixture.

Cook onions and garlic in 1 tablespoon (15 mL) of butter. When soft, combine with the feta-and-greens mix.

To assemble
Butter a 9 in. x 13 in. (23 cm x 33 cm) baking pan.

Place a phyllo leaf in the pan (it's okay to let the edges climb the sides). Brush generously with butter. Continue stacking and buttering layers, until stacked 8 layers high.

Spread on half of the filling.

Continue with another stack of 8 leaves, spread on the other half of the filling.

Apply the remaining phyllo in buttered layers.

Bake uncovered at 375°F (190°C) for 45 minutes or until golden.

---

* Inspired by Mollie Katzen's *The Moosewood Cookbook*.

Part III: *Getting Wild in the Kitchen*

Healthful Weed Pie*

This yummy pie is made in layers in a standard-sized casserole dish.

Layer one (base layer)
8 cups (2 L or ½ brown paper grocery bag) mixed young leaves (nettle shoots or leaves, chickweed, fireweed, lamb's quarters, shepherd's purse, or dandelion. If you don't have enough wild greens, add spinach and/or kale).

Layer two
3 tablespoons (45 mL) butter
2 tablespoons (30 mL) onion, chopped
2 garlic cloves, chopped
2 tablespoons (30 mL) flour
1 cup (250 mL) milk
Bay leaf (optional) or 1 large Labrador-tea leaf

Layer three
1 cup (250 mL) cheese, grated

Layer four
2 eggs, beaten
¾ cup (175 mL) flour
⅓ cup (80 mL) milk
⅓ cup (80 mL) water (or leftover water from steaming greens)
Pinch sea salt
Dash nutmeg

Layer five (top layer)
½ cup (125 mL) Parmesan cheese, freshly grated

To make
Preheat oven to 400°F (200°C)

Layer one
Steam greens until tender. Use them to line the bottom of a large, oiled casserole dish.

---

*This recipe was created and given to me by Marie-France Campagna, a Yukoner and a passionate organic gardener.

Main Courses and Sides

Layer two
In a frying pan, melt butter. Add onion and garlic, cooking till onion is transparent or limp.

Stir in flour, milk, and bay leaf or Labrador-tea leaf. Cook over low heat, stirring till thickened.

Pour mixture over greens.

Layer three
Spread grated cheese over mixture.

Layer four
In a bowl, combine eggs, flour, milk, water, salt, and nutmeg.

Pour over grated cheese.

Layer Five
Sprinkle Parmesan cheese on top.

To bake
Bake for 20 minutes, or until set and golden. Bon appétit!

Serves up to 4 people.

> "The act of putting into your mouth what the earth has grown is perhaps your most direct interaction with the earth."
>
> —Frances Moore Lappé, *Diet For A Small Planet*

ADDITIONAL RECIPE

Lamb's Quarters Omelette, page 114

Wild onions spice up any meal. CA

Wild-Greens Stir-Fry

1 tablespoon (15 mL) sunflower or olive oil
1 large bunch wild onion chives, chopped
3 cloves garlic, chopped
½ onion, chopped
¼ cup (60 mL) tamari
4 cups (1 L) mixed vegetables (such as chopped zucchini, chopped peppers, shredded carrot, peas, beans, etc.)
1 can (400 mL) coconut milk
2 cups (500 mL) mixed greens (nettle leaves, chickweed, fireweed leaves, lamb's quarters, plantain leaf, or dandelion leaves all work great alone or as a combination)
Sprinkle of nuts or seeds (I like almonds and hemp seeds)

On medium heat, warm oil in frying pan. Add chives, garlic, and onions, cooking till tender.

Stir in 1 tablespoon of the tamari.

Add vegetables, then coconut milk. Cover and steam for a few minutes until vegetables are just slightly tender.

Add greens, cover, and steam until just wilted.

Stir in nuts or seeds.

Serve on its own or over rice, couscous, fried tofu, meat, or pasta. Top with goat cheese, if desired.

Serves up to 4 people.

DESSERTS AND SWEET TREATS

USING WILD INGREDIENTS in the creation of tasty desserts lends an interesting and local twist to any sweet treat. It seems throughout the summer months every dessert I make has wild plants incorporated—even if it's just a sprig of mint accompanying a bowl of ice cream, bright yellow dandelion petals decorating a cake, or juicy wild berries bursting with flavour in frozen fruitsicles!

Dandelion-Petal Cake*

The dandelion flower's beauty shines through in this moist and delicious cake.

The cake
2 cups (500 mL) all-purpose flour
2 teaspoons (10 mL) baking powder
1½ teaspoons (7 mL) baking soda
1 teaspoon (5 mL) cinnamon
1 teaspoon (5 mL) salt
1 cup (250 mL) cane sugar
1 cup (250 mL) Dandelion-Petal Syrup (see page 355)
1½ cups (375 mL) melted butter or sunflower oil
4 eggs, beaten
1 can (approximately 500 mL) crushed pineapple, drained
½ cup (125 mL) coconut, shredded
2 cups (500 mL) dandelion petals

The icing (This cake is great with or without icing)
2 cups (500 mL) cream cheese, room temperature
1 cup (250 mL) honey, maple, birch or dandelion syrup
Dandelion petals sprinkled on top

*Dandelion cake recipe inspired by the late Rose Barlow.

Part III: *Getting Wild in the Kitchen*

To make the cake
Mix dry ingredients in a bowl (except for the sugar).

In a separate bowl, mix sugar, dandelion syrup, butter, and eggs till creamy.

Fold in pineapple and coconut.

Stir in dry ingredients till well blended.

Fold dandelion petals into the batter.

Pour batter into an oiled 11" x 13" (28 cm x 33 cm) cake pan and bake at 350°F (180°C) for about 40 minutes.

To ice
Pour sweetener over cream cheese and blend together with a fork.

Ice the cake once it has cooled.

Decorate with dandelion petals.

Dandelion petals ready to be made into cake. BG

Desserts and Sweet Treats

Wild-Blueberry Fruitsicles

Growing up, my children loved running to the freezer and finding homemade fruitsicles. To this day they still prefer these to store-bought Popsicles.

- 2 cups (500 mL) wild blueberries, fresh or frozen
- 1 cup (250 mL) yogourt
- ½ cup (125 mL) honey
- 1 teaspoon (5 mL) vanilla extract
- 1 teaspoon (5 mL) lemon juice
- 1 teaspoon (5 mL) lemon peel, grated
- 1 teaspoon (5 mL) rosewater
- ⅓ cup (80 mL) whipping cream

In a blender, combine all ingredients (except the whipping cream) and blend until smooth.

Whip the cream till soft peaks form, fold in fruit mixture.

Pour into ice pop moulds and insert sticks.

Makes approximately 12 fruitsicles in standard-sized moulds. You could also freeze this recipe in small cups or dessert dishes, and serve with a sprig of wild mint or sprinkled with rose petals.

Creamy Cranberry-Vanilla Ice

This easy-to-make, creamy dessert is great on its own or with fresh fruit.

- 1 cup (250 mL) wild cranberries
- 1 cup (250 mL) vanilla yogourt
- 1 cup (250 mL) honey or cane sugar

Place all ingredients in the blender, blend on high speed until smooth.

Pour into a sealed container and freeze. You can also pour into ice pop moulds for great creamy cransicles!

Let thaw until soft before serving.

Part III: *Getting Wild in the Kitchen*

Roasted Dandelion-Root Ice Cream*

Dandelion-root ice cream has a sweet mocha-goodness taste!

2½ cups (625 mL) heavy cream
1½ cups (375 mL) half-and-half cream
1¼ cups (300 mL) cane sugar
5 egg yolks
2 tablespoons (30 mL) roasted dandelion root

In a coffee mill, grind dandelion root into a powder, making sure there are no chunks. (If there are chunks, run the powder through a flour sifter or a fine sieve to remove them.)

Place both creams and sugar in a medium-sized pot or double boiler. Warm on a medium-low heat, until it just reaches a simmer, stirring to dissolve sugar.

Add dandelion-root powder and continue cooking, barely simmering, for 10 minutes making sure not to let it boil.

Remove from heat and let the root powder infuse in the cream mixture for 45 minutes.

In another pot, whisk egg yolks. Slowly add dandelion-root cream, stirring gently. Heat slowly on medium-low heat, stirring until the mixture coats the back of a spoon.

Chill in the fridge. Freeze in an ice-cream maker.

---

*Recipe inspired by the late Rose Barlow.

Desserts and Sweet Treats

Flower-Delight Tempura*

A delightful way to end a feast! This recipe is fun to eat in a group and is a great way of sharing what you are grateful for! This can also be served as an appetizer or main course with a savoury sauce.

1 cup (250 mL) water
½ cup (125 mL) flour
1 teaspoon (5 mL) cornstarch
¼ cup (60 mL) sunflower oil for frying (you may need more if you have a large frying pan)
Approximately 4 cups (1 L) of edible flowers (freshly bloomed fireweed flowers, dandelion flowers, goldenrod flowers, and strawberry blite)
Powdered sugar, honey or honey thinned with water

In a bowl, mix water, flour, and cornstarch to make a batter.

Heat oil in large frying pan, wok, or stainless steel pot until a drop of water sizzles when dropped in it.

Roll flowers in batter. Place in hot oil until crisp (watch them carefully, they cook quickly!), remove with tongs.

Drain and pat dry with a paper towel.

Dust with sugar, drizzle with honey, or serve with a honey-thinned-with-water dipping sauce and serve immediately.

Serves up to 4 people.

Strawberry blite. BG

---

*This recipe was created and given to me by Marie-France Campagna, a Yukoner and a passionate organic gardener.

Part III: *Getting Wild in the Kitchen*

FRUIT LEATHER

Roll up your sleeves and get ready for some fun! Fruit leathers make tasty, healthy snacks the whole family can enjoy.

Great for eating on the trail, on long river trips, or as a treat in the school lunchbox, fruit leather packs a nutritious punch. When making fruit leathers I have been known to slip in a few medicinal herbs as well, and why not? The berries generally disguise any bitter flavours that some people may not find palatable!

Cranberries, blueberries, rhubarb, raspberries, and rosehips are some of my boreal-forest favourites for making fruit leather, but you can also use currants, Saskatoon berries, strawberries, and mossberries, on their own or blended together.

Apples, apricots, peaches, and pears work great in combination with wild boreal fruits. Wild herbs like nettle, lamb's quarters, or mint can be ground and added to the purée, and spices such as ginger, cinnamon, and cardamom can also be added. Rice syrup, birch syrup, maple syrup, or honey can be added as a sweetener.

Fruit leather can be made from the purée left over from jelly making.

Raw Uncooked Fruit Leather Process

Uncooked fruit leather is perfect for raw foodies! Not cooking the wild fruits helps preserve the many enzymes, vitamins, and minerals. This makes raw fruit leather a delicious and healthy raw-food sweet snack.

Making fruit leather is easy. Simply purée the fruit in a blender and then dehydrate it in a dehydrator or oven. BG

Simply purée fruit in a blender, adding enough water (or herbal tea/infusion) to make a thick smooth purée (you don't want it too thin or it will take too long to dry).

Dehydrator method
Pour mixture onto dehydrator trays lined with a Teflex sheet or parchment paper.

With the back of a large spoon, spread out to ⅛–¼ in. (3–6 mm) thickness.

Dehydrate for approximately 10–12 hours at 95°F (35°C), or until fruit peels away easily from the Teflex sheet or parchment paper.

With a sharp knife, cut leather into strips.

Roll pieces up and store in plastic wrap, wax paper, or parchment paper, or in a sealed container, away from excess heat and moisture.

Oven method
Cover a cookie sheet with parchment paper.

Desserts and Sweet Treats

Spread purée on parchment paper so that it's ⅛–¼ in. (3–6 mm) thick.

Dry on low, 95°F (35°C). It may take a couple of days to dry, depending on your stove.

Living Fruit Leather
This raw fruit leather is rich in antioxidants.

1 part wild cranberries
1 part wild blueberries
⅒ part powdered nettle leaf (a pinch)
⅒ part powdered wild chamomile (a pinch)
½ part rosehip juice, syrup, or decoction
20 drops rhodiola tincture
A small amount of water to blend

Cooked Process
The cooked process is good to use if you are introducing fruits that brown easily, such as apples or pears.

4 parts berries and/or fruit
1 part water
¼ part fresh or ⅛ dry herbs (optional)*

In a stainless-steel or glass cooking pot, simmer over low heat to reduce the liquid down by half, while mashing the berries.

If you want a smooth consistency, run the purée through a blender once it has cooled.

Pour cooled purée onto dehydrator trays or a cookie sheet lined with parchment paper. (See drying instructions above.)

*Note: If you are going to add fresh herbs to your leather, purée herbs in the blender with a small amount of water and add to the fruit purée halfway through the cooking process. If you are using dried herbs that are already powdered pour in at the beginning of the cooking process.

ADDITIONAL RECIPE

Saskatoon-Berry Pie, page 224

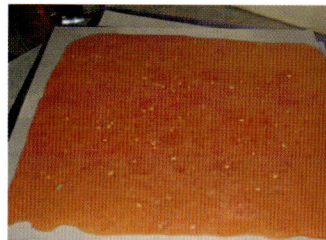

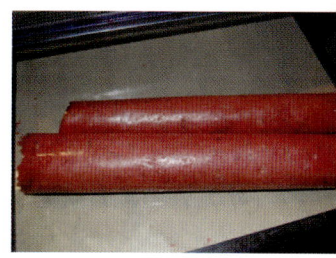

Top: Fruit leather purée, ready to be dehydrated.

Middle: Dehydrated and ready to be rolled up.

Above: Rolled up and ready to be enjoyed. BG

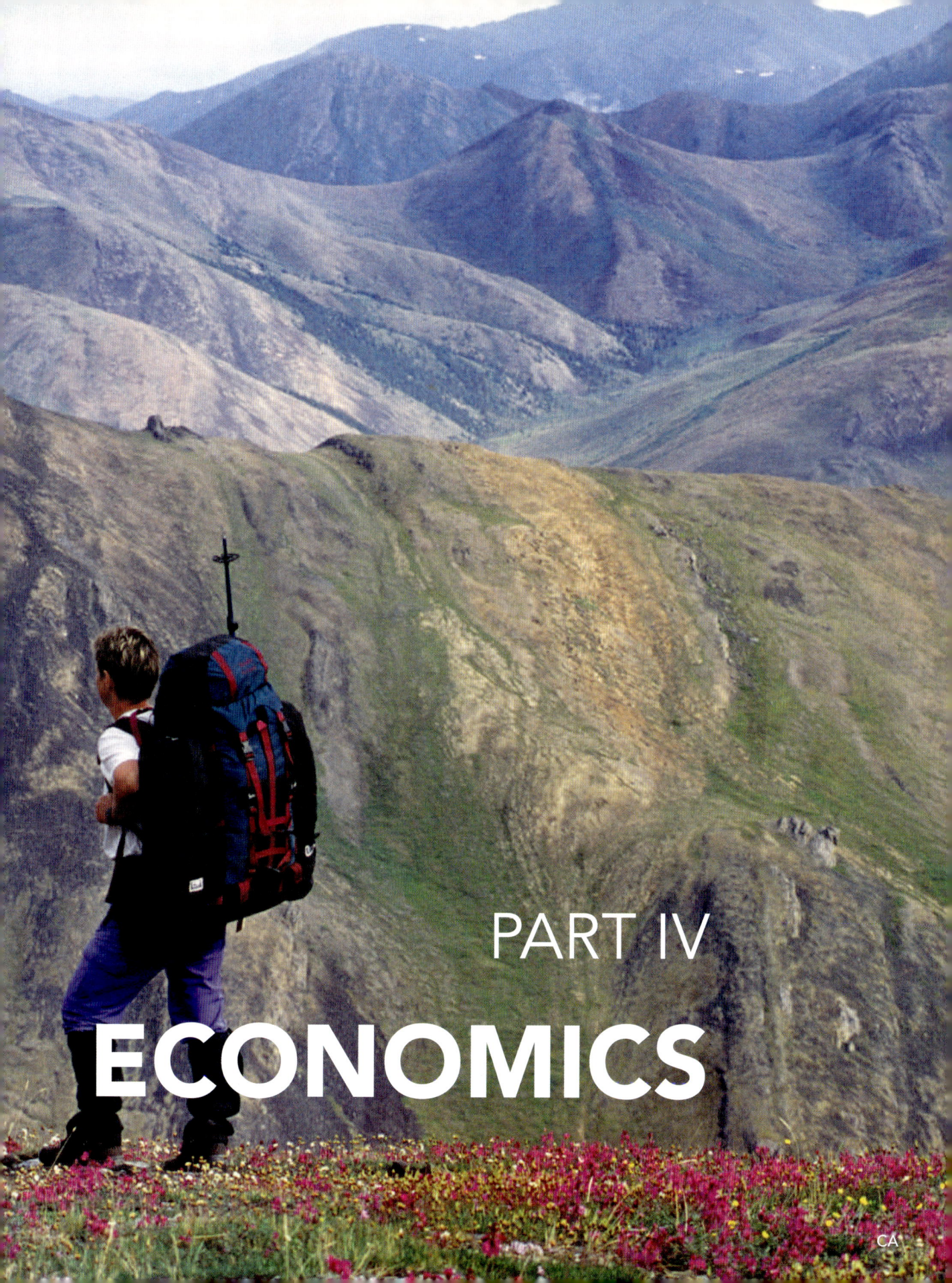

PART IV
ECONOMICS

Above: Sylvia Frisch hauls buckets full of birch sap that will be made into Uncle Berwyn's Birch Syrup. BL

Previous: For people who love the outdoors, starting up a non-timber forest-products business lets them make money while pursuing their passion for being outside. CA

STARTING UP A NON-TIMBER FOREST-PRODUCTS BUSINESS

OVER THE PAST FIFTEEN YEARS, I have turned my passions for plants, care for the environment, my love of people and our unique northern region into an original and successful business adventure.

The Aroma Borealis Herb Shop began as a home-based business in 1995. Using one room in my house, I began creating herbal products to sell at craft fairs and stores throughout the Yukon. Three years later, in February 1998, my partner Mike and I opened a store in the heart of downtown Whitehorse, Yukon.

Since those early days, I have formulated and developed hundreds of herb and aromatherapy products for the natural health market. Many of my products combine wild plants from the subarctic/boreal forest region with organically grown herbs and essential oils from all over the world, and are what I refer to as "healing collages from around the globe."

Many of these products are available in stores across North America, and the Aroma Borealis store in Whitehorse is still going strong. Not only has the business helped support

Above: Autumn rain settling on an aspen leaf. FM

Left: The Aroma Borealis Herb Shop began as a home-based business and grew into a storefront, online business, and wholesaler. CA

Part IV: *Economics*

my family, but we maintain full-time staff throughout the year, and we buy local products and botanicals from wild harvesters. Along the way, I was recognized as the "Business Person of the Year" by the Yukon Chamber of Commerce, in 2006, and was awarded a Gold Award of Excellence by *Alive* magazine for the Aroma Borealis Green-Aid Ointment.

In "business speak" you would say that Aroma Borealis has a strong competitive advantage because it spans many levels of the natural health industry: we wild-harvest herbs, manufacture products, sell retail via our storefront and our website, sell wholesale to businesses in the Canadian health food, aromatherapy, gift, and esthetic sectors, and provide education through workshops, newspaper and magazine articles, and now this book.

I've shown with the Aroma Borealis store that boreal-forest-plant products are viable alternatives to consumer products from "outside" the Yukon, and their production supports our local economy. While Aroma Borealis may be a unique enterprise, there is a long history of sustaining families and communities with what's found in the boreal forest.

First Nation peoples have lived off the land of the northern boreal forest for millennia, often trading inland plants and pelts with coastal tribal groups. Early non-First Nations people also relied on the forest for the food, shelter, and medicine they needed to survive.

In the last fifty years, however, the northern economy has relied very heavily on raw materials being extracted from the North and shipped long distances for processing and manufacturing in other regions. During this period, with the advent of airplanes and roads, northerners have become reliant on food, medicine, clothing, and shelter shipped to us from far away. Along the way we have begun to lose touch with the world at our doorstep and the riches the forest offers us.

Fortunately, in the last decade a local products and local-processing movement has been growing, and has created what are known as value-added products: furniture, home decor, artwork, herbal medicines, teas, beer, food, beauty and health-care products.

The average consumer has also become aware of the environmental and social problems associated with a globalized economy. The result is more and more people are looking for ways to shop locally so as to reduce their carbon footprints and support their local economies.

At the moment there are many opportunities for people to respond to current and emerging issues and needs by developing commercial applications for what are widely known as "non-timber forest products" (NTFP). By definition, an NTFP can be any product or service other than timber that's produced from the forest. This includes fruits, foods, medicinal plants, and a range of bark and wood products (such as carvings and baskets).

Starting a business with wild boreal botanicals can be a wonderful way to share the bounty of the land and do work you love, while making a bit of extra money.

Many Yukoners are successfully doing this on a small-scale level. For example, vendors at the vibrant Fireweed Community Market in Whitehorse—including local farmers, artisans, cooks, medicine makers—sell products as diverse as beautiful goat cheeses, fresh produce, wild jams, jellies, and chutneys, dried mushrooms, herbal teas, honey, willow and spruce-root baskets, jewellery, furniture, and even coffins manufactured from local wood.

In the Yukon we are fortunate to have working farms that produce crops, feeds, greens, vegetables, poultry, and livestock, including elk and bison. While many people are deterred from choosing this way of life because of the soil conditions and cool temperatures in the

Starting Up a Non-timber Forest-products Business

Lyndsey Berwyn Larson, the creator of Uncle Berwyn's Yukon Birch Syrup, is another example of someone who has created a successful non-timber forest-product business. His birch syrup, made from the sap of birch trees near Mayo, Yukon, has become a coveted commodity in the territory, and is even used by the Yukon Brewing Company to create the McQuestin Birch Beer, a specialty beer which is only available at the brewery in Whitehorse.

Left: Two of "Uncle Berwyn's" buckets filling with birch sap near Mayo, Yukon. BL

North, we need more northern farmers. Farmers contribute to our local economies and our food-production needs. As this becomes more recognized, hopefully our local agricultural sector will grow.

Currently, Shiela Alexandrovich of Wheaton River Garden, 50 km southwest of Whitehorse, is a great example of a farmer who uses the bounty of the land to support herself. In the summer months, Shiela and her partner Poul, grow organic vegetables that they sell at the Fireweed Market. The couple's farming is also sustained in part by Community Supported Agriculture (CSA), a popular way for consumers to invest in locally grown food and buy directly from a farmer: investors purchase a certain number of shares in return for a portion of the produce; the farmer receives guaranteed customers and income throughout the growing season. If Shiela and Poul have the time, they also make jams and jellies. During the winter months, Shiela uses wild materials to create beautiful and functional willow baskets, willow statues, jewellery, and much more.

At Aurora Mountain Farms, north of Whitehorse, Simone and Tom Rudge grow certified organic crops, use the milk from their goats and local botanicals to make soap, and produce an array of preserves from locally harvested herbs and berries. They sell cranberry jam, Saskatoon-berry jam, yarrow jelly, dandelion jelly, fireweed jelly, highbush-cranberry jelly, raspberry jelly, and rhubarb marmalade.

The Rudges also feed many of the "volunteer" wild weeds that grow on the farm—chickweed, lamb's quarters, and dandelion—to their poultry and livestock. This saves them money and gives their animals enzyme-rich, fresh, and nutritious greens throughout the summer seasons.

If you're thinking about starting a "wild herb business," it's important to consider that while there's a greater demand for natural products in the marketplace today, there are also more-stringent laws for natural-product producers.

A budding econony—in recent years, the local-processing movement has been growing, using boreal forest resources to create furniture, art, herbal medicines, teas, beer, food, beauty and health-care products. BG

Under the Natural Health Products Regulations that came into effect on January 1, 2004, natural health products (NHPs) are defined as: vitamins and minerals, herbal remedies, homeopathic medicines, traditional medicines such as traditional Chinese medicines, probiotics, and "other products" such as amino acids and essential fatty acids.

If you are manufacturing herbal remedies for sale in Canada that would fall under natural health products (including tinctures, capsules, salves, and creams), and, for example, were to make a health claim such as "heals wounds and prevents scarring," the law requires you to have site and product licences for each remedy you sell.

Luckily, all the information you need, including application guides and forms, is available on the Health Canada website, *www.hc-sc.gc.ca*.

Similarly, if you are making cosmetics such as creams, lotions, salves, deodorants, powders, sprays, and body or facial-cleansing products, as a manufacturer you must register the products with The Cosmetic Program of Health Canada. Be aware that you're forbidden to make health claims for these products. The word "healing," for example, is considered a health claim. For more information on this, refer to Health Canada's publication, *Guidelines for Cosmetic Advertising and Labelling Claims*.

Even though there are a number of rules and regulations you'll need to understand when starting up a non-timber forest-product business, there are also many government agencies and non-governmental organizations that can assist you.

I believe that as more people develop and sell new and innovative value-added products, not only will we see our local economies flourish, but we will reap social, environmental, and cultural benefits. By promoting the sustainable use of the forest, we are supporting the social and cultural values of the communities the producers live in.

Many of the people who wild harvest botanicals for Aroma Borealis do it because they love the outdoors. When they deliver their botanicals to us, I often hear them say, "Why not get paid for doing something you love to do?"

Starting Up a Non-timber Forest-products Business

This is a big part of my business philosophy, too: do what you love, and the money and opportunities will follow.

When it comes to creating top-quality herbal medicines, foods, and cosmetic products from northern boreal forest botanicals, I believe the possibilities are endless. I hope this book helps guide you to a place that inspires you to get out and learn about the plants that grow wild, and that you'll find it to be a useful tool for learning good practices of plant identification, and sustainable harvesting of medicinal and food plants.

If you've been inspired to use the information in this book to start your own business, below is a list of government agencies and non-profit organizations that can help you realize your dreams.

Federal Government Agencies

Health Canada
Address Locator 0900C2,
Ottawa, ON K1A 0K9
E-mail: publications@hc-sc.gc.ca
Telephone: (613) 957-2991
Toll free: 1-866-225-0709
Website: www.hc-sc.gc.ca

Natural Health Products Directorate
Health Products and Food Branch
Health Canada
Qualicum Tower A
2936 Baseline Rd., A.L. 3302A,
Ottawa, ON K1A 0K9
Website: www.hc-sc.gc.ca

Consumer Product Safety
Cosmetics Program
Health Canada
MacDonald Building
123 Slater St., 4th Floor A.L. 3504D,
Ottawa, ON K1A 0K9
E-mail: cosmetics@hc-sc.gc.ca
Telephone: (613) 946-6452
Fax: (613) 952-3039
Website: www.healthcanada.gc.ca/cosmetics

Cosmetics Program
British Columbia and Yukon Regional Product Safety Office
Health Canada
Suite 400-4595 Canada Way
Burnaby, BC V5G 1J9
E-mail: Bby.ProdSafe@hc-sc.gc.ca
Telephone: (604) 666-5003
Toll free: 1-866-662-0666

Canadian Food Inspection Agency (Yukon)
Box 2703,
Whitehorse, YT Y1A 2C6
Contact: Valerie Whelan
E-mail: Valerie.Whelan@gov.yk.ca
Telephone: (867) 667-5272
Fax: (867) 393-6222

Canadian Food Inspection Agency (BC)
BC Mainland / Interior (includes Yukon)
4321 Still Creek Dr., Suite 400,
Burnaby, BC V5C 6S7
Telephone: (604) 666-6513
Fax: (604) 666-1261

Yukon Government Agencies
Agriculture Branch
E-mail: agriculture@gov.yk.ca
Telephone: (867) 667-5838
Toll free: 1-800-661-0408 ext. 5838
Fax: (867) 393-6222
Website: www.agriculture@gov.yk.ca

Non-Governmental Organizations (NGOs)

National Non-Timber Forest Products Network of Canada
The network is a partnership of organizations, agencies, businesses, and people concerned with the sustainable and ethical development of the non-timber forest-products sector in Canada. The group's aim is to enhance the ability of communities, researchers, resource managers, policy-makers, economic-development specialists, and other groups and individuals to develop expertise, share knowledge, make informed decisions, and ultimately, work together more effectively to develop and manage these resources for the benefit of rural and remote communities across Canada.
Website: www.ntfpnetwork.ca

Part IV: *Economics*

Royal Roads University, Centre for Non-Timber Resources
The centre provides: education, training, curriculum development in the non-timber sector; community-economic development, First Nations and non-timber resources; wild foods, medicinal and floral plants, fungi and non-timber forest-product inventory and policy development.
Website: www.royalroads.ca

The Boreal Centre for Conservation Enterprise
A non-profit organization of community-based entrepreneurs, and rural development practitioners who are focused on development of ecologically and socially responsible enterprise in the boreal forest bio-region.
P.O. Box 285
Moberly Lake, BC V0C 1X0
E-mail: theborealcentre@gmail.com
Telephone: (250) 788-9635
Fax: (250) 788-9636
Website: www.peaceriverwatershed.ca/borealenterprise/education.htm

Canadian Herb, Spice and Natural Health Products Coalition (CHSNC)
The CHSNC's mission is to build a viable national herb, spice, and natural health product industry that highlights regional strengths and expertise and optimizes national synergies.
Website: www.saskherbspice.org/CHSNC

Yukon Agricultural Association
203-302 Steele St.,
Whitehorse, YT Y1A 2E5
E-mail: admin@yukonag.ca
Telephone: (867) 668-6864
Fax: (867) 393-9566
Website: www.yukonag.ca

Growers of Organic Food Yukon (GOOFY)
Website: www.organic.yukonfood.com

Canadian Organic Growers (COG)
Website: www.cog.ca

Fireweed Community Market
On its website the Fireweed Community Market has posted guidelines for home-prepared food, as well as the forms you'll need if you plan to sell food at the market.
E-mail: fireweedmarket@yahoo.ca
Telephone: (867) 393-2255
Website: www.fireweedmarket.yukonfood.com

The Canadian NTFP Business Companion: Ideas, Techniques and Resources for Small Businesses in Non-Timber Forest Products & Services
is a CD that covers about sixty-five NTFP business areas.
Contact: Gina H. Mohammed, PhD, Research Director
P&M Technologies
66 Millwood St.,
Sault Ste. Marie, ON P6A 6S7
E-mail: gm@pmtech.ca
Telephone: (705) 946-2882
Website: *www.pmtech.ca*

FOR REFERENCE

Herbalists through the Ages 389
The Celtic Tree Alphabet 392
Botany Basics 393
Botanical Terms 395
Understanding Herbal Constituents and Phytochemistry 401
Medicinal Actions of Plants 403
Herbal Usage Chart 406
Medicinal Actions Chart 416
Vitamin Chart 420
Bibliography 424
Index 428
Contributing Authors and Photographers 440

HERBALISTS THROUGH THE AGES

OVER THE AGES a number of luminaries have made their mark in the botanical world by cataloguing and classifying plants, noting their edible and medicinal properties, and creating and recommending herbal remedies, making it easier for the rest of us to go forth and forage. Throughout *The Boreal Herbal*, the work of these early herbalists is referred to often, and I feel they deserve a more formal introduction here.

Pliny the Elder (AD 23–79)

Pliny the Elder, a Roman statesman, historian, author, naturalist, and herbalist, was born in Como, in what is now Italy, in AD 23. After years of travelling, writing, and serving in the military, Pliny published *Natural History*, a thirty-seven-volume encyclopedia of ancient scientific knowledge. It included eight volumes dealing specifically with healing herbs, as well as folk remedies and pharmacopoeias passed down from the Greeks and earlier cultures.

Among his staggering number of medical recommendations, Pliny made some wild assertions, such as "kissing the hairy muzzle of a mouse" can remedy a cold. But many of Pliny's recommendations guided physicians for centuries and continue to benefit us today.

Pliny was a champion of herbal medicine who beseeched physicians to grow and use their own medicinal plants. He lamented the great cost of imported medicines while herbal remedies were available in people's backyards. He is quoted as saying, "For a tiny sore a medicine is imported from the Red Sea, though genuine remedies form the daily dinner of even the very poorest. But if remedies were sought in the kitchen garden…none of the arts would become cheaper than medicine."

Pliny the Elder died in 79 AD while attempting to save people from the eruption of Mount Vesuvius. He was quoted well into the Middle Ages, and for more than one and a half millennia, *Natural History* was venerated as a pillar of human knowledge.

Hildegard of Bingen (1098–1179)

Hildegard of Bingen was one of the most celebrated herbal authorities of her time. The German mystic, poet, composer, playwright, and abbess was born in 1098 and lived to be eighty-two—an astounding age in an era where average life expectancy was around thirty.

Hildegard was well known for her healing powers and use of herbs, tinctures, and precious stones. She wrote her *Physica* in 1160, the earliest book on natural history penned in German. In it, she outlined her nine healing systems as Plants, Elements, Trees, Stones, Fish, Birds, Animals, Reptiles, and Metals; describes a number of herbs and their medicinal uses, and offers simple home remedies. *Physica* was widely published during the Renaissance and was an influential text well into the sixteenth century.

It is fortunate that Hildegard did not live between 1300 and 1650. If she had been practising herbalism at that time, she probably would have been burned at the stake as a witch. Instead, she is venerated as a saint and "the light of her people and of her time."

Paracelsus (1493–1541)

Considered the father of modern chemistry, Paracelsus, or Philippus Theophrastus Aureolus Bombastus von Hohenheim, was a Swiss physician, alchemist, botanist, and philosopher.

Born in Einsiedeln, Switzerland, he began studying medicine at the University of Basel at the age of sixteen.

Part V: *For Reference*

It is believed that he received his doctorate degree from the University of Ferrara, in Italy, and then became an itinerant physician (and an occasional journeyman miner) whose wanderings took him through Germany, France, Spain, the Netherlands, Denmark, Sweden, and Russia. He is also reported to have travelled as far as China, Egypt, and Turkey in search of knowledge.

Like many university-trained doctors of his time, Paracelsus was a practising astrologer, and he concerned himself with finding astrological talismans for curing disease. He is also credited for being the first western physician to use chemicals and minerals in medicine and for inventing laudanum, an opium tincture used for treating pain. During his lifetime, he did much to popularize the Doctrine of Signatures in its medical application.

Unfortunately Paracelsus, whose name means "equal to or greater than Celsus," in reference to the famous Roman encyclopedist, was also known to have been rather arrogant. This quality put him at odds with many of his contemporaries, particularly as he was also known for challenging many of the medical tenets of the day—once going so far as to burn traditional medical texts during what became a short-lived stint as chair of medicine at the University of Basel.

John Gerard (1545–1611)

Herbalist and surgeon John Gerard was a famous gardener and author of one of the most well-known herbals ever written.

Born in Nantwich, England, he became a Barber-Surgeon's apprentice in London in 1562. It was during his studies that he first developed a passion for plants and started his own medicinal herb garden. After completing his apprenticeship, and as a means of expanding his plant collection, Gerard became a ship's surgeon, collecting medicinal plants and seeds from ports he visited in Denmark, Sweden, and Russia.

Back in England, his garden became quite famous, and he was asked to oversee several illustrious gardens of the time, including the "physic garden" at the College of Physicians, where medical students were trained in the medicinal properties of plants and herbs. He also oversaw the creation of a botanical garden, also for training physicians, at Cambridge University.

Gerard is most well known for publishing *The Herball or Generall Historie of Plantes* in 1597. In 1633, *The Herball* was revised and expanded by botanist Thomas Johnson. The revised edition, which was republished in 1975, contains descriptions of 2,850 plants and about 2,700 illustrations. Selections and adaptations of Gerard's work continue to be published today.

Nicholas Culpeper (1616–1654)

The English herbalist, physician, botanist, political radical, and astrologer Nicholas Culpeper spent much of his life outdoors cataloguing medicinal herbs. Like Pliny, he believed medicine should be readily available to the masses and lambasted physicians for charging such high fees for their medicines when healing plants were universally available.

Culpeper devoted himself to the sick and the poor. He opened an apothecary shop in London in 1644, where he translated scores of medical books from Latin into English. The results were twofold: he made the texts accessible to laypeople, while threatening the monopolies over medicine maintained by university-trained physicians.

In 1653, Culpeper published his celebrated *The English Physician Enlarged,* now known as *Culpeper's Complete Herbal.* It included descriptions of more than 350 common herbs of England and recipes for making traditional medicines. In *Culpeper's Complete Herbal,* he draws on the wisdom of the ancient astrological herbalists, who linked

Yukon biologist Bruce Bennett, who has a collection of 7,000 Yukon plants in his herbarium, owns this copy of John Gerard's *The Herbal*. A reprint of the 1633 edition, *The Herbal,* is 1,678 pages long and continues to be a rich source of botanical knowledge.

Herbalists Through The Ages

herbs to different signs of the zodiac and made recommendations based on what sign and planet ruled over the part of the body that was ailing. *Culpeper's Complete Herbal* sold as far afield as colonial America and has never gone out of print.

Carolus Linnaeus (1707–1778)

The Swedish botanist and explorer Carolus Linnaeus (also known as Carl Linnaeus) is most famous for his classification of all natural things. He adopted a system of binomial nomenclature for naming plants, animals, and minerals. His reasoning behind this system, which assigned two Latin names (genus and species) to each thing, is that if people learned the Latin names, they would not need to know the names in any other language. He was fond of saying, "God created, Linnaeus organized."

Linnaeus loved plants and studied them—and all life—with passion. He was a prolific author, penning more than 180 works throughout his illustrious career. In *Genera Plantarum* (1737), he outlined his system for classifying plants, which was based largely on the number of stamens and pistils of the flower. Though more than 230 years have passed since his death, Linnaeus's system has remained the basis for modern taxonomy.

In *Species Plantarum* (1753), he endeavored to name and characterize all known plants—in a time when explorers were constantly discovering new species. Many of his names for flowering plants survive with little or no change. The twinflower, one of Linnaeus's favourite plants, was given its Latin name in his honour. He took *Linnaea borealis* as his personal symbol when he was knighted in 1757.

After Linnaeus's death, his priceless herbarium—which purportedly contains 13,832 sheets of preserved plant specimens—was moved from Sweden to England, and can be found at London's Linnaean Society.

Carl Linnaeus's favourite flower, the twinflower, was named *Linnaea borealis* in his honour.

Part V: *For Reference*

THE CELTIC TREE ALPHABET

THE ANCIENT CELTS, like the aboriginal people of North America, believed that all things in nature were interconnected. Trees were venerated as sacred beings and identified as ancestors of human beings that were either inhabited by spirits or possessed their own. Different trees had different traits and magical energies including the power to heal.

Celtic priests (Druids) used the Celtic Tree Alphabet, or Ogham (pronounced o'um), a particular and secret method for communicating with one another and memorizing their abundant tree knowledge. The origins of Ogham have been lost in time, but legend says the alphabet was named after Ogma, the Celtic god of eloquence.

Ogham contains twenty main letters. The letters are subdivided into four groups of five letters each. Six more letters were added later. The original twenty are represented by lines (consonants) and notches (vowels). The six additional letters are represented by distinctive symbols. Each letter is named for a tree and is associated with a certain month of the Celtic calendar, as well as a particular colour, class, deity, planet, stone, and meaning.

The first letter "B," named for Birch, carries the meanings of "new beginnings; changes; purification." It's said the first message written in Ogham was seven Bs carved into a birch tree, a warning from Ogma to the deity Lugh that his "wife will be carried away seven times to the otherworld unless the birch protects her." Other trees listed in the alphabet include Willow ("S"), which means "gaining balance in your life"; and Aspen ("E"), meaning "problems; doubts; fears."

The Druids used the Ogham for divination. The letters were painted on or carved into flat pieces of wood that were selected and read much like Tarot cards. There is also mention of "omen sticks," sticks marked with runes and Ogham letters. Similar to the I Ching, the sticks were cast to the ground and predictions were made based on the way the sticks fell.

Celtic shamans believed that certain herbs held healing powers and they used the Ogham in their preparations. While concocting their remedies, healers drew the Ogham sign of the month in the air over the mixture. If that particular month's energies were not deemed appropriate, the healer chose another sign. With its energy of "new beginnings," Birch was a popular and effective choice.

Ogham fell into disuse after the first centuries of the Christian era in Britain but the letters and their meanings are still known, including letters for the Boreal trees below.

Beithe: Birch
Letter: B
Meaning: new beginnings; changes; purification.

Luis: Rowan
Letter: L
Meaning: controlling your life; protection against control by others.

Fearn: Alder
Letter: F, V
Meaning: help in making choices; spiritual guidance and protection.

Saille: Willow
Letter: S
Meaning: gaining balance in your life

Ailim: Silver fir
Letter: A
Meaning: learning from past mistakes; take care in choices.

Eadha: White poplar or aspen
Letter: E
Meaning: problems; doubts; fears.

BOTANY BASICS
A brief overview of key botanical terms

Types of Plants
Plants have adopted varying life strategies.

Annual plants complete their life cycle in a single growing season: from seed germination to seed production and dispersal.

Biennial plants complete their life cycle in two years: growing leaves and storing energy in their roots in the first season, then producing flowers and seeds in the following season.

Perennial plants persist for more than two years: producing seed when conditions are good; many perennials produce woody tissue.

Flowers
Plants such as dandelions and arnica produce showy flowers to attract pollinators: bees and butterflies. Plants such as birch trees that produce inconspicuous flowers rely on wind for pollen transfer. Flowers can contain the following general components:

Flowers can be radially symmetrical or bilaterally symmetrical.

When flowers contain both stamens and pistils, they are said to be, in botanical terms, perfect. If flowers are unisexual, bearing either stamens or pistils but not both, they are said to be imperfect. Individual plants that produce both male and female imperfect flowers are called monoecious. Species that produce male and female flowers on separate plants ("male plants" and "female plants") are called dioecious.

Flowers can be solitary on a peduncle (flowering stalk), or can be arranged in clusters of various shapes known as inflorescences.

Illustrations by Tanya Handley

Part V: *For Reference*

Inflorescences within the sunflower family (Asteraceae) are unique. These daisy-like flowers are in fact a compact cluster of many miniature flowers ("florets") that usually form a disc and rays. Together, the disc and rays form a capitulum. Disc florets are perfect flowers with inconspicuous, tubular petals. Together, they form the button-like centre of the inflorescence. The disc is usually bordered by a series of ray florets. Ray florets are imperfect (with pistils), with each having a modified, enlarged petal.

Note: See Botanical Terms (next page) for definitions.

Inflorescences in the sunflower family

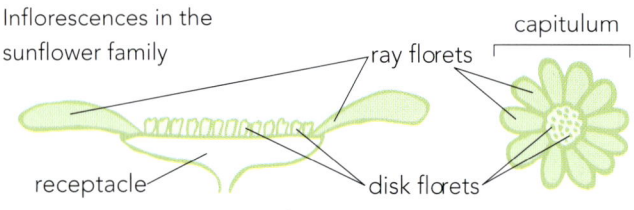

Leaves

Parts of a leaf:

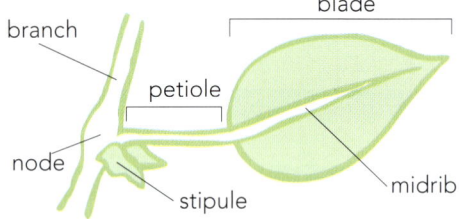

Leaf Placement: Leaves can be simple or compound. They can be opposite, alternate, whorled or a basal rosette on the stems.

Position Terms:

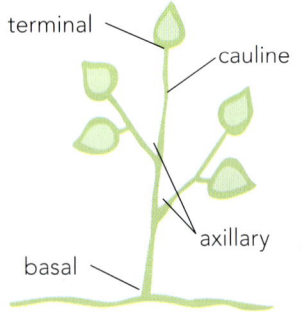

Leaf margins:

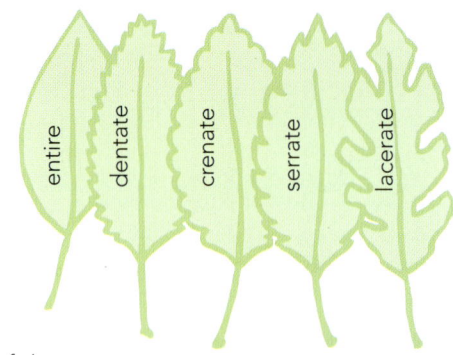

Leaf shapes:

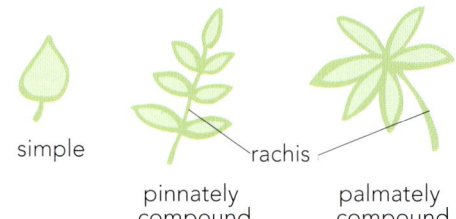

Leaf placement:

Botanical shapes (leaves, petals, fruit):

BOTANICAL TERMS

Achene: A dry single-seeded fruit that does not open to release its seed. Dandelions have achenes.

Aerial: The part of the plant growing above the ground.

Alternate: Leaf or flower arrangements along the stem in which the leaves are not opposite to each other or whorled. (See leaf placement illustration.)

Amphi-Atlantic: Native to both sides of the Atlantic Ocean, Europe, and North America.

Amphi-Beringian: Occurring on both sides of the Bering Strait, but lacking on the Atlantic side of the Earth.

Annual: Plants complete their life cycle in a single growing season or year, from seed germination to seed production and dispersal.

Anther: A male flower part forming the top part of a stamen and bearing the pollen in sacs. (See parts of a flower illustration.)

Arborescent: Treelike.

Ascending: (Of stems) growing obliquely upward; (of parts of a plant) directed obliquely forward with respect to the organ to which they are attached.

Astringent: When tasting tends to pucker the tissues of the mouth or when used topically causes contraction of tissues.

Awl-shaped: Leaf gradually tapering from the base to a slender or stiff point. (See botanical shapes illustrations.)

Axil: The angle between two structures such as a leaf and the stem from which it grows.

Axillary: Located in, or arising from, an axil. (See leaf position illustration.)

Axis: The main stem.

Basal: Located at the base of either a plant or leaves where attached to the plant. (See leaf position illustration.)

Bearded: Bearing long and/or stiff hairs.

Berry: A fruit developed from a single ovary, fleshy, pulpy or juicy throughout, containing one or many seeds; loosely, any pulpy or juicy fruit.

Biennial: Plants complete their life cycle in two growing seasons or years, growing leaves and storing energy in their roots in the first season, then producing flowers and seeds in the second.

Blade: The expanded terminal part of a flat organ, such as a leaf, petal, or sepal, in contrast to the narrowed basal portion.

Bract: A specialized leaf from the axil of which a flower arises; differing from foliage leaves in size, shape, or texture, but in some species gradually modified from them; sometimes applied to a specialized leaf that usually subtends a flower or an inflorescence.

Branchlet: The ultimate division of a branch.

Bud: A developing stem, leaf, or flower.

Bulb: A short, vertical, underground plant part that's purpose is food storage or reproduction on which specialized leaves are prominently developed.

Calyx: The outer series of floral leaves forming the perianth of a flower, often green, frequently enclosing the rest of the flower in bud, occasionally coloured or petal-like; in some groups of plants greatly reduced or completely lacking. (See parts of a flower illustration.)

Cambium: A layer, usually one cell thick, of persistent meristematic tissue. Meristem is the tissue in all plants consisting of undifferentiated cells, found in zones of the plant where growth can take place.

Capillary: Hair-like and slender.

Capsule: A dry fruit developed from a compound ovary that bursts open, releasing the seeds; almost always containing two or more seeds.

Catkin: A dense spike or raceme bearing many small inconspicuous or petal-free flowers. (See types of flowers illustration.)

Part V: *For Reference*

Cauline: Situated on, or pertaining to, the stem, used to describe leaf position. (See leaf position illustration.)

Chaff: The thin scales or bracts underneath individual flowers of many of the Asteraceae (Compositae) family.

Chlorophyll: The substance that gives plants their green colour.

Cilia: Marginal hairs.

Claw: The narrow basal portion of some sepals and petals.

Compound: Made up of two or more similar parts, as in a leaf which has leaflets. (See leaf shapes illustration.)

Cone: A globose to cylindrical arrangement of crowded bracts or scales subtending reproductive organs and usually hard or woody or long persistent; a structure of similar appearance although possibly of different morphological nature.

Contiguous: Connected.

Cordate: Heart-shaped; applied sometimes to whole organs but more often to only the base of leaves, petals and fruits. (See botanical shapes illustrations.)

Cordilleran: Pertaining to the mountains of western North America.

Corolla: The inner whorl of the perianth, between the calyx and the stamens; a collective term for the petals of a flower. (See parts of a flower illustration.)

Corymb: A broad, flat-topped inflorescence flower cluster in which the flower stalks arise from different points on the main stem and the marginal flowers are the first to open. (See types of flower clusters illustration.)

Creeping: Growing along the surface of the ground and emitting roots at intervals, usually from the nodes.

Crenate: Leaf with shallow roundish teeth on the margin, scalloped. (See leaf margin illustration.)

Crisp: Curled.

Cyme: A broad, flat-topped inflorescence flower cluster in which the central flower is usually the first to open. (See types of flower clusters illustration.)

Deciduous: Trees or shrubs shedding their leaves annually in the autumn.

Decumbent: Prostrate at base, erect or ascending elsewhere.

Dehiscent: The process or act of opening, usually in a fruit.

Deltoid: Broadly triangular leaf shape. (See botanical shapes illustrations.)

Dentate: Sharp, outward-pointing teeth on the leaf margin. (See leaf margins illustrations.)

Depressed: Flattened.

Diffuse: Loosely spreading.

Dioecious: Having male and female flowers on different plants of the same species.

Disc: The central portion of composite flowers, made up of a cluster of disc flowers. (See inflorescence in the sunflower family illustration.)

Divided: Cut into distinct parts, usually describing a leaf cut to the mid-rib or to the base.

Downy: Pubescent; covered with soft, fine hairs.

Drupe: A fleshy indehiscent fruit enclosing a nut or hard stone containing generally a single seed such as highbush cranberry.

Drupelet: A small drupe.

Embryo: The rudimentary plant formed in a seed.

Endemic: Confined geographically to a single area.

Entire: A leaf having a continuous, unbroken margin. (See leaf margin illustrations.)

Epidermis: The outer cell layer of a plant.

Erect: (Of whole plant) growing essentially in a vertical position; (of part of a plant) describing the position of a structure that extends in the same direction as the organ that bears it.

Evergreen: Remaining green throughout the winter.

Exfoliate: To come off in scales or flakes.

Botanical Terms

Fertile: Capable of normal reproduction functions; e.g., a fertile stamen produces pollen, a fertile pistil produces ovules, and a fertile flower normally produces fruit although it may lack stamens.

Fibrous: Resembling fibres.

Filament: The stalk that supports the pollen-bearing anther in the male reproductive organ stamen of a flower. (See parts of a flower illustration.)

Fleshy: Thick and juicy; succulent.

Floret: A small flower, usually one of several in a cluster. (See illustration.)

Floriferous: Flower bearing.

Follicle: A dry dehiscent fruit developed from a simple ovary, and dehiscent usually along one suture only.

Forked: Divided into equal branches.

Fruit: A ripened ovary together with such other parts of the plant as are regularly associated with it.

Glabrous: Smooth, without hairs.

Gland: A secreting organ in plants usually producing nectar or volatile oil, and either internal or external.

Glandular: Bearing glands that produce tiny globules of sticky or oily substance.

Glaucous: Grey, greyish green, or bluish green with a thin coat of fine, removable particles often waxy in nature.

Globose: Globe shaped; spherical.

Globular: Spherical. (See botanical shapes illustrations.)

Habitat: The kind of place in which a plant grows, such as bogs and woods.

Hair: An epidermal appendage, usually slender, simple, or variously branched.

Head: A dense flower cluster composed of sessile or nearly sessile flowers crowded on a short axis.

Herb: A plant—annual, biennial, or perennial—with stems dying back to the ground at the end of the growing season.

Herbaceous: (Of plant) dying back to the ground at the end of the growing season; (of part of a plant) leaf-like in colour or texture.

Hybrid: A cross between two species.

Indehiscent: Describes a fruit that does not open up to release seeds when ripe.

Indigenous: Native; not introduced.

Inflorescence: A complete flower cluster, including the axis and bracts. (See types of flower clusters illustrations.)

Innovation: An offshoot, usually from the base of a stem.

Internode: The portion of a stem between one node and the next.

Irregular: Describes a flower in which the members of one or more sets of organs (usually the corolla) differ among themselves in size, shape, or structure.

Involucre: A set of bracts subtending a flower or an inflorescence.

Keel: A sharp or conspicuous longitudinal ridge.

Lacerate: Torn; describing a flat organ, such as a leaf or petal, with an irregularly jagged margin. (See leaf margins illustrations.)

Lance-attenuate: Lanceolate leaf shape, with the tip tapering. (See botanical shapes illustrations.)

Lance-oblong: Lanceolate leaf shape and oblong. (See botanical shapes illustrations.)

Lanceolate: Leaf shaped like a lance head; much longer than wide and widest below the middle. (See botanical shapes illustrations.)

Lateral: Situated on, or arising from, the side of an organ, such as a lateral inflorescence.

Latex: The milky juice of some plants.

Leaflet: A single segment of a compound leaf.

Legume: A dry dehiscent fruit derived from a simple ovary and usually dehiscing along two sutures.

Linear: Narrow and elongate with essentially parallel sides.

Part V: *For Reference*

Lobe: A partial division of an organ, such as a leaf. The term generally applies to a division less than halfway to the base of the mid-rib.

Lunate: Crescent-shaped leaf. (See botanical shapes illustrations.)

Lustrous: Glossy and shiny.

Mealy: Covered with meal or with fine granules.

Median: Pertaining to the middle.

Mid-rib: The median or central rib of a leaf. (See parts of a leaf illustrations.)

Monoecious: Bearing both staminate and pistillate flowers but not perfect ones.

Mucilaginous: Slimy and moist.

Nerve: A prominent longitudinal vein of a leaf or other organ. The adjective **nerved** is often used as a suffix; e.g., three-nerved.

Nodal: Located at or pertaining to a node.

Node: A point on the stem from which leaves or branches arise, characterized internally by certain anatomical features. (See parts of a leaf illustrations.)

Nutlet: A small nut or one of the sections of the mature ovary of some members of the **Boraginaceae**, or **Lamiaceae** families.

Oblong: Describes a flat organ broader than linear but maintaining its width with little change for a considerable part of its length. Also describing a solid object, such as a fruit or seed, which is essentially cylindrical or prismatic and therefore appears oblong when viewed from the side. (See botanical shapes illustrations.)

Obovate: Leaves are reversed ovate, having the end of the leaf broader. (See botanical shapes illustrations.)

Obtuse: Blunt.

Opposite: (Of leaves, flowers, branches, etc.) situated diametrically opposite each other at the same node; situated directly in front of another organ; e.g., stamens opposite the petals. (See leaf placement illustration.)

Oval: Broadly elliptic. (See botanical shapes illustrations.)

Ovary: The basal, usually expanded portion of a pistil within which the ovules are borne. (See parts of a flower illustration.)

Ovate: Egg-shaped leaf, wider below the middle. (See botanical shapes illustrations.)

Ovoid: Egg-shaped, solid. (See botanical shapes illustrations.)

Ovule: A reproductive organ within the ovary in which the female cell is produced and which, after further development, becomes a seed.

Palmate: Radiating from a single point like the spreading fingers of an outstretched hand. (See leaf shapes illustrations.)

Panicle: Compound or branched inflorescence flower cluster of the racemose type; often applied to any compound inflorescence that is loosely branched and longer than thick. (See types of flower cluster illustrations.)

Pappus: An outgrowth of hairs, scales, or bristles from the summit of the achene of many species of Asteraceae family.

Parasite: A plant that derives its food and water wholly or chiefly from another plant to which it is attached.

Pedicel: The stalk of a single flower in an inflorescence.

Perennial: Plants that persist for more than two years, producing seed when conditions are good; many perennials produce woody tissue.

Perianth: The outer structure of a flower, made up of the corolla, the calyx, or both.

Persistent: A plant part such as a scale on a pine cone that lasts beyond maturity without falling off.

Perfect: Describes a flower containing both stamens and pistils.

Petal: A separate segment of the corolla. (See parts of a flower illustration.)

Petiole: The basal stalk-like portion of an ordinary leaf, in contrast with the expanded blade. (See parts of a leaf illustration.)

Petiolate: With a petiole.

Botanical Terms

Pinnate: Having branches or lobes or leaflets or veins arranged on two sides of a rachis or stem. (See leaf shapes illustrations.)

Pistil: The innermost or central organ or organs of a flower, composed typically of ovary, style, and stigma. (See parts of a flower illustration.)

Pistillate: Having a pistil; usually applied to flowers lacking stamens.

Pod: Strictly the fruit of a legume; loosely often a synonym of capsule.

Pollen: Spores borne within the anther, which produce the male cells.

Pollinate: To transfer pollen from a stamen to a stigma.

Pollination: The act or process of pollinating.

Procumbent: Prostrate or trailing, but not rooting at the nodes.

Prostrate: Flat on the ground.

Pubescent: Bearing hairs on the surface.

Pyramidal: Pyramid shaped.

Raceme: A common type of inflorescence having an elongated, unbranched axis and lateral flowers, with the lowest opening first. (See types of flower clusters illustrations.)

Racemose: A general type of inflorescence in which all flowers are axillary and lateral; the axis therefore theoretically capable of indefinite prolongation.

Rachis: The main stem of a flower cluster or a compound leaf. (See leaf shapes illustrations.)

Radiant, radiate: Spreading from, or arranged around, a common centre.

Radical: Belonging to the root.

Ray: The strap-like marginal flower in Asteraceae family. (See illustration.)

Recumbent: Leaning or reposing upon the ground.

Reflexed: Bent backward.

Regular: Describing a flower in which the members of each circle of parts are similar in size and shape.

Resinous: Having resin.

Rhizome: An underground stem, usually horizontal in direction, usually emitting roots from the lower side and leafy stems from the upper.

Rib: A primary and prominent vein of a leaf. (See parts of a leaf illustration.)

Riparian: Growing by rivers or streams.

Rosette: A cluster of leaves in a circular arrangement at the base of a plant, often called the basal rosette. (See leaf placement illustrations.)

Runner: An elongated, slender, prostrate branch or stem that grows roots from nodes or the tip.

Sagittate: Arrow-shaped leaf; lanceolate or triangular in outline with two basal lobes turned back or down. (See botanical shapes illustrations.)

Scale: A small, thin, or flat structure on a plant or other outgrowth on a plant surface.

Seed: A ripened ovule.

Sepal: One separate segment of the calyx. (See parts of a flower illustration.)

Seriate: Arranged in a series of rows, either transverse or longitudinal.

Serrate: Leaf margin toothed with the apex of each tooth sharp. (See leaf margins illustrations.)

Sessile: Leaves or flowers attached directly and without a petiole, pedicel or other type of stalk.

Sheath: An organ wholly or partly surrounding another organ at the base, such as the sheathing leaf of a grass.

Simple: A leaf that has one part, not subdivided into leaflets. (See leaf shapes illustrations.)

Spatulate: Leaf shaped like a spatula, maintaining its width or somewhat broadened toward the rounded summit. (See botanical shapes illustrations.)

Spike: An elongate inflorescence of the racemose type with sessile or subsessile flowers; the term is often loosely applied to an inflorescence of different morphological nature but of similar superficial appearance. (See types of flower clusters illustration.)

Part V: *For Reference*

Spikelet: A small or secondary spike subtended by a common pair of bracts, as in grasses.

Spine: A sharp-pointed rigid structure, usually a highly modified leaf or stipule.

Spiral: Describing the arrangement of like organs, such as leaves, at regular angular intervals.

Sporadic: Occurring here and there without continuous range.

Spore: A one-celled asexual reproductive organ. The term is used almost exclusively in flowerless plants.

Stamen: A member of the third set of floral organs, typically composed of anther and filament. (See parts of a flower illustration.)

Stellate: Star-like, with radiating branches and often referring to the pattern of hairs on the surface of a leaf. (See botanical shapes illustrations.)

Stem: The main upward-growing axis of a plant that bears the leaves and flowers.

Sterile: Not producing seeds or fruit.

Stigma: The terminal portion of a pistil, which receives the pollen. (See parts of a flower illustration.)

Stipule: An appendage at the base of a petiole, usually in pairs. (See parts of a leaf illustration.)

Stolon: A horizontal branch arising at or near the base of a plant and taking root and developing new plants at the nodes or apex.

Style: The attenuated part of a pistil connecting the stigma to the ovary. (See parts of a flower illustration.)

Sub: A prefix to many adjectives, meaning more or less, or somewhat.

Subtending: To lie underneath something so as to surround or enclose it.

Succulent: Fleshy, thickened and juicy.

Suture: The seam (or line of opening) of a pod or capsule.

Talus: Rock debris at the base of a cliff or slope.

Terminal: At the end of the branch or stem. (See position terms illustration.)

Terrestrial: Growing in the soil as distinct from growing in water or in other habitats.

Trailing: Prostrate but not rooting.

Tuber: A thickened portion of a rhizome or root, serving for food storage and often also for propagation.

Umbel: A racemose type of inflorescence with greatly abbreviated axis and elongate pedicels. In a compound umbel the branches are again umbellately branched at the summit. (See types of flower clusters illustrations.)

Umbellate: Arranged in umbels.

Unisexual: Bearing either stamens or pistils.

Vein: Any of the vascular bundles externally visible, such as in a leaf.

Whorl: A circle of three or more leaves or branches or pedicels arising from one node. (See leaf placement illustration.)

Wing: A thin, paperlike flat margin bordering or extending from a seed capsule, stem or flower.

Winter bud: A shortened and crowded, hibernating vegetative shoot.

Wintergreen: Remaining green throughout the winter.

UNDERSTANDING HERBAL CONSTITUENTS AND PHYTOCHEMISTRY

by Robert Rogers

> As the soul of a human being can never be understood from its chemistry or grammar, so cannot plant purpose, intelligence, or soul. Plants are much more than the sum of their parts.
> —Stephen Harrod Buhner

THE FOLLOWING SUMMARIES of plant constituents explain how they contribute to human health.

Sugars, fatty acids, proteins, and sterols are involved in the primary metabolism of plants. They are the basic building blocks of life. By and large, the medicinally important parts of plants are their secondary metabolites, which include phenols, terpenes, and nitrogen-derived compounds.

Anthraquinones, found in rhubarb root, have laxative activity, while **hypericin** and **pseudohypericin** from St. John's Wort oil contribute to the plant's bright red colour. **Emodin** is found in various Rumex species and contributes to laxative activity, but also to anti-cancer activity.

Coumarins give a vanilla-like scent to bedstraw, sweet grass, and sweet clover. For example, **Melitoside**, from sweet clover, reduces edema, varicose inflammation, and lymphatic congestion. **Xanthotoxin**, a furanocoumarin produced by cow parsnip, can cause phototoxicity.

Glycosides are glycoside-yielding glucoses. They are secondary metabolites that upon hydrolysis yield sugars. Glycosides are poorly digested and, when consumed by healthy bacteria in the colon, are absorbed into the bloodstream. The benefit of medicinal plants is often dependent upon healthy intestinal flora for breakdown, absorption, and use in the body.

Cyanogenic glycosides are found in chokecherry and pin cherry bark, as well as mountain ash pits. They are poisonous. However, after the bark is dried and reconstituted, hydrolysis in the digestive system gives a slow release, is detoxified and removed from the body, while giving respiratory relief in irritated conditions.

Bitter glycosides found in gentian root help promote digestion and are antidepressive.

Bitter compounds are also found in wormwood leaf and dandelion root.

Ketones are found in Artemisia species and are neurotoxic in large doses. Alcohol, such as menthol found in the mint family, is a major constituent in many essential oils. **Methone**, a ketone, is a major constituent of wild mint and thus makes it unsuitable to consume during pregnancy.

Phenols are basic building blocks. Phenolic acids, which include benzoic and cinnamic derivates, form salicin, a phenolic glycoside found in poplar, alder and willow bark. Rosmarinic acid is a phenolic acid ester found in comfrey, and caffeic acid (found in dandelion root, yarrow, and tree resins) is antimicrobial, antioxidant, and analgesic. Usnic acid, a benzofuran found in *Usnea* species, is antibacterial. Arbutin, found in bearberry and blueberry leaf, is a urinary disinfectant and antiseptic. Benzoic acid is found in cloudberry and is a natural preservative.

Saponins are large, soap-like molecules. One end of the molecule is hydrophilic (water-loving) and the other end is lipophilic (fat-loving). Saponins are found in alfalfa, chickweed, buffaloberry, wild sarsaparilla, baneberry, and wild licorice. They influence steroid hormone production and modulate immune function. Sterols and their glucosides, found in nettle root, influence hormonal homeostasis.

Terpenes, including monoterpenes, diterpenes, and sesquiterpenes, compose the majority of constituents in essential oils. **Monoterpenes** such as pinene, from the pine family, are antibacterial, antifungal, and support adrenal health. **Sesquiterpenes**, from Labrador tea, aid liver and kidney detoxification. Volatile or essential oils are important

Part V: *For Reference*

components of many medicinal herbs; care should be taken to preserve and retain them when possible.

Flavonoids such as quercitin are common in foods and medicinal herbs such as goldenrod, and possess antihistamine, anti-inflammatory, anti-fatigue, and antidepressant activity.

Anthocyanins are red-to-blue pigments found in blueberries and Saskatoon berries, and have antioxidant and anti-inflammatory benefits. Rosehips contain bioflavonoids that offer immune-enhancing and anti-inflammatory support.

Isoflavonoids, found in red and white clover, and in wild licorice root, have affinity for estrogen receptors and influence excess or deficient hormonal conditions in both female and male bodies.

Tannins precipitate protein, and thus were long used to cure animal hides by First Nation peoples. The tannins react and cross link proteins on mucus membranes, and by astringing or tightening tissue, making it less permeable. In diarrhea, the tannins create a protective level of precipitated protein on the mucosa of the intestinal wall, numbing nerve response and reducing peristalsis. They also inhibit bacterial growth, reduce excessive secretions, and neutralize inflammatory proteins.

They are irritating, poorly absorbed, and are either condensed or hydrolysable. The latter are found in bearberry leaves, while related **ellagitannins** are found in geranium species. Ellagic acid is plentiful in raspberries and strawberries, and possesses anti-cancer properties. Condensed tannins are found in pine bark, roseroot, fireweed, and the leaves of red raspberry. They aid in the treatment of diarrhea, sore throats, etc.

Alkaloids are compounds that contain nitrogen and are able to pass the blood-brain barrier, thus affecting the nervous system and neurotransmitters. Alkaloids in nature are often combined with salts. Morphine from opium poppies is a central nervous-system depressant, while caffeine is a stimulant. Alkaloids in poison hemlock and water hemlock cause progressive paralysis and death. Alkaloids in valerian promote relaxation and alleviate insomnia, while alkaloids from buffaloberry leaves are monoamine oxidase inhibitors.

Pyrrolizidine alkaloids found in coltsfoot may cause hepatotoxicity over time.

Gums, oleoresins, and balsams—such as spruce gum, pine pitch, or balsam poplar bud resin, often collected by bees to form propolis—are sticky, water-insoluble substances produced by plants in response to injury. These compounds, which harden with time, are nutritive, soothing, and disinfecting. Oleoresins, such as the turpentine from conifers, are insoluble in water.

Mucilages are hydrophilic polysaccharides that form gels and make herbs such as plantain and Iceland moss soothing and protective to skin and gastrointestinal tissue.

Carbohydrates are an assortment of simple and complex sugars. Beta-D-glucans are complex chains of polysaccharides found in oats and barley, as well as in medicinal mushrooms. They help regulate blood sugar and immune health.

Inulins, or linear fructans, are found in the root of chicory and dandelion. They are not easily digested except by healthy intestinal bacteria that in turn help keep our immune system healthy.

Arabinogalactans are water-soluble polysaccharides found in tamarack and larch inner bark. They have been well studied for their role in immune health.

The mineral content of plants helps alleviate various medical conditions. Organic **silica**, found in horsetail, helps repair and maintain connective tissue. **Chromium** in blueberry leaves supports blood-sugar regulation, and the **sodium-potassium** ratio of many herbs can be useful for issues of water retention and kidney/bladder health.

All this said, it would be too reductionist and biochemical to suggest that plants are healers based on their constituents. If we looked at the human body as a collection of minerals, vitamins, and chemicals, we would miss the personality, the spiritual nature, and other intangibles that comprise a living being.

Good herbalists understand the biochemistry of plants—they just don't let that get in the way of *knowing* the plants.

MEDICINAL ACTIONS OF PLANTS

See also Medicinal Actions Chart, page 416.

Adaptogen herbs' main purposes are to assist the body in adapting to stress. They help us increase our resistance to fatigue and depression, enhancing our mood and increasing energy. *Devil's club, rhodiola root.*

Alterative herbs, also known as blood cleansers, gradually restore the proper functions of the body and increase health and vitality. *Crampbark, juniper, bedstraw, yellow dock, Labrador tea, mint, nettle, plantain, red clover, shepherd's purse, poplar.*

Analgesic (or anodyne) herbs reduce pain and are either applied externally or taken internally. *Aspen, devil's club, gentian root, juniper, Labrador tea, mint, pine, poplar, spruce, valerian, willow, yarrow.*

Anesthetic (local) topically numbs an area to achieve pain relief. *Wild mint.* Anthelmintic or **vermafuge** herbs expel worms from the digestive system. Can be toxic in high doses. *Gentian root, wild onion, wild sage (wormwood).*

Antibacterial plants prevent the growth of bacteria. *Devil's club, juniper, mossberry branches, usnea. yarrow.*

Antibilious herbs help the body remove excess bile from the liver; used in cases of jaundice. *Dandelion, wild sage (wormwood).*

Anti-catarrhal herbs help break up and remove mucus, phlegm from the lungs, sinuses and/or other parts of the body. *Bearberry leaf, coltsfoot, eyebright, goldenrod, mint, yarrow.*

Anticoagulant is a substance that prevents coagulation; it stops blood from clotting. *Aspen, poplar.*

Anti-diabetic herbs lower glucose levels in the blood. *Rhodiola.*

Antidiarrheal herbs provide symptomatic relief of diarrhea. *Raspberry leaf, strawberry leaf.*

Antidepressant herbs help to lessen the symptoms of depression. *Rhodiola.*

Antiemetic herbs relieve nausea or vomiting. *Mint, raspberry leaves.*

Antifungal herbs help to treat fungal infections such as athlete's foot, ringworm and candida. *Dandelion, devil's club, mossberry branches, pine, spruce.*

Antihistamine herbs are used as treatment and prevention of the symptoms of allergies. *Mossberry, nettle, plantain.*

Anti-inflammatory plants bring down inflammation in or on the body. *Arnica, alder, aspen, bedstraw, birch, blueberry, wild chamomile, chickweed, cranberry, currants, fireweed, gentian, goldenrod, juniper, mint, plantain, poplar, rhodiola, Saskatoon berry, shepherd's purse, sorrel (sheep), tamarack larch, valerian, willow, yarrow.*

Antilithic herbs prevent the formation and removal of stones and gravel in the urinary system. *Bearberry leaves.*

Antimicrobial herbs help destroy or resist pathogenic micro-organisms. *Bearberry leaf, fireweed, juniper, mint, pine, plantain, rosehips, wild sage (wormwood), spruce, yellow dock.*

Antioxidant plants reduce the damage caused by oxidation, such as the harm caused by free radicals. *Bedstraw, blueberry, cloudberry, cranberry, currants, fir, horsetail, mossberry, raspberry, red clover, rhodiola, Saskatoon berry, sorrel (sheep), strawberry and leaf, willow.*

Antipyretic herbs help to reduce a fever. *Chickweed.*

Anti-rheumatic herbs help to reduce the pain and inflammation of rheumatoid arthritis. *Aspen, birch, chickweed, dandelion, juniper, mint, poplar.*

Part V: *For Reference*

Antiscorbutic plants cure and prevent scurvy because they are high in vitamin C. *Aspen, bearberry leaf, birch syrup, blueberry, chickweed, cloudberry, currants, cranberry, dandelion, dock roots, fir, fireweed, gentian root, highbush cranberry, horsetail, juniper, Labrador tea, lamb's quarters, mint, mossberry, mountain ash berries, nettle, wild onion, pine, plantain, poplar catkins, raspberry and leaf, red clover, rhubarb, rosehip, shepherd's purse, soapberry, sorrel, sorrel (sheep), spruce, strawberry and leaf, tamarack larch, willow catkins, yarrow.*

Antispasmodic herbs ease spasms or cramps. *Wild chamomile, coltsfoot, crampbark, dandelion, mint, nettle, raspberry leaf, red clover, rosehips, valerian.*

Antiseptic herbs are antimicrobial substances that are applied to the skin to reduce the possibility of infection. *Aspen, birch, cranberry, currants, fir, fireweed, gentian, goldenrod, juniper, mint, wild onion, pine, plantain, poplar, raspberry, spruce, strawberry and leaf, sorrel (sheep), willow, yarrow.*

Antiviral plants are used for preventing or treating viral infections. *Devil's club, Saskatoon berry.*

Aphrodisiac herbs are used to arouse sexual desire. The word aphrodisiac is derived from the name Aphrodite, the Greek goddess of sensuality and love. *Rhodiola.*

Aromatic herbs have a strong odour that stimulates the limbic and digestive systems. *Wild chamomile, juniper, mint, valerian, wild sage (wormwood).*

Astringent herbs generally contain tannins that help reduce excessive secretions and discharges as in the case of diarrhea. They cleanse the body and assist it in discharging unnecessary morbid matter and help to achieve density and firmness of tissue. *Alder, aspen, birch, cloudberry, cranberry, crampbark, currants, bearberry leaf, bedstraw, dandelion, eyebright, fireweed, horsetail, lamb's quarters, mountain ash, nettle, plantain, poplar, raspberry, red clover, river beauty, rose leaves, rosehips, Saskatoon berry, shepherd's purse, sorrel (mountain), sorrel (sheep), strawberry and leaf, sweetgrass, willow, yarrow, yellow dock.*

Bitter herbs stimulate the digestive system. *Aspen, bearberry leaf, wild chamomile, chickweed, dandelion, eyebright, gentian, mint, red clover, wild sage (wormwood), poplar, willow, yarrow.*

Bacteriostatic herbs inhibit the growth of bacteria. *Dandelion.*

Cardiac-tonic herbs affect and strengthen the heart. *Raspberry leaf, rhodiola.*

Carminative herbs help relax the stomach, thereby supporting digestion and help against bloating and gas in the digestive tract. *Wild chamomile, goldenrod, juniper, mint, valerian, yarrow.*

Cathartic herbs can purge the bowels. *Devil's club, soapberry.*

Cholagogue herbs stimulate the secretion of bile from the gall bladder. They also have a laxative effect on the digestive system. *Aspen, dandelion, gentian, poplar, red clover.*

Demulcent herbs soothe and protect irritated or inflamed tissue. *Bearberry leaf, chickweed, coltsfoot, lungwort, plantain.*

Diaphoretic herbs promote perspiration that in turn helps the skin eliminate toxins. *Birch, chamomile, coltsfoot, dandelion, devil's club, fir, goldenrod, juniper, Labrador tea, mint, yarrow.*

Digestive herbs promote digestion. *Currants, juniper, wild onion, river beauty, yarrow.*

Disinfectant herbs kill microorganisms that might carry disease. *Juniper.*

Diuretic herbs increase the output of urine. *Alder, aspen, bearberry leaf, bedstraw, birch, chickweed, coltsfoot, cranberry, crampbark, currants, dandelion, fir, fireweed, goldenrod, horsetail, juniper, Labrador tea, nettle, plantain, poplar, red clover, rosehips, shepherd's purse, sorrel (sheep), willow, yarrow, yellow dock.*

Emetic herbs cause vomiting in high doses. *Alder, devil's club, fir.*

Medicinal Actions of Plants

Emmenagogue herbs both stimulate and balance menstrual flow. *Gentian, juniper berry, mint, raspberry leaf, wild sage (wormwood), yarrow.*

Emollient herbs are applied to the skin to soften, soothe or protect it from the elements. *Chickweed, coltsfoot, plantain, rose flowers.*

Enuresis herbs help to stop bedwetting. *Bearberry leaves.*

Expectorant herbs help remove excess amounts of mucus from the respiratory system. *Coltsfoot, devil's club, fir, gentian, lungwort, plantain, poplar, red clover, yarrow.*

Febrifuge or **Antipyretic** herbs help bring down fevers. *Aspen, chickweed, gentian, mint, poplar, raspberry leaf, yarrow.*

Galactagogue herbs help promote and increase the flow of breast milk. *Dandelion, mint, nettle, raspberry leaf, red clover, twinflower.*

Haemostatic herbs stop bleeding and are also considered as styptics. *Alder, birch, mountain ash, nettle, shepherd's purse, soapberry.*

Hepatic herbs tone and strengthen the liver to increase the flow of bile. *Bedstraw, dandelion, sorrel (sheep), wild sage (wormwood), yarrow, yellow dock root.*

Hypnotic herbs induce sleep. *Wild chamomile, valerian.*

Hypotensive herbs reduce blood pressure. *Wild onion, valerian, yarrow.*

Immune-stimulant herbs stimulate the immune system to help fight disease. *Plantain, rhodiola, tamarack larch.*

Laxative herbs promote evacuation of the bowels. *Bedstraw, chickweed, currants, dandelion, devil's club, fir, soapberry, sorrel (sheep), yellow dock root.*

Mucolytic herbs act as an expectorant to dissolve thick mucus and help relieve respiratory difficulties. *Fir.*

Nervine herbs tone and strengthen the nerves. Some act as stimulants, some relaxants. *Arnica (topically or homeopathically), crampbark, wild chamomile, mint, red clover, valerian.*

Pectoral herbs have a general strengthening and healing effect on the respiratory system. *Coltsfoot, fir bark, Labrador tea, lungwort, nettle, plantain, poplar buds.*

Refrigerant herbs lower the body temperature. *Cranberry, gentian, mint, raspberry.*

Rubefacient herbs are applied to the skin to increase circulation. *Fir, juniper, stinging nettle.*

Sedative herbs calm the nervous system and reduce stress. *Aspen, coltsfoot, crampbark, wild chamomile, poplar, red clover, valerian.*

Stimulant herbs enliven and stimulate the physiological functions of the body. *Fir, juniper, mint, raspberry and leaf, wild sage (wormwood), valerian, yarrow.*

Stomachic herbs relieve stomach aches, help heal ulcers and eliminate heartburn. *Wild chamomile, chickweed, dandelion, devil's club, gentian, mint.*

Styptic herbs reduce or stop external bleeding. *Bearberry leaf, eyebright, goldenrod, horsetail, lungwort, plantain, raspberry leaf, shepherd's purse, yarrow.*

Tonic herbs strengthen and tone either a specific organ or the whole body. *Aspen, bearberry leaf, bedstraw, birch, coltsfoot, cranberry, dandelion, eyebright, fir, gentian, juniper, Labrador tea, nettle, poplar, red clover, raspberry, strawberry and leaf, valerian, wild sage (wormwood), willow, yarrow, yellow dock root.*

Urinary antiseptic herbs inhibit the growth and development of unbeneficial microorganisms in the urinary system. *Bearberry leaf, birch, cranberries, juniper.*

Uterine-stimulant herbs stimulate the uterus. *Juniper, shepherd's purse.*

Vulnerary herbs are applied externally and help heal cuts and wounds. *Arnica (do not use on broken skin), bedstraw, chickweed, horsetail, moss, plantain, poplar buds, wild sage, shepherd's purse, willow, yarrow.*

Part V: *For Reference*

HERBAL USAGE CHART

This chart allows you to quickly reference which plants can be used to treat a condition (i.e., allergies) or have a desirable quality (i.e., antioxidant). You can also use the chart to see what the healing qualities are of the different plants. Always remember to consult the herb's plant profile for any contraindications before use and to learn what part of the plant should be used (i.e., rose petals or rosehips).

USAGE / HERB	ALDER	ARNICA	ASPEN	BEARBERRY	BEDSTRAW	BIRCH	BLUEBERRY	CHICKWEED	CLOUDBERRY	COLTSFOOT	CRANBERRY	CURRANTS	DANDELION	DEVIL'S CLUB	DOCK	FIR	FIREWEED	GENTIAN	GOLDENROD	HIGHBUSH CRANBERRY	HORSETAIL	JUNIPER	LABRADOR TEA	LAMB'S QUARTERS	LUNGWORT
ACNE	✦					✦		✦					✦	✦		✦	✦	✦		✦		✦	✦		
ADDICTIONS																							✦		
ADRENAL NOURISHMENT																							✦		
ALLERGIES						✦		✦			✦								✦						
ALTITUDE SICKNESS																									
ANEMIA															✦						✦				
ANTIBIOTIC					✦																	✦	✦		
ANTI-INFLAMMATORY	✦	✦	✦		✦	✦	✦				✦	✦	✦			✦	✦	✦	✦		✦		✦	✦	
ANTIOXIDANT					✦		✦	✦	✦		✦	✦				✦	✦				✦		✦		
ANXIETY																	✦				✦				
APHRODISIAC																									
APPETITE STIMULANT													✦					✦							
ARTHRITIS	✦	✦	✦										✦		✦	✦			✦				✦		
ASTHMA										✦					✦										✦
BANDAGE								✦									✦				✦				✦
BEDWETTING				✦																					
BENIGN PROSTATIC HYPERPLASIA (BPH)		✦															✦		✦						
BLADDER INFECTION				✦		✦					✦	✦	✦								✦				
BLOOD CIRCULATION						✦																			
BLOOD CLEANSER					✦											✦					✦		✦	✦	
BLOOD PRESSURE—LOW						✦	✦												✦						
BLOOD PRESSURE—HIGH							✦						✦												
BOILS						✦		✦								✦	✦								
BREAST INFECTIONS								✦					✦												
BREASTFEEDING—CRACKED NIPPLES								✦																	
BREASTFEEDING TONIC													✦												

406 | The Boreal Herbal

Herbal Usage Chart

MINT	MOSSBERRY	MOUNTAIN SORREL	NETTLE	PLANTAIN	PINE	POPLAR	RASPBERRY	RED CLOVER	RHODIOLA	RIVER BEAUTY	ROSE	SASKATOON BERRY	SHEEP SORREL	SHEPHERD'S PURSE	SOAPBERRY	SPRUCE	STRAWBERRY	STRAWBERRY BLITE	SWEETGRASS	TAMARACK	TWINFLOWER	VALERIAN	WILD CHAMOMILE	WILD ONION	WILD SAGE	WILLOW	YARROW	
♦					♦		♦		♦						♦	♦	♦					♦			♦			ACNE
									♦																			ADDICTIONS
					♦				♦							♦												ADRENAL NOURISHMENT
	♦		♦	♦						♦																		ALLERGIES
									♦																			ALTITUDE SICKNESS
			♦				♦	♦			♦					♦												ANEMIA
																												ANTIBIOTIC
♦	♦		♦		♦		♦	♦	♦		♦	♦	♦	♦		♦	♦		♦	♦		♦	♦			♦	♦	ANTI-INFLAMMATORY
	♦	♦			♦	♦	♦	♦			♦	♦	♦			♦										♦		ANTIOXIDANT
																						♦	♦					ANXIETY
									♦																			APHRODISIAC
																											♦	APPETITE STIMULANT
♦	♦		♦			♦		♦									♦			♦		♦	♦			♦		ARTHRITIS
			♦	♦																								ASTHMA
			♦								♦					♦										♦	♦	BANDAGE
																											♦	BEDWETTING
			♦																									BENIGN PROSTATIC HYPERPLASIA (BPH)
				♦							♦																♦	BLADDER INFECTION
							♦				♦														♦			BLOOD CIRCULATION
♦			♦	♦			♦				♦		♦	♦														BLOOD CLEANSER
																												BLOOD PRESSURE—LOW
														♦	♦												♦	BLOOD PRESSURE—HIGH
				♦													♦	♦							♦			BOILS
																												BREAST INFECTIONS
			♦																									BREASTFEEDING—CRACKED NIPPLES
							♦	♦								♦					♦			♦				BREASTFEEDING TONIC

Wild Food and Medicine Plants of the North

Part V: *For Reference*

USAGE	ALDER	ARNICA	ASPEN	BEARBERRY	BEDSTRAW	BIRCH	BLUEBERRY	CHICKWEED	CLOUDBERRY	COLTSFOOT	CRANBERRY	CURRANTS	DANDELION	DEVIL'S CLUB	DOCK	FIR	FIREWEED	GENTIAN	GOLDENROD	HIGHBUSH CRANBERRY	HORSETAIL	JUNIPER	LABRADOR TEA	LAMB'S QUARTERS	LUNGWORT
BROKEN BONES																					✦				✦
BRUISES		✦																							
BURNS			✦													✦	✦		✦		✦		✦		
BURSITIS																									
CANDIDA											✦									✦		✦	✦		
CANKERS								✦																	
CHICKEN POX	✦						✦																		
CHOLESTEROL							✦	✦	✦				✦												
CIRCULATION	✦	✦					✦															✦	✦		
COLDS AND FLUS			✦								✦			✦		✦							✦		
COLIC																									
CONSTIPATION					✦								✦		✦										
COUGHS			✦	✦				✦		✦						✦	✦	✦		✦	✦		✦	✦	✦
CRAMPING									✦		✦								✦						
CUTS			✦								✦					✦									✦
CYSTITIS			✦	✦	✦	✦													✦			✦			
DANDRUFF						✦															✦	✦			
DEODORANT					✦											✦					✦	✦			
DEPRESSION																			✦						
DERMATITIS								✦													✦				
DETOXIFICATION					✦								✦	✦	✦						✦		✦		
DIABETES MANAGEMENT							✦						✦	✦									✦		
DIARRHEA		✦	✦		✦	✦		✦											✦				✦	✦	✦
DIGESTION	✦		✦		✦		✦			✦		✦						✦	✦		✦				
DIURETIC	✦		✦	✦	✦		✦			✦	✦	✦	✦			✦			✦	✦	✦	✦	✦	✦	
DRY SKIN							✦									✦						✦			
EARACHE					✦																				
ECZEMA				✦	✦		✦		✦		✦	✦		✦		✦					✦				
EDEMA (SEE DIURETICS)					✦								✦												
ENDOMETRIOSIS					✦								✦						✦						
ENERGY INCREASE																							✦		
EXCESS MUCUS AND PHLEGM													✦		✦						✦				
EXHAUSTION																									

Herbal Usage Chart

MINT	MOSSBERRY	MOUNTAIN SORREL	NETTLE	PLANTAIN	PINE	POPLAR	RASPBERRY	RED CLOVER	RHODIOLA	RIVER BEAUTY	ROSE	SASKATOON BERRY	SHEEP SORREL	SHEPHERD'S PURSE	SOAPBERRY	SPRUCE	STRAWBERRY	STRAWBERRY BLITE	SWEETGRASS	TAMARACK	TWINFLOWER	VALERIAN	WILD CHAMOMILE	WILD ONION	WILD SAGE	WILLOW	YARROW	
				✦																								BROKEN BONES
				✦																								BRUISES
			✦	✦		✦					✦																	BURNS
																										✦		BURSITIS
	✦																						✦					CANDIDA
	✦			✦			✦						✦															CANKERS
			✦																							✦		CHICKEN POX
				✦				✦													✦							CHOLESTEROL
✦											✦														✦		✦	CIRCULATION
✦	✦		✦		✦		✦	✦								✦					✦	✦	✦	✦		✦	✦	COLDS AND FLUS
✦																							✦					COLIC
			✦										✦	✦														CONSTIPATION
✦			✦	✦	✦		✦									✦				✦							✦	COUGHS
					✦	✦					✦									✦	✦				✦			CRAMPING
			✦	✦	✦	✦					✦			✦	✦	✦				✦					✦	✦	✦	CUTS
				✦																								CYSTITIS
✦			✦																				✦					DANDRUFF
✦					✦																		✦		✦		✦	DEODORANT
						✦	✦																					DEPRESSION
										✦																		DERMATITIS
			✦				✦						✦			✦				✦			✦				✦	DETOXIFICATION
																												DIABETES MANAGEMENT
	✦	✦		✦		✦	✦				✦	✦				✦							✦	✦				DIARRHEA
✦							✦		✦							✦							✦	✦	✦		✦	DIGESTION
			✦	✦			✦				✦			✦	✦	✦				✦						✦	✦	DIURETIC
✦					✦		✦				✦						✦					✦					✦	DRY SKIN
																									✦			EARACHE
			✦	✦	✦		✦										✦	✦		✦								ECZEMA
																												EDEMA (SEE DIURETICS)
													✦							✦							✦	ENDOMETRIOSIS
						✦			✦				✦							✦								ENERGY INCREASE
✦			✦								✦																	EXCESS MUCUS AND PHLEGM
✦				✦				✦	✦														✦					EXHAUSTION

Part V: *For Reference*

USAGE / HERB	Alder	Arnica	Aspen	Bearberry	Bedstraw	Birch	Blueberry	Chickweed	Cloudberry	Coltsfoot	Cranberry	Currants	Dandelion	Devil's Club	Dock	Fir	Fireweed	Gentian	Goldenrod	Highbush Cranberry	Horsetail	Juniper	Labrador Tea	Lamb's Quarters	Lungwort
EXPEL WORMS																♦									
EYE FATIGUE							♦																		
EYE IRRITATIONS								♦			♦				♦										
FATIGUE														♦		♦			♦						
FEVER			♦					♦			♦					♦							♦		
FLATULENCE			♦															♦	♦			♦		♦	
FOOT ODOUR																♦									
FRACTURED BONES		♦																			♦				♦
GINGIVITIS											♦										♦				
GOUT						♦					♦	♦										♦			
GUM INFLAMMATION														♦											
HAIR RINSE																			♦		♦				
HAIR GROWTH STIMULANT						♦																			
HEADACHES			♦			♦					♦								♦		♦				
HEART HEALTH							♦				♦														
HEARTBURN			♦								♦		♦						♦						
HEMORRHAGES—BLOOD CLOTTER																♦									
HEMORRHOIDS	♦																								♦
HEPATITIS						♦							♦	♦											
HIGH CHOLESTEROL								♦	♦	♦															
HORMONE BALANCER														♦											
HORMONE BALANCER—HOT FLASHES																									
IMMUNE SYSTEM							♦				♦	♦					♦						♦	♦	
INDIGESTION			♦								♦		♦					♦	♦				♦		
INFECTION—TOPICAL			♦					♦	♦						♦	♦			♦				♦	♦	
INFLAMMATION		♦	♦			♦	♦	♦		♦				♦			♦		♦						
INSECT BITES AND STINGS								♦								♦	♦		♦			♦	♦	♦	
INSECT REPELLENT																							♦		
INTESTINAL FLORA REGULATION								♦																	
INTESTINAL PAIN																	♦								
IRRITABILITY																									
IRRITABLE BOWEL SYNDROME (IBS)																									

Herbal Usage Chart

Usage	MINT	MOSSBERRY	MOUNTAIN SORREL	NETTLE	PLANTAIN	PINE	POPLAR	RASPBERRY	RED CLOVER	RHODIOLA	RIVER BEAUTY	ROSE	SASKATOON BERRY	SHEEP SORREL	SHEPHERD'S PURSE	SOAPBERRY	SPRUCE	STRAWBERRY	STRAWBERRY BLITE	SWEETGRASS	TAMARACK	TWINFLOWER	VALERIAN	WILD CHAMOMILE	WILD ONION	WILD SAGE	WILLOW	YARROW
EXPEL WORMS												♦		♦										♦	♦	♦		
EYE FATIGUE		♦																						♦				
EYE IRRITATIONS				♦						♦											♦			♦				
FATIGUE		♦				♦				♦														♦				
FEVER	♦			♦		♦		♦																			♦	♦
FLATULENCE	♦							♦																♦	♦			♦
FOOT ODOUR	♦					♦											♦							♦		♦		
FRACTURED BONES					♦																						♦	
GINGIVITIS	♦										♦																	
GOUT				♦	♦			♦	♦									♦			♦			♦			♦	
GUM INFLAMMATION									♦													♦						♦
HAIR RINSE				♦										♦														
HAIR GROWTH STIMULANT																												
HEADACHES	♦							♦															♦				♦	
HEART HEALTH	♦							♦	♦			♦																
HEARTBURN	♦																							♦				♦
HEMORRHAGES—BLOOD CLOTTER															♦													♦
HEMORRHOIDS				♦				♦				♦								♦							♦	♦
HEPATITIS																												
HIGH CHOLESTEROL				♦				♦													♦							
HORMONE BALANCER			♦					♦						♦														♦
HORMONE BALANCER—HOT FLASHES	♦			♦				♦																				
IMMUNE SYSTEM											♦							♦			♦							♦
INDIGESTION	♦																							♦				♦
INFECTION—TOPICAL				♦									♦								♦				♦			♦
INFLAMMATION				♦	♦		♦	♦	♦					♦						♦	♦			♦			♦	♦
INSECT BITES AND STINGS	♦		♦	♦	♦				♦																	♦		♦
INSECT REPELLENT	♦																							♦				♦
INTESTINAL FLORA REGULATION																						♦						
INTESTINAL PAIN											♦												♦	♦				
IRRITABILITY			♦						♦														♦					
IRRITABLE BOWEL SYNDROME (IBS)				♦					♦																			

Wild Food and Medicine Plants of the North

Part V: *For Reference*

USAGE / HERB	ALDER	ARNICA	ASPEN	BEARBERRY	BEDSTRAW	BIRCH	BLUEBERRY	CHICKWEED	CLOUDBERRY	COLTSFOOT	CRANBERRY	CURRANTS	DANDELION	DEVIL'S CLUB	DOCK	FIR	FIREWEED	GENTIAN	GOLDENROD	HIGHBUSH CRANBERRY	HORSETAIL	JUNIPER	LABRADOR TEA	LAMB'S QUARTERS	LUNGWORT
IRRITATED SKIN								♦								♦	♦								♦
ITCHY SKIN	♦		♦		♦			♦								♦							♦		
JAUNDICE													♦		♦										
JOINT PAIN		♦												♦		♦						♦	♦		
KIDNEY INFECTION																									
KIDNEY IRRITATIONS								♦																	
KIDNEY STONES						♦					♦		♦								♦	♦			
LARYNGITIS										♦											♦				
LAXATIVE					♦			♦					♦	♦	♦	♦	♦						♦		
LICE														♦									♦		
LIVER CLEANSE													♦										♦		
LOWER BLOOD SUGAR							♦				♦											♦			
LUNG CONGESTION										♦						♦									♦
LYMPH HEALTH					♦										♦								♦		
MEASLES								♦									♦								
MENOPAUSE																									
MENSTRUAL HEALTH					♦				♦		♦				♦					♦	♦				
MENTAL ALERTNESS																									
MOOD ENHANCEMENT																									
MORNING SICKNESS																									
MOTION SICKNESS																									
MOUTH HEALTH					♦		♦			♦	♦	♦				♦	♦		♦					♦	
MUSCLE PAIN AND INFLAMMATION	♦	♦	♦		♦									♦			♦	♦			♦	♦	♦		
NAUSEA		♦									♦														
NERVOUSNESS														♦											
NOSEBLEEDS																									
OSTEOPOROSIS																					♦				
OVARIAN CYSTS								♦																	
PAIN RELIEF		♦	♦			♦					♦			♦		♦	♦		♦	♦		♦	♦		
PMS												♦	♦												
PNEUMONIA										♦															♦
POSTPARTUM BLEEDING																									
POSTPARTUM UTERINE CRAMPING																					♦				
PREMATURE AGING					♦		♦		♦				♦												

412 | The Boreal Herbal

Herbal Usage Chart

	MINT	MOSSBERRY	MOUNTAIN SORREL	NETTLE	PLANTAIN	PINE	POPLAR	RASPBERRY	RED CLOVER	RHODIOLA	RIVER BEAUTY	ROSE	SASKATOON BERRY	SHEEP SORREL	SHEPHERD'S PURSE	SOAPBERRY	SPRUCE	STRAWBERRY	STRAWBERRY BLITE	SWEETGRASS	TAMARACK	TWINFLOWER	VALERIAN	WILD CHAMOMILE	WILD ONION	WILD SAGE	WILLOW	YARROW
IRRITATED SKIN				✦	✦							✦												✦				
ITCHY SKIN	✦		✦	✦			✦					✦					✦							✦			✦	
JAUNDICE																												
JOINT PAIN	✦																							✦		✦	✦	
KIDNEY INFECTION					✦																							
KIDNEY IRRITATIONS				✦						✦						✦												
KIDNEY STONES																												
LARYNGITIS					✦																							
LAXATIVE				✦	✦									✦	✦						✦		✦					
LICE																												
LIVER CLEANSE				✦																								✦
LOWER BLOOD SUGAR																												
LUNG CONGESTION	✦			✦	✦		✦										✦											
LYMPH HEALTH							✦																					
MEASLES				✦								✦																✦
MENOPAUSE	✦			✦			✦					✦			✦													✦
MENSTRUAL HEALTH	✦		✦			✦	✦	✦		✦		✦	✦	✦	✦			✦				✦	✦		✦	✦		✦
MENTAL ALERTNESS	✦						✦										✦											
MOOD ENHANCEMENT							✦																					
MORNING SICKNESS								✦										✦										
MOTION SICKNESS	✦																											
MOUTH HEALTH				✦		✦	✦					✦					✦	✦	✦		✦							✦
MUSCLE PAIN AND INFLAMMATION	✦					✦	✦	✦									✦						✦	✦		✦	✦	✦
NAUSEA	✦							✦																✦				✦
NERVOUSNESS				✦		✦				✦							✦						✦	✦				
NOSEBLEEDS															✦													✦
OSTEOPOROSIS				✦				✦				✦																
OVARIAN CYSTS																												
PAIN RELIEF	✦	✦		✦		✦	✦					✦					✦						✦	✦	✦	✦	✦	✦
PMS				✦				✦						✦														✦
PNEUMONIA						✦											✦											
POSTPARTUM BLEEDING															✦													
POSTPARTUM UTERINE CRAMPING																												
PREMATURE AGING				✦						✦		✦																

Wild Food and Medicine Plants of the North

Part V: *For Reference*

USAGE \ HERB	ALDER	ARNICA	ASPEN	BEARBERRY	BEDSTRAW	BIRCH	BLUEBERRY	CHICKWEED	CLOUDBERRY	COLTSFOOT	CRANBERRY	CURRANTS	DANDELION	DEVIL'S CLUB	DOCK	FIR	FIREWEED	GENTIAN	GOLDENROD	HIGHBUSH CRANBERRY	HORSETAIL	JUNIPER	LABRADOR TEA	LAMB'S QUARTERS	LUNGWORT
PROSTATE HEALTH																	✦		✦			✦			
PROSTATITIS			✦																			✦			
PSORIASIS						✦		✦			✦				✦		✦								
RASHES			✦			✦		✦							✦									✦	
RESTLESS-LEG SYNDROME																				✦					
RHEUMATIC PAIN						✦		✦			✦	✦	✦		✦				✦	✦	✦	✦	✦		
RINGWORM								✦							✦										
ROSACEA								✦																	
SCABIES																						✦			
SCIATIC PAIN										✦									✦	✦		✦			
SEASONAL AFFECTED DISORDER (SAD)														✦											
SINUS CONGESTION						✦											✦					✦	✦		
SKIN HEALTH	✦		✦	✦	✦			✦					✦		✦	✦					✦		✦	✦	✦
SLEEP PROBLEMS																							✦		
SMOKING CESSATION																									
SORE THROAT			✦					✦		✦						✦	✦		✦	✦		✦			
SPRAINS		✦				✦		✦												✦	✦				✦
STOMACH HEALTH	✦			✦			✦	✦					✦	✦					✦	✦	✦		✦		
STOP BLEEDING	✦				✦														✦		✦				✦
STRESS								✦					✦	✦									✦		
SUNSCREEN				✦																					
TEETH — HEALTH			✦								✦						✦		✦	✦					
TEETHING																									
TENDONITIS																			✦			✦			
THROAT IRRITATIONS							✦		✦				✦	✦				✦				✦		✦	
THYROID BALANCER																				✦					
TOBACCO SUBSTITUTE				✦						✦													✦		
TUBERCULOSIS														✦		✦					✦				
ULCERS — STOMACH								✦		✦				✦							✦				
URINARY HEALTH	l		✦	✦		✦	✦	✦			✦		✦						✦		✦	l			
VAGINAL HEALTH				✦															✦	✦		✦			
VARICOSE VEINS																									
WEIGHT LOSS					✦			✦					✦												
WOUNDS	✦		✦			✦			✦		✦			✦	✦	✦			✦		✦	✦	✦	✦	✦

Herbal Usage Chart

	MINT	MOSSBERRY	MOUNTAIN SORREL	NETTLE	PLANTAIN	PINE	POPLAR	RASPBERRY	RED CLOVER	RHODIOLA	RIVER BEAUTY	ROSE	SASKATOON BERRY	SHEEP SORREL	SHEPHERD'S PURSE	SOAPBERRY	SPRUCE	STRAWBERRY	STRAWBERRY BLITE	SWEETGRASS	TAMARACK	TWINFLOWER	VALERIAN	WILD CHAMOMILE	WILD ONION	WILD SAGE	WILLOW	YARROW	
		♦		♦	♦			♦	♦								♦	♦											PROSTATE HEALTH
																													PROSTATITIS
				♦		♦		♦	♦									♦			♦								PSORIASIS
	♦		♦	♦	♦		♦			♦							♦				♦								RASHES
																													RESTLESS-LEG SYNDROME
	♦					♦	♦										♦							♦	♦	♦	♦		RHEUMATIC PAIN
					♦																								RINGWORM
																													ROSACEA
			♦																										SCABIES
																											♦		SCIATIC PAIN
										♦																			SEASONAL AFFECTED DISORDER (SAD)
					♦												♦											♦	SINUS CONGESTION
	♦			♦	♦	♦	♦		♦			♦					♦	♦							♦	♦		♦	SKIN HEALTH
				♦					♦														♦	♦					SLEEP PROBLEMS
				♦																									SMOKING CESSATION
							♦	♦	♦								♦			♦	♦	♦		♦		♦		♦	SORE THROAT
				♦													♦						♦	♦		♦			SPRAINS
	♦	♦		♦													♦							♦		♦	♦	♦	STOMACH HEALTH
				♦	♦			♦							♦													♦	STOP BLEEDING
	♦			♦				♦	♦			♦											♦	♦					STRESS
																													SUNSCREEN
				♦														♦						♦				♦	TEETH — HEALTH
																								♦					TEETHING
																	♦										♦		TENDONITIS
				♦								♦												♦					THROAT IRRITATIONS
																				♦									THYROID BALANCER
	♦																									♦			TOBACCO SUBSTITUTE
		♦				♦										♦													TUBERCULOSIS
	♦			♦													♦												ULCERS — STOMACH
			♦	♦																						♦			URINARY HEALTH
																													VAGINAL HEALTH
								♦				♦																	VARICOSE VEINS
																											♦		WEIGHT LOSS
				♦		♦	♦										♦	♦			♦					♦	♦	♦	WOUNDS

Wild Food and Medicine Plants of the North

Part V: *For Reference*

MEDICINAL ACTIONS CHART

This chart allows you to quickly reference which plants have specific medicinal actions. For an understanding of the medicinal actions, consult the Medicinal Actions of Plants glossary (page 403). Always consult the herb's plant profile for more detailed information, including any contraindications, before use.

ACTION / HERB	Alder	Arnica	Aspen	Bearberry	Bedstraw	Birch	Blueberry	Chickweed	Cloudberry	Coltsfoot	Cranberry	Currants	Dandelion	Devil's Club	Dock	Fir	Fireweed	Gentian	Goldenrod	Highbush Cranberry	Horsetail	Juniper	Labrador Tea	Lamb's Quarters	Lungwort
ADAPTOGEN														✦											
ALTERATIVE					✦										✦				✦			✦	✦		
ANALGESIC			✦												✦							✦	✦		
ANESTHETIC (LOCAL)																									
ANODYNE			✦																✦						
ANTHELMINTIC																			✦						
ANTIBACTERIAL															✦							✦			
ANTIBILIOUS													✦												
ANTI-CATARRHAL					✦					✦									✦						
ANTICOAGULANT		✦																							
ANTIDEPRESSANT																									
ANTI-DIABETIC																									
ANTIDIARRHEAL																									
ANTIEMETIC																									
ANTIFUNGAL													✦	✦											
ANTIHISTAMINE																									
ANTI-INFLAMMATORY	✦	✦	✦			✦	✦	✦	✦				✦	✦			✦	✦	✦			✦			
ANTILITHIC				✦																					
ANTIMICROBIAL				✦											✦			✦				✦			
ANTIOXIDANT						✦	✦		✦		✦	✦							✦		✦				
ANTIPYRETIC							✦																		
ANTI-RHEUMATIC		✦				✦	✦						✦									✦			
ANTISCORBUTIC																									
ANTISEPTIC				✦		✦							✦	✦			✦	✦	✦			✦			
ANTISPASMODIC								✦						✦							✦				
ANTIVIRAL													✦												
APHRODISIAC																									
AROMATIC																						✦			
ASTRINGENT				✦	✦	✦			✦		✦	✦				✦	✦				✦	✦		✦	
BACTERIOSTATIC													✦												
BITTER		✦	✦					✦					✦					✦							

Medicinal Actions Chart

Action	MINT	MOSSBERRY	MOUNTAIN SORREL	NETTLE	PLANTAIN	PINE	POPLAR	RASPBERRY	RED CLOVER	RHODIOLA	RIVER BEAUTY	ROSE	SASKATOON BERRY	SHEEP SORREL	SHEPHERD'S PURSE	SOAPBERRY	SPRUCE	STRAWBERRY	STRAWBERRY BLITE	SWEETGRASS	TAMARACK	TWINFLOWER	VALERIAN	WILD CHAMOMILE	WILD ONION	WILD SAGE	WILLOW	YARROW
ADAPTOGEN										♦																		
ALTERATIVE	♦			♦	♦		♦		♦						♦													
ANALGESIC	♦					♦	♦			♦													♦				♦	♦
ANESTHETIC (LOCAL)	♦																											
ANODYNE						♦																				♦		
ANTHELMINTIC																								♦	♦			
ANTIBACTERIAL		♦																										♦
ANTIBILIOUS																									♦			
ANTI-CATARRHAL																												♦
ANTICOAGULANT							♦																					
ANTIDEPRESSANT										♦																		
ANTI-DIABETIC										♦																		
ANTIDIARRHEAL								♦									♦											
ANTIEMETIC	♦							♦																				
ANTIFUNGAL		♦				♦								♦														
ANTIHISTAMINE		♦		♦	♦																							
ANTI-INFLAMMATORY	♦			♦		♦		♦		♦	♦	♦											♦	♦	♦		♦	♦
ANTILITHIC																												
ANTIMICROBIAL	♦			♦	♦					♦							♦									♦		
ANTIOXIDANT		♦				♦	♦	♦				♦	♦					♦									♦	
ANTIPYRETIC																												
ANTI-RHEUMATIC	♦						♦																					
ANTISCORBUTIC			♦											♦														
ANTISEPTIC	♦			♦		♦	♦	♦							♦		♦	♦							♦		♦	♦
ANTISPASMODIC	♦			♦				♦	♦						♦								♦	♦				
ANTIVIRAL												♦																
APHRODISIAC									♦																			
AROMATIC	♦																							♦	♦	♦		
ASTRINGENT			♦	♦		♦	♦	♦				♦	♦	♦	♦			♦		♦							♦	♦
BACTERIOSTATIC																												
BITTER	♦					♦	♦																	♦		♦		♦

Wild Food and Medicine Plants of the North

Part V: *For Reference*

ACTION	ALDER	ARNICA	ASPEN	BEARBERRY	BEDSTRAW	BIRCH	BLUEBERRY	CHICKWEED	CLOUDBERRY	COLTSFOOT	CRANBERRY	CURRANTS	DANDELION	DEVIL'S CLUB	DOCK	FIR	FIREWEED	GENTIAN	GOLDENROD	HIGHBUSH CRANBERRY	HORSETAIL	JUNIPER	LABRADOR TEA	LAMB'S QUARTERS	LUNGWORT
CARDIAC TONIC																									
CARMINATIVE																			✦			✦			
CATHARTIC														✦											
CHOLAGOGUE			✦										✦					✦							
DEMULCENT				✦				✦		✦															
DIAPHORETIC							✦			✦			✦	✦		✦			✦			✦	✦		
DIGESTIVE											✦	✦					✦					✦			
DISINFECTANT														✦								✦			
DIURETIC	✦		✦	✦	✦	✦		✦	✦	✦	✦	✦	✦		✦	✦	✦		✦	✦	✦	✦	✦		
EMETIC	✦													✦	✦										
EMMENAGOGUE																			✦			✦			
EMOLLIENT								✦		✦															
ENURESIS				✦																					
EXPECTORANT										✦				✦		✦			✦						✦
FEBRIFUGE OR ANTIPYRETIC			✦			✦												✦							
GALACTAGOGUE														✦											
HAEMOSTATIC	✦				✦														✦						
HEPATIC					✦								✦		✦										
HYPNOTIC																									
HYPOTENSIVE																									
IMMUNE STIMULANT																									
LAXATIVE						✦		✦					✦	✦	✦	✦									
MUCOLYTIC																✦									
NERVINE		✦																	✦						
PECTORAL										✦							✦						✦		✦
REFRIGERANT									✦										✦						
RUBEFACIENT																✦						✦			
SEDATIVE			✦							✦									✦						
STIMULANT		✦														✦						✦			
STOMACHIC								✦					✦	✦				✦							
STYPTIC					✦														✦		✦				✦
TONIC				✦	✦	✦					✦	✦	✦	✦		✦		✦				✦	✦		
URINARY ANTISEPTIC				✦		✦					✦											✦			
UTERINE STIMULANT																						✦			
VERMAFUGE																									
VULNERARY					✦			✦													✦				

Medicinal Actions Chart

	MINT	MOSSBERRY	MOUNTAIN SORREL	NETTLE	PLANTAIN	PINE	POPLAR	RASPBERRY	RED CLOVER	RHODIOLA	RIVER BEAUTY	ROSE	SASKATOON BERRY	SHEEP SORREL	SHEPHERD'S PURSE	SOAPBERRY	SPRUCE	STRAWBERRY	STRAWBERRY BLITE	SWEETGRASS	TAMARACK	TWINFLOWER	VALERIAN	WILD CHAMOMILE	WILD ONION	WILD SAGE	WILLOW	YARROW
CARDIAC TONIC								✦		✦																		
CARMINATIVE	✦																						✦	✦				✦
CATHARTIC																✦												
CHOLAGOGUE							✦	✦																				
DEMULCENT					✦																							
DIAPHORETIC	✦																							✦				✦
DIGESTIVE	✦										✦													✦				✦
DISINFECTANT						✦											✦											
DIURETIC				✦	✦		✦	✦				✦	✦	✦													✦	✦
EMETIC																												
EMMENAGOGUE	✦							✦																		✦		✦
EMOLLIENT					✦							✦																
ENURESIS																												
EXPECTORANT				✦	✦		✦	✦																				✦
FEBRIFUGE OR ANTIPYRETIC	✦						✦	✦																				✦
GALACTAGOGUE	✦			✦			✦	✦														✦						✦
HAEMOSTATIC				✦										✦	✦													✦
HEPATIC													✦													✦		✦
HYPNOTIC																							✦	✦				
HYPOTENSIVE																							✦		✦			✦
IMMUNE STIMULANT				✦		✦														✦								
LAXATIVE														✦		✦												
MUCOLYTIC																												
NERVINE	✦							✦															✦	✦				
PECTORAL				✦	✦	✦																						
REFRIGERANT	✦							✦																				
RUBEFACIENT				✦																								
SEDATIVE							✦	✦															✦	✦				
STIMULANT	✦					✦																	✦			✦		✦
STOMACHIC	✦																						✦					
STYPTIC					✦		✦								✦													✦
TONIC				✦				✦	✦									✦						✦		✦	✦	✦
URINARY ANTISEPTIC																												
UTERINE STIMULANT															✦													
VERMAFUGE																	✦									✦		
VULNERARY				✦	✦											✦										✦	✦	✦

Wild Food and Medicine Plants of the North

Part V: *For Reference*

VITAMIN CHART

There is not much information available on the vitamin and mineral content of wild boreal plants, so this list is not as comprehensive as it could be. Soil conditions also affect the nutrient content of plants.

HERB	ALDER	ARNICA	ASPEN	BEARBERRY	BEDSTRAW	BIRCH	BLUEBERRY	CHICKWEED	CLOUDBERRY	COLTSFOOT	CRANBERRY	CURRANTS	DANDELION	DEVIL'S CLUB	DOCK	FIR	FIREWEED	GENTIAN	GOLDENROD	HIGHBUSH CRANBERRY	HORSETAIL	JUNIPER	LABRADOR TEA	LAMB'S QUARTERS	LUNGWORT
VITAMINS																									
VITAMIN A				♦			♦	♦	♦		♦	♦	♦		♦		♦	♦					♦	♦	
BETA CAROTENE																								♦	
THE B VITAMINS												♦	♦												
VITAMIN B1 (THIAMIN)						♦						♦									♦	♦			
VITAMIN B12 (CYANOCOBALAMIN)																									
VITAMIN B2 (RIBOFLAVIN)						♦	♦					♦									♦	♦			
VITAMIN B3 (NIACIN)							♦	♦			♦	♦									♦	♦			
VITAMIN B5 (PANTOTHENIC ACID)							♦					♦													
VITAMIN B6							♦																		
VITAMIN B9 (FOLIC ACID)							♦																		
CHOLINE																									
INOSITOL																									
VITAMIN C (ASCORBIC ACID)		♦	♦			♦	♦	♦			♦	♦	♦			♦	♦	♦	♦		♦	♦	♦	♦	
VITAMIN D													♦												
VITAMIN E							♦																		
VITAMIN K							♦														♦				
MINERALS																									
ALUMINUM																									
BARIUM																									
CALCIUM				♦		♦	♦	♦			♦	♦	♦		♦						♦	♦		♦	
CHROMIUM																					♦	♦			
COBALT																					♦	♦			
COPPER						♦									♦										
IRON				♦		♦	♦	♦			♦	♦	♦		♦						♦	♦	♦	♦	
MAGNESIUM				♦		♦	♦	♦				♦	♦								♦	♦			
MANGANESE				♦		♦	♦	♦				♦	♦		♦						♦	♦			
PHOSPHORUS							♦	♦			♦	♦	♦		♦						♦	♦		♦	

Vitamin Chart

	MINT	MOSSBERRY	MOUNTAIN SORREL	NETTLE	PLANTAIN	PINE	POPLAR	RASPBERRY	RED CLOVER	RHODIOLA	RIVER BEAUTY	ROSE	SASKATOON BERRY	SHEEP SORREL	SHEPHERD'S PURSE	SOAPBERRY	SPRUCE	STRAWBERRY	STRAWBERRY BLITE	SWEETGRASS	TAMARACK	TWINFLOWER	VALERIAN	WILD CHAMOMILE	WILD ONION	WILD SAGE	WILLOW	YARROW	
VITAMINS																													
	♦		♦	♦	♦			♦	♦			♦		♦	♦			♦							♦	♦		♦	VITAMIN A
				♦		♦								♦	♦		♦												BETA CAROTENE
			♦	♦										♦															THE B VITAMINS
								♦	♦					♦	♦													♦	VITAMIN B1 (THIAMIN)
									♦																				VITAMIN B12 (CYANOCOBALAMIN)
								♦						♦	♦			♦										♦	VITAMIN B2 (RIBOFLAVIN)
				♦				♦	♦			♦		♦	♦													♦	VITAMIN B3 (NIACIN)
																		♦											VITAMIN B5 (PANTOTHENIC ACID)
																		♦											VITAMIN B6
																		♦											VITAMIN B9 (FOLIC ACID)
														♦	♦														CHOLINE
															♦														INOSITOL
	♦	♦		♦	♦	♦	♦	♦	♦		♦	♦		♦	♦	♦	♦	♦			♦				♦		♦	♦	VITAMIN C (ASCORBIC ACID)
				♦										♦															VITAMIN D
						♦		♦				♦		♦											♦				VITAMIN E
	♦			♦	♦			♦						♦	♦			♦										♦	VITAMIN K
MINERALS																													
													♦																ALUMINUM
													♦																BARIUM
	♦		♦					♦	♦				♦	♦	♦	♦		♦							♦			♦	CALCIUM
				♦				♦	♦																			♦	CHROMIUM
				♦				♦																				♦	COBALT
													♦	♦		♦								♦					COPPER
	♦		♦	♦				♦	♦			♦	♦	♦	♦	♦		♦						♦				♦	IRON
	♦			♦				♦	♦						♦									♦				♦	MAGNESIUM
				♦				♦	♦					♦	♦			♦										♦	MANGANESE
				♦				♦	♦					♦	♦										♦			♦	PHOSPHORUS

Part V: *For Reference*

HERB	ALDER	ARNICA	ASPEN	BEARBERRY	BEDSTRAW	BIRCH	BLUEBERRY	CHICKWEED	CLOUDBERRY	COLTSFOOT	CRANBERRY	CURRANTS	DANDELION	DEVIL'S CLUB	DOCK	FIR	FIREWEED	GENTIAN	GOLDENROD	HIGHBUSH CRANBERRY	HORSETAIL	JUNIPER	LABRADOR TEA	LAMB'S QUARTERS	LUNGWORT
POTASSIUM							♦	♦			♦	♦	♦		♦					♦		♦		♦	
SELENIUM													♦		♦						♦	♦			
SILICA																					♦				
SILICON													♦									♦			
SODIUM							♦															♦			
SULFUR																									
TIN																					♦	♦			
ZINC				♦		♦	♦	♦					♦						♦	♦					
OTHER																									
PROTEIN								♦			♦				♦							♦		♦	

Vitamin Chart

	MINT	MOSSBERRY	MOUNTAIN SORREL	NETTLE	PLANTAIN	PINE	POPLAR	RASPBERRY	RED CLOVER	RHODIOLA	RIVER BEAUTY	ROSE	SASKATOON BERRY	SHEEP SORREL	SHEPHERD'S PURSE	SOAPBERRY	SPRUCE	STRAWBERRY	STRAWBERRY BLITE	SWEETGRASS	TAMARACK	TWINFLOWER	VALERIAN	WILD CHAMOMILE	WILD ONION	WILD SAGE	WILLOW	YARROW	
				♦	♦			♦	♦			♦		♦	♦			♦							♦			♦	POTASSIUM
								♦	♦			♦													♦			♦	SELENIUM
				♦																									SILICA
								♦	♦																			♦	SILICON
				♦				♦						♦														♦	SODIUM
													♦												♦				SULFUR
								♦																				♦	TIN
	♦			♦				♦						♦											♦			♦	ZINC
																													OTHER
				♦				♦	♦			♦	♦															♦	PROTEIN

Part V: *For Reference*

BIBLIOGRAPHY

Andre, Alestine and Alan Fehr. *Gwich'in Ethnobotany: Plants Used by the Gwich'in for Food, Medicine, Shelter and Tools*. Inuvik, NWT: Gwich'in Social and Cultural Institute and Aurora Research Institute, 2001.

Andrews, Ted. *Nature-Speak: Signs, Omens & Messages in Nature*. Jackson, TN: Dragonhawk Publishing, 2004.

Anodea, Judith. *Eastern Body, Western Mind: Psychology and the Chakra System as a Path to the Self*. Berkeley, CA: Celestial Arts, 1996.

Barker, Cicely Mary. *Flower Fairies of the Wayside*. London, UK: Blackie & Son Ltd., 1948.

Black, Martha Louise. *Yukon Wild Flowers*. Vancouver, British Columbia: Price, Templeton Syndicate, 1940.

Blumenthal Mark. *The Complete German Commission E Monographs: Therapeutic Guide to Herbal Medicines*. Austin, TX: The American Botanical Council, 1998.

Blumenthal Mark. *The ABC Clinical Guide to Herbs*. Austin, TX: The American Botanical Council, 2003.

Boon, Heather and Michael Smith. *The Botanical Pharmacy: The Pharmacology of 47 Common Herbs*. Kingston, ON: Quarry Press, 1999.

Boreal Centre for Conservation Enterprise Society. *Boreal Forest Bounty: A Botanical Species Resource Guide for Conservation Enterprise Development*. Bloomington, IN: Trafford Publishing, 2007.

Boutenko, Victoria. *Green Smoothie Revolution: The Radical Leap Towards Natural Health*. Berkeley, CA: North Atlantic Books, 2009.

Bradley P.R., ed. *British Herbal Compendium: A Handbook of Scientific Information on Widely Used Plant Drugs, Volume 2*. Bournemouth, UK: British Herbal Medicine Association, 2006.

Catty, Suzanne. *Hydrosols: The Next Aromatherapy*. Rochester, NY: Healing Arts Press, 2001.

Cody, William J., ed. *Flora of the Yukon Territory*. National Research Council of Canada Monograph Publishing Program, Ottawa, ON: NRC Research Press, 1996.

Cooperative Extension Service, University of Fairbanks. *Collecting and Using Alaska's Wild Berries & Other Wild Products*. Fairbanks, AK: University of Alaska Fairbanks, 2007.

Council of Yukon First Nations. *Land of My Ancestors: Plants as food and medicine: Yukon First Nations perspective on our environment*. Dec 1993.

Council of Yukon First Nations. *Land of My Ancestors: Trees and Forests: Yukon First Nations perspective on our environment*. Dec 1993.

Crewe, Jodi and Jill Johnstone. *Plant Use in Vuntut Gwitchin Territory*. Old Crow, YT: Vuntut Gwitchin First Nation, 2008.

Englehart, Matthew and Terces. *Sacred Commerce: Business as a Path of Awakening*. Berkeley, CA: North Atlantic Books, 2008.

Germano, Carl and Zakir Ramazanov. *Arctic Root: The Powerful New Ginseng Alternative*. New York, NY: Kensington Publishing Corp., 1999.

Gladstar, Rosemary. *Herbal Recipes for Vibrant Health*. North Adams, MA: Story Publishing, 2008.

Gladstar, Rosemary. *Herbal Healing for Women*. New York, NY: Simon & Schuster, 1993.

Gladstar, Rosemary. *The Science and the Art of Herbalism: Sage Mountain home study course*. East Barre, VT.

Gordon, Lesley. *Green Magic—Flowers, Plants, Lore and Legend*. New York, NY: Viking Press, 1977.

Green, James. *The Herbal Medicine-Makers Handbook: A Home Manual*. Freedom, California: The Crossing Press, 2000.

Grieve M. *A Modern Herbal, Volume 1*. New York, NY: Dover Publications, 1971. [Reprint of 1931 Harcourt, Brace & Company publication]

Gurudas. *The Spiritual Properties of Herbs*. San Rafael, CA: Cassandra Press, 1988.

Hammond, Herb. *Maintaining Whole Systems on Earth's Crown: Ecosystem-Based Conservation Planning for the Boreal Forest*. Slocan Park, BC: Silva Forest Foundation, 2009.

Henry, J. David. *Canada's Boreal Forest*. Washington, DC: Smithsonian Institution Press, 2002.

Hobbs, Christopher. *Medicinal Mushrooms*. Loveland, CO: Botanica Press, 1986.

Hoffmann, David. *The Holistic Herbal: A Safe and Practical Guide to Making and Using Herbal Remedies*. Shaftesbury, Dorset: Element Book Limited, 1996.

Hoffmann, David. *Medical Herbalism: The Science and Practice of Herbal Medicine*. Rochester, NY: Healing Arts Press, 2003.

Bibliography

Johnson, D., L. Kershaw, A. MacKinnon and J. Pojar, eds. *Plants of the Western Boreal Forest and Aspen Parkland*. Calgary, AB: Lone Pine Publishing and the Canadian Forest Service, 1995.

Johnson, Steve. *The Essence of Healing: A Guide to the Alaskan Flower, Gem and Environmental Essences*. Homer, AK: Alaskan Flower Essence Project, 1996.

Kallas, John. *Edible Wild Plants: Wild Foods from Dirt to Plate*. Layton, UT: Gibbs Smith, 2010.

Kaminski, Patricia and Richard Katz. *Flower Essence Repertory: A Comprehensive Guide to North American and English Flower Essences for Emotional and Spiritual Well Being*. Nevada City, CA: The Flower Essence Society and Earth Spirit Inc. 1994 Edition.

Kaur, Sat Dharam, N.D. *The Complete Natural Medicine Guide to Breast Cancer: A Practical Manual for Understanding, Prevention, and Care*. Toronto, ON: Robert Rose Inc., 2005.

Lammers, Mickey. *Flowers of the Yukon*. Vancouver, BC: Evergreen Press Ltd., 1979.

MacKinnon, Andy, et al. *Edible & Medicinal Plants of Canada*. Edmonton, AB: Lone Pine Publishing, 2009.

Marles, Robin J., et al. *Aboriginal Plant Use in Canada's Northwest Boreal Forest*. Vancouver, BC: UBC Press, 2000.

Meyer, Joseph E. *The Herbalist*. Glenwood, IL: Meyerbooks, 1960.

McCutcheon, A. "Quality Control and Product Standards: an exploration of current issues in botanical quality, draft." Vancouver, 2002.

Montgomery, Pam. *Plant Spirit Healing: A Guide to Working with Plant Consciousness*. Rochester, Vermont, Bear & Company. 2008.

Moore, Michael. *Medicinal Plants of the Pacific Northwest*. Santa Fe, NM: Red Crane Books, 1993.

Pedersen, Mark. *Nutritional Herbology: A Reference Guide to Herbs*. Warsaw, IN: Wendell W. Whitman Company, 1987.

Pojar, J. and A. MacKinnon, eds. *Plants of Coastal British Columbia*. Calgary, AB: B.C. Ministry of Forests and Lone Pine Publishing, 1994.

Pollan, Michael. *The Botany of Desire: A Plant's-Eye View of the World*. Toronto, ON: Random House Publishing Group, 2001.

Rogers, Dale Robert. *Rogers' Herbal Manual*. Edmonton, AB: Karamat Wilderness Ways, 2000.

Rose, Jeanne. *Distillation—A how to book*. San Francisco, CA: Institute of Aromatic Studies, 2001.

Rose, Jeanne. *Herbs & Aromatherapy for the Reproductive System*. San Francisco, CA: Jeanne Rose Earth Medicine Books, 1994.

Schnaubelt, Kurt. *Medical Aromatherapy: Healing with Essential Oils*. Berkeley, CA: North Atlantic Books, 1999.

Schofield, Janice. *Discovering Wild Plants: Alaska, Western Canada, The Northwest*. Bothell, WA: Alaska Northwest Books, 1989.

Schroeder, Lori. *Reading Yukon Forests*. Whitehorse, YT: Yukon Conservation Society, 1997.

Scott, Julian. *Natural Medicine for Children*. New York, NY: Avon Books, 1990.

Sheppard-Hanger, Sylla. *The Aromatherapy Practitioner Reference Manual*. Tampa, FL: Atlantic Institute of Aromatherapy, 1995.

Sherry, Erin and Vuntut Gwitchin First Nation. *The Land Still Speaks: Gwich'in Words About Life in Dempster Country*. Old Crow, YT: Vuntut Gwitchin First Nation and Erin Sherry, 1999.

Small, Ernest and Paul Catling. *Canadian Medicinal Crops*. Ottawa, ON: NRC Research Press, 1999.

Tierra, Michael. *The Spirit of Herbs: A Guide to the Herbal Tarot*. Stamford, CT: U.S. Games Systems, Inc. 1993.

Tierra, Michael. *Planetary Herbology*. Twin Lakes, WI: Lotus Press, 1988.

Tigner, Daniel. *Canadian Forest Tree Essences: Vibrational Healing Through the Natural Resonance of Trees*. Ottawa, ON: Canadian Forest Tree Essences Inc., 1998.

Tompkins, Peter and Christopher Bird. *The Secret Life of Plants*. New York, NY: Harper & Row, 1989.

Turner, Nancy J. *Food Plants of Coastal First Peoples*. Vancouver, BC: UBC Press, 1995.

Turner, Nancy J. *Food Plants of Interior First Peoples*. Vancouver, BC: UBC Press, 1997.

Vuntut Gwitchin First Nation. *Plant Use in Vuntut Gwitchin Territorry*. Old Crow, YT: Vuntut Gwitchin First Nation, 2008.

Vuntut Gwitchin First Nation and Shirleen Smith. *People of the Lakes: Stories of Our Van Tat Gwich'in Elders*. Edmonton, AB: University of Alberta Press, 2009.

Walker, Marilyn. *Harvesting the Northern Wild*. Yellowknife, NWT: Outcrop Ltd., 1984.

Weed, Susun S. *Wise Woma Herbal for the Childbearing Year*. Woodstock, NY: Ash Tree Publishing, 1986.

Weed, Susun. *Healing Wise*. Woodstock, NY: Ash Tree Publishing, 1989.

Part V: *For Reference*

Weed, Susun. *Menopausal Years: The Wise Woman Way.* Woodstock, NY: Ash Tree Publishing, 1992.

Wildlife Viewing Program, Environment Yukon. *Common Yukon Roadside Flowers.* Whitehorse, YT: Government of Yukon, 2010.

Willard, Terry. *Edible and Medicinal Plants of the Rocky Mountains and Neighbouring Territories.* Calgary, AB: Wild Rose College, 1992.

Wood, Matthew. *The Book of Herbal Wisdom: Using Plants As Medicine.* Berkeley, CA: North Atlantic Books, 1997.

Wood, Matthew. *The Earthwise Herbal: A Complete Guide to New World Medicinal Plants.* Berkeley, CA: North Atlantic Books, 2009.

Wood, Matthew. *The Earthwise Herbal: A Complete Guide to Old World Medicinal Plants.* Berkeley, CA: North Atlantic Books, 2008.

Zevin, Igor Vilevich, et al. *A Russia Herbal.* Rochester, NY: Healing Arts Press, 1997.

Suggested Reading

Davis, Wade. *The Wayfinders: Why Ancient Wisdom Matters in the Modern World.* Toronto, ON: House of Anansi Press, 2009.

Fireweed Community Market Society. *Celebrate Yukon Food: Seasonal Recipes.* Whitehorse, YT: Fireweed Community Market Society, 2006.

Genest, Miche. *The Boreal Gourmet: Adventures in Northern Cooking.* Madeira Park, BC: Lost Moose, 2010.

MacKinnon, J.B. and Alisa Smith. *The 100-Mile Diet: A Year of Local Eating.* Toronto, ON: Vintage Canada, 2007.

Wildlife Viewing Program, Environment Yukon. *Common Yukon Roadside Flowers.* Whitehorse, YT: Yukon Government, 2010.

On-Line References

Natural Health Products Regulations, Part 3, "Good manufacturing Practices, valid from 1 January 2004," http://www.hc-sc.gc.ca/hpfb-dgpsa/nhpd-dpsn/gmp_e.html.

World Health Organization (WHO). 2004, "WHO Guidelines for Good Manufacturing and Collection Practices (GACP) for medicinal plants," http://www.who.int/medicines/library/trm/medicinalplants/agricultural.pdf.

Studies Referenced

Sun, Shi. Oplopanax horridus; anti-proliferative effect; human breast cancer MCF-7 cells; 43 non-small cell lung cancer (NSCLC) cells; apoptosis; cell cycle. *Phytotherapy Research.* Fitoterapia 81 (2010).

Research and development in the area of wild berries. University of Kuopio, www.kuopioinnovation.fi/upload/files/OSKE_A4_berrieshealth_final.pdf.

Publications and Websites on the Subject of Flower Essences

Canadian Forest Tree Essences, Vibrational Healing through the Natural Resonance of Trees by Daniel Tigner, or visit www.essences.ca

Dr. Bach has many books about the Bach flower essences. Visit the Bach Centre website, www.bachcentre.com

Findhorn Flower Essences by Marion Leigh, or visit the website at www.findhornessences.com

Flower Essence Repertory—A Comprehensive Guide to North American and English Flower Essences for Emotional and Spiritual Well Being by Patricia Kaminski and Richard Katz. Or visit the Flower Essence Society website at www.flowersociety.org

For more information on the different properties of each type of northern flower essence, visit the Alaskan Flower Essence website at www.alaskanessences.com

For more information on the Bailey Flower Essences visit their website at www.baileyessences.com

For more information on Woodland Essences visit their website at www.woodlandessence.com

Other Websites

This website is an "e-flora" for central Yukon. It's part of a project called Central Yukon Species Inventory Project (CYSIP). The larger CYSIP includes lichens, birds, mammals, and fungi as well. The goal of CYSIP is to document all the species endemic (growing wild) to Central Yukon. www.flora.dempstercountry.org

Tombstone Park Information www.environmentyukon.gov.yk.ca/parksconservation/tombstonepark

Northern Birch Syrup www.birchboy.com or www.yukonbirch.ca

The Boreal Gourmet: Adventures in Northern Cooking www.borealgourmet.com

Bibliography

Herbal Supplies
  www.aromaborealis.com
  www.richters.com
  www.clefdeschamps.net

Herbal Organizations
  Canadian Council of Herbalist Associations www.herbalccha.org

Canadian Herbalist's Association of BC
  www.chaofbc.ca

Rose Barlow's website
  www.rosesprodigalgarden.org

Ontario Herbalists Association (publishers of the *Canadian Journal of Herbalism*)
  www.herbalists.on.ca

America Herbalist Guild
  www.americanherbalistsguild.com

American Botanical Council (publishers of *HerbalGram*)
  www.abc.herbalgram.org

North American Institute of Medical Herbalism
  www.naimh.com

Canadian Herb, Spice and Natural Health Products Coalition www.saskherbspice.org/CHSNC

HerbMed
  www.herbmed.org

Herbal Education
  Dominion Herbal College www.dominionherbal.com

Wild Rose College of Natural Healing www.wrc.net

Boreal Forest Conservation Groups

Canadian Environmental Network
  www.cen-rce.org

Canadian Parks and Wilderness Association Yukon Chapter
  www.cpawsyukon.org

Canadian Boreal Initiative
  www.borealcanada.ca

Forest Ethics
  www.forestethics.org

WWF–Canada
  www.wwf.ca

Yukon Conservation Society
  www.yukonconservation.org

Silva Forest Foundation
  www.silvafor.org

Taiga Rescue Network
  www.taigarescue.org

Lakehead University Faculty of Natural Resources Management
  www.borealforest.org

David Suzuki Foundation
  www.davidsuzuki.org

Northern Agriculture

Growers of Organic Food Yukon (Goofy)
  www.organic-yukonfood.com

Yukon Agriculture Association
  www.yukonag.ca

Yukon Master Gardener Program
  www.yukoncollege.yk.ca

Part V: *For Reference*

INDEX

Aaron's rod 98
Abies lasiocarpa 247
abrasion: *See* skin: wound
abuse 93, 143, 258
Achillea borealis 179, 182
Achillea millefolium 178
Achillea millefolium ssp. *borealis* 179
acids
    acid-alkaline ratio 126, 206
    amino 111, 113, 140
    ascorbic: *See* vitamins: C
    benzoic 193, 198
    citric 219
    ellagic 219, 230
    fatty 80, 114, 198
        gamma linolenic acid (GLA) 206
        omega-3 114, 198, 231
    folic: *See* vitamins: B-9
    formic 124, 127
    fumaric 156
    hydrochloric 79
    lactic 139
    oxalic (oxalates) 27, 90, 103, 114, 159, 162, 166, 198
    removal 279
    silicic 116
    stomach 135
    tannic 235
    uric 100, 210
    usnic 264
acne: *See* skin: inflammation
adaptogen 85, 139
addictions 263, 109, 135
aging: premature 192
Alaska birch 241
Alaskan Brewing Company 262
Alaskan Flower Essence Project 61, 81, 86, 93, 105, 109, 114, 116, 127, 143, 147, 153, 170, 173, 177, 182, 193, 216, 228, 237, 245, 258, 263, 268, 272
Alaskan ginseng 83
Alaskan larch 265
Alaska white spruce 260
Alberta Natural Health Agricultural Network (ANHAN) 140
alcohol
    ethanol 108
    ethyl 291, **292**
Alder 235–7
Alexandrovich, Shiela 272, 383
Algonquin 265
alkaloids 76, 218
allantoin 116, 135
allergies 62, 81, 99, 101, 125–6, 133, 153, 183, 215, 240, 264
    capsules for 301
    eye 65
    hay fever 197
    infusion for 281
    to cats 99
    to trees 250, 258
Allium schoenoprasum 129
Alnus crispa 235
Alnus incana 235
Alnus viridus 235

Alnus viridus ssp. *fruiticosa* 235
alpine arnica 47
alpine bearberry 51–2
alpine bilberry 187
alpine cranberry 195
alpine fir 247
alpine holy grass 167
alpine sorrel 157
alsike clover 72
alterative 56, 68, 88–9, 108, 125, 133, 155, 201, 209, 257, 266
altitude sickness 140
Alzheimer's disease 188
Amelanchier alnifolia 221
amenorrhea: *See* reproductive system: menstruation
amphi-Atlantic 165
analgesic 85, 96, 108, 119, 175, 179–80, 209, 239, 252, 257, 261, 270, 294
Ancient Aromatic Panax Oil 86
anemia 89, 104, 125, 140, 146
anesthetic: local 119
angelica 127
anger 153
anodyne 239, 270
anthelmintic 96, 130, 152
anthocyanidins 197, 230
anthocyanins 188, 192, 215, 219
Anthoxanthum hirtum 167
Anthoxanthum monticola 167
anthraquinones 161
anti-aging 139
antibacterial 53, 57–8, 85, 108, 146–7, 179–80, 197, 209, 211, 215, 243, 245, 264, 321
antibilious 79
antibiotic: *See* antibacterial
anti-cancer: *See* anti-carcinogenic
anti-carcinogenic 70, 139, 219, 230
anti-catarrhal 52–3, 74, 99, 179
anticoagulant 239, 257
antidepressant 139
anti-diabetic 139
antidiarrheal 116, 218, 230
antiemetic 119, 218
antifungal 85, 89, 99–100, 211, 215, 245, 252–3, 261, 264
anti-haemorrhagic 227
antihistamine 98–100, 116, 125, 133, 143, 215
anti-inflammatory 49, 55–7, 60, 64–5, 79, 91, 93, 96, 98–100, 108, 119, 126, 133, 143, 155, 161, 175, 179–80, 182, 188, 196, 205, 209–10, 215, 221–2, 236, 239–40, 243, 245, 257, 266, 270, 294, 307
anti-irritant 91
antilithic 52–3
antimetastasis 181
antimicrobial 52–3, 88, 91, 119, 133, 145, 152, 209–10, 212, 252–3, 261, 300, 313
anti-mutagenic 139, 219, 230
antioxidant 56–7, 66, 68–9, 98, 100, 104, 108, 139, 143, 157–8, 161, 188, 190, 192, 196–7, 205–6, 215, 218–9, 221–2, 230, 245, 248, 270
antipyretic 64
anti-rheumatic 64–5, 79, 119, 209, 239, 243, 257
antiscorbutic 145, 158
antiseptic 52–3, 91, 96, 99–100, 119, 130, 133, 161, 179–80, 196–7, 205, 209–10, 212, 218, 230, 239–40, 243, 248, 250, 252–3, 257, 261–2, 270

antispasmodic 60, 68, 74–5, 79, 100, 119–20, 125, 145–6, 175–6, 182, 201–2, 218, 294, 300
anti-tumour 245
antiviral 85, 211, 222–3, 264, 321
anxiety 71, 100, 176, 216, 301
aphrodisiac 139–40
apothecary, tools for the kitchen 36
appetite: *See* digestive system
Araliaceae (ginseng family) 83
arbutin 53, 197
Arctic butterbur 73
Arctic coltsfoot 75
Arctic dock 87
Arctic root 137
Arctic sweet coltsfoot 73
Arctostaphylos uva-ursi 51–2, 197
Arctous alpina 51–2
Arctous rubra 51–2
Arnica 47–50, 306
Arnica angustifolia 47–8
Arnica cordifolia 47–8
Arnica Ease 245
Aroma Borealis 64, 105, 121, 147, 151, 179, 211–2, 240, 245, 257, 272, 301, 314
aromatherapy 60, 109, 148, 182, 212, 253, 263
aromatic 60, 119, 152, 175, 209
arrow-leaved coltsfoot 73
Artemisia absinthium 151
Artemisia frigida 149–51
Artemisia tilesii 149–51
arthritis: *See* skeletal system: joints
Ashthorn, Heather 219
Aspen: *See* **Trembling Aspen**
Aspirin 170, 239, 270, 273
Asteraceae (aster, daisy, or sunflower family) 47, 59–60, 62, 73, 77, 81, 98, 149–50, 178
asthenia 139
asthma 69, 75, 88
astringent 52, 56–7, 60, 68, 79, 88, 104, 112, 116, 125–6, 133, 143, 145, 147, 155–6, 158, 161–2, 168, 172, 179, 181–3, 192, 196, 201, 205, 207, 211–2, 218, 222, 230–1, 236, 239, 243, 245, 250, 257, 266, 270, 300
atherosclerosis: *See* cardiovascular system
Atkinson, Bob 272
aura, sacred tree 30
Aura Borealis 119
Aurora Mountain Farms 383
avron 191
azulene 182
Bach, Dr. Edward 97, 147, 240, 254, 268, 272, 301–2
back: pain 271
bacteria 298
    airborne 212
    Bacillus subtilis 108
    E. coli 108, 197
    Mycobacterium tuberculosis 215
    Staphylococcus aureus bacterium 108, 264
    Streptococcus 202, 264
    See also antibacterial; prebiotic
bacteriostatic 79
Bailey Flower Essences 76
Bailie, Ceilidh-Anne Gray 39, 40, 44–5
Bailie, Markie-May Gray 146, 272
bain-marie method 313
bake(d) apple 191
balancer 69, 71, 88, 101, 109, 114, 121, 126, 146–7, 180, 206

Index

"balm of Gilead" 257
balms 257, **309–14**
　See also **Recipes**
Balsam Poplar 243, **256–8**
Barker, Cicely Mary
　Flower Fairies of the Wayside 69, 94
Barlow, Rose 344, 347, 355, 359, 362, 371, 374
Basic Trifolium Compound 70
baths 99, 104, 108, 133, 152, 179, 201, 209–10, 236, 239, 243, 249–50, 253, 257, 271, **324–9**
　footbaths 49, 61, 81, 103, 121, 146–7, 227, 236, 248, 250, 261, 266–7, 324, **327**
　full bath 60–1, 64, 81, 101, 150, 152, 177, 183, 211, 227, 324
　hand bath 324
　sitz baths 52–3, 69, 181, 324, **328**
　steam bath 249, **328**
　See also **Recipes**
Bearberry 51–4, 151, 197, 210, 292
bears, safety and awareness 21–2
Bedstraw 55–8, 210, 322
bedwetting 53
bees 79, 91, 271, 288
Bennett, Bruce 59, 175, 235, 390
Betula alba 245
beta carotene: See vitamins: A
Betulaceae (birch family) 235, 242
Betula glandulosa 241
Betula neoalaskana 241
Betula papyrifera 241
beverage: See food and beverage
bilberry 197
binomial plant names, system of 26
bioflavonoids: See phytochemicals: flavonoids
biome 24
Birch 107, 161, **241–6,** 321, 329, 346
"birch-canker polypore" 245
birds 271
bird's-eye berry 51
birthing: See reproductive system
bite: See skin: inflammation; skin: wound
bitter gentian 95
bitter 52, 60, 64, 68, 79, 96–7, 119, 152, 179, 180, 239, 257, 270,
bitters 297
Black, George 189
Black, Martha Louise, 189
　Yukon Wild Flowers 57, 97, 150, **189,** 197, 206, 212
blackberry 214
black crowberry 214
Blackfoot 112
black spruce 259
bladder: See urinary system
bleeding: See body part, system or condition; cardiovascular system: blood; skin: wound
blessed thistle 70
blindness: snow 61
bloating 96
blood: See cardiovascular system
blue bell 115
Blueberry 187–90, 197, 216, 223, 289
blue mountain tea 98
bog bilberry 187
bog blueberry 187
boils: See skin: infection; skin: inflammation
bones: See skeletal system
Boon, Heather 79
Boraginaceae (borage family) 115
Borago officinalis 116

Boreal 19
The Boreal Centre for Conservation Enterprise 386
the boreal forest 24
The Boreal Gourmet: Adventures in Northern Cooking 227, 244, 262
Boreal Summer-Solstice Tea Ceremony 288
botanical terms 395–400
botany, basics of 393–4
BPH: See glandular system: prostate
brandy 297–8
Brassicaceae (mustard or cabbage family) 154
breasts
　breastfeeding 61, 64, 69, 76, 85, 97, 173, 219, 230
　cancer 70–1, 80, 219, 230
　cysts 65, 80, 308
　infection 65
　inflammation
　　mastitis 69
　nipples
　　inflammation 64–5
breathing: See respiratory system
broad-leaf fireweed 141
broad-leaved plantain 132
"broad-leaved willow herb" 141
brook mint 118
bruises: See skin
buffaloberry 225, 227
burdock 70, 161
burning nettle 123
bursa: bursitis 271
"bush bandage" 145–6, 179
bush cranberry 200
caffeine 107
Caisse, Rene 70, 161
cambium 258, 261
Camellia sinensis 279
Campagna, Marie-France 356, 363, 368, 375
camphor 152
camphoraceous 253, 325
Canada goldenrod 98
Canada gooseberry 204
Canadian balsam 247
Canadian Forest Tree Essences 253
Canadian Herb, Spice and Natural Health Products Coalition (CHSNC) 386
Canadian mint 118
The Canadian NTFP Business Companion (compact disc) 386
Canadian Organic Growers (COG) 386
Canadian Pharmacy Journal 161
cancer 57, 140, 267
　estrogenic 70
　hormone-sensitive 71, 85
　prevention 188, 197
　risk factors 70
　therapy drugs 70
　treatment 245
　See also anti-carcinogenic; body part, system or condition
canning tips 349
Caprifoliaceae (honeysuckle family) 171, 200
Capsella bursa-pastoris 154
capsules 104, 125, 176, 197, 218, 271, **301**
　See also **Recipes**
cardiac tonic 218
cardio protective 139

cardiovascular system
　aids and remedies 197
　atherosclerosis 188
　blood
　　anemia 230
　　bleeding 57, 103–4, 126, 180
　　hemorrhages 155–6
　　hemorrhoids 69, 116, 135, 146, 165, 181
　　internal 75, 104, 179, 236
　　blood-sugar levels 85, 136, 189, 197, 210
　　cholesterol 243
　　circulation 120, 130, 180, 213, 319
　　cleanser (purifier) 57, 68, 79, 88–9, 108, 119, 125, 133, 146, 161, 210
　　clotting (coagulation) 104, 156
　　pressure 57, 79–80, 96–7, 109, 156, 180, 227
　　red cells 146
　　strengthener 219
　　thinner 71, 170, 273
　　tonic 218, 223, 261
　blood vessels
　　infusion for 281
　heart
　　aids and remedies 69–70, 146–7, 188
　　ailments 105
　　cholesterol 69, 79, 135, 188, 267
　　irregularity 49
　　pain 250
　　palpitations 120, 176
　　rate 139
　　stimulant 237
carminative 60, 99–100, 119, 175–6, 179, 209
Carcross 291
Caryophyllaceae (pink family) 63
cataracts 85
catarrh 99
cathartic 85, 109, 226–7
Catty, Suzanne 100, 109, 182, 212, 250, 263, 322
cedar 169
cell-proliferant 116
cell-regenerating 161–2
cellulite 211
Celtic tree alphabet 392
　See also Ogham
chaga 245
chakra
　ajna (third-eye; sixth) 183, 253
　crown (seventh) 121, 212
　fifth 66
　fourth 66
　heart 66, 117, 147, 303
　root (first) 105, 212
　solar plexus (third) 49, 66, 81
Chamerion angustifolium 91
Chamerion latifolia 141
Chamomile: See false chamomile; German chamomile; **Wild Chamomile**
ch'at àn dagàii 178
Chenopodiaceae (goosefoot family) 111, 164
Chenopodium album 111
Chenopodium ambrosiodes 113
Chenopodium capitatum 112
Chenopodium capitatum 164
chest
　congestion 117, 130
　pain 53, 88, 250
Chickweed 63–6, 190, 292, 319, 337–40, 356, 359–61, 364–5, 367–8, 370, 383, 401, 403–5, 410

Part V: *For Reference*

childbirth: *See* reproductive system: birthing
children
    dosages for 40–1
        herbs for 59–61, 69, 151
        herbs to avoid 76, 153, 156
    at tea ceremony 288
chiming bells 115–6
chives 129, 131
chlorophyll 114, 125, 127
cholagogue 68, 79, 96, 239, 257
cholerectic 79
cholesterol: *See* cardiovascular system: heart
choline: *See* vitamins: B complex
Christopher, Dr. John 70
chronic fatigue syndrome 139
cicatrizant 182
circulatory system: *See* cardiovascular system; lymphatic system
Clark's Rule 40
cleanser 61, 71, 114, 125, 147, 156, 161–2, 168–9, 210, 249
    household 253–4
    See also body part, system or condition
cleavers 55–6
Cloudberry 191–4
clover: *See* alsike clover; cow clover; **Red Clover;** wild clover
cobalt 127, 182, 203
Cocos nucifera 310
Cody, William J.: *The Flora of the Yukon* 138, 235
colds: *See* respiratory system
colic 176: *See* digestive system
Coltsfoot 73–6, 202, 210, 328
comfrey 116, 135
common bearberry 51–2
common juniper 208
common nettle 123
common sheep sorrel 160
common starwort 63
The Complete Natural Medicine Guide to Breast Cancer 80, 308
Compositae 77
compost 127
compresses 56, 64–6, 119–21, 165, 192, 201, 206, 236, 243, 245, 250, 263, 267, 271, 320
    cold 320
    hot 57
concentration: aids and remedies 121
conjunctivitis: *See* eye: infection
constipation: *See* digestive system
contaminants 81
    See also poisons; toxins
Cooper, Wendy 224
cosmetics, Health Canada guidelines for 384
cottonwood 256
coughs: *See* respiratory system
coughwort 73
coumarin 57, 170
Council of Yukon First Nations 261, 263
cowberry 195
cow clover 67
crampbark 69, 155, 176, 200, 271
    See also **Highbush Cranberry (Crampbark)**
Cranberry 153, 172, **195–9,** 289
    See also Highbush Cranberry; upland cranberry
Crassulaceae (stonecrop family) 137
creams 49, 60–1, 64–6, 85–6, 91, 93, 99, 101, 108, 119, 133, 168, 170, 179–80, 201, 209–10, 239, 243, 245, 248, 250, 252–3, 257, 261, 271, 304, 307–8, **315–17**

contamination 315
creatinine 126
Cree 75, 22
creeping juniper 208
crocus 226
Crohn's disease 96
crowberry 143, 214
Crown Chakra Vibrational Essence 121
Culpeper, Nicholas 75, 130, 134, 162, 390
    Culpeper's Complete Herbal 176, 390
Cupressaceae (cypress family) 208
curled dock 87
curlewberry 214
Currants 204–7
cut: *See* skin: wound
cystitis: *See* urinary system: infection
cystoliths 127
cysts
    dermoid 65
    ovarian 65
Dandelion 77–82, 126, 190, 236, 285–7, 289, 292, 297, 302, 308, 322, 336–8, 345, 347, 350, 355, 359–62, 364–6, 368, 370–2, 374–5, 383, 393, 395, 401–5, 410
dandruff: *See* skin: inflammation
dandy-lioness 77
decoctions 53, 58, 60, 79, 81, 85, 105, 108, 139–40, 146, 165, 176, 201–2, 211, 215, 222–3, 226–7, 236, 243, 245, 249, 252, 267, 271, 280, **286**
    See also **Recipes**
DEET 183
deetru'jak 208
demulcent 52, 64–6, 74, 116, 133, 210, 257
Dena'ina 61, 75, 83, 171
Dene (Slavey) 164
dent-de-lion 77
deodorants 119, 121
depression 70, 97, 100, 117, 139, 301
detoxifier 81, 109, 161–2, 168, 182, 212–3, 249
Devil's Club 83–6, 236
diabetes 80, 85, 90, 136
    Type 2 108, 189
diaphoretic 74, 79, 85, 99, 108, 119–20, 179–80, 209, 243, 248–9
diarrhea: *See* digestive system
Diet for a Small Planet 369
digestive 69, 130, 143, 179, 205, 209
digestive system (tract)
    aids and remedies 125, 130, 135, 143, 152, 182, 241
        bitters for 297
        elixir for 297
    ailments 60, 79, 91, 223, 271
    appetite 79, 96, 130, 152–3, 210
    bowels
        inflammation 69
        irritable-bowel syndrome 135
        movements 267, 297
        weakness 100
    cleanser 113
    colic 60–1, 120
    colon
        cancer 108, 219, 230
        diverticulosis 220
    constipation 53, 65, 79, 88–9, 135, 162, 226–7
    detoxifier 182
    diarrhea 53, 66, 96, 108–9, 113, 116, 135, 146, 156, 159, 162, 189, 192, 207, 215, 219, 222, 224, 226, 240, 243, 257

digestion (and indigestion) 65, 80, 89, 96, 108, 119–20, 180, 182, 197, 202, 210, 225–7, 236, 240, 297
duodenum
    ulcer 65
dysentery 240
enzymes 197–8, 267
esophagus
    cancer 219, 230
gallbladder 79, 210, 240
gas (flatulence) 60, 69, 113, 120, 176, 180, 210, 240
    pain 100
gastritis 96, 146
good bacteria
    Lactobacillus 267
heartburn 60, 79, 96, 120, 182, 197, 240
hemorrhoids 69, 116, 135, 146, 165, 181, 236, 328
irritation 90, 96, 148
liver
    ailments 79, 88
    bile production 79, 96
    cleanser 109, 180, 79
    disease 267
    function 80, 125, 245
    hepatitis 57, 79, 88
    regenerator 109
    stimulant 88
nausea 97, 120, 180, 240
    infusion for 284
pancreas
    cancer 219, 230
parasites
    pinworms 61, 151
    roundworms 151
    tapeworms 52
    worms 91, 130, 148, 152, 161–2
stimulant 69, 243, 297
stomach
    acid 135
    ailments 53, 131, 135, 215, 219, 222, 224, 240, 261
    bleeding 273
    bloating 180
    cramps 153, 202, 210
    hyperacidity 96
    infection 53
    irritable 202
    irritation 273
    pain 93, 96, 113, 236
    tonic 227
    ulcer 65, 75, 85, 104, 135, 261
tonic 152–3, 243
vomiting 49, 61, 86, 97, 237
dineech'üh 214
Dioscorides: *De Materia Medica* 138
Discovering Wild Plants 76, 237, 323
disinfectant 209, 212, 248, 252, 266, 317
diuretic 52, 56–7, 64–5, 68, 74, 79–80, 88, 99–100, 104, 108, 125–6, 133, 135, 145, 155, 161–2, 179–80, 196, 201–2, 205, 209–10, 236, 239–40, 243, 248–9, 257, 266, 270
divining 101
dizziness 49
Dock 87–90, 124
The Doctrine of Signatures 25, 171, 173
dogs 180
    poisons to 245
dogwood 54

Index

dosage 38–41
dosage, for children 40
    Clark's Rule 40
    Cowling's Rule 41
    Young's Rule 41
double-boiler method (infused oils) 306
double-boiler method (ointments) 312
dream pillow 56, 59–61
Dupuytren's contracture 57
dwarf birch 241
dwarf fireweed 141
dye 54, 56, 58, 81, 90, 101, 166, 177, 203, 216, 227–8
dysmenorrhea: *See* reproductive system: menstruation
dzih kò 259
dzih tl'ùu 259
ear
    earache 58
    ear-drop treatment 223
    inner
        infection 267
    oil 58
Earle, Teresa 42-43, 172
echinacea 190
Eclectic medicine 202
ecosystem 271
eczema 125: *See* skin: inflammation
edema 80, 105, 243
Edible and Medicinal Plants of the Rocky Mountains and Neighbouring Territories 104
Edwards, Nicole: *Sage and Wild Roses* (CD) 152
Einstein, Albert 31
Elaeagnacae (oleaster family) 225
electrolytes
    abnormalities 90
    potassium 80
    sodium 80
elixirs 230, **297**
emetic 85–6, 248, 250
emmenagogue 96, 119, 152, 179, 181, 218, 209
emollient 64, 74, 116, 133, 145, 257
emotions 80, 127, 212, 245, 250, 288–9, 301
Empetraceae (crowberry family) 214
Empetrum nigrum 214, 216
Empetrum nigrum ssp. *hermaphroditum* 214
emphysema 88
endurance 138
energy 93, 138–9, 183, 206, 223, 288–9, 325
    infusion for 281
    of plants 28
    tincture for 294
Englishman's foot 132
enuresis 52
entire-leaved stonecrop 137
enzymes 136, 189, 197
epazote 113
epilepsy 134
Epilobium angustifolium 91
Epilobium latifolium 141
Equisetaceae (horsetail family) 102
Equisetum arvense 102
Equisetum palustre 105
Ericaceae (health family) 51, 106, 187, 200
ESCOP: *See* European Scientific Cooperative on Phytotherapy
The Essence of Healing 61, 81, 86, 93, 105, 109, 114, 116, 127, 143, 147, 153, 170, 173, 177, 182, 193, 216, 228, 237, 245, 258, 263, 268, 272

essences
    flower and plant 49, 54, 60–1, 68, 71, 73, 76, 79, 81, 93, 104–5, 109, 121, 140, 147, 156, 173, 183, 207, 212, 222–3, 240, 254, 263, 268, 272, **301–3**
    "mother essence" 173
    See also **Recipes**
Essiac Tea (or Formula) 70, 161–2
exhaustion 69, 96, 253
expectorant 64–5, 68–9, 74, 85, 96, 108, 116, 133, 179–80, 182, 210, 248, 253, 257, 266, 300
extraction: *See* maceration
eyes
    cataracts 215
    compress 65
    drops 135
    eye-drop treatment 223
    infection
        conjunctivitis (pink eye) 65, 135
    inflammation
        allergies 65
        blepharitis 135
        wound 65
    irritation 64, 146, 169
        itching 99, 125
        watering 65, 99,125
    soreness 215
    vision 197
    wash 61, 64–5, 85, 133, 146, 169
eyewash: *See* eye: wash
Fabaceae (pea family) 67
face: *See* skin
fairy clock 77
false chamomile 59
fatigue 61, 89, 100, 138–9, 216, 249, 253
    See also chronic fatigue syndrome
fats: plant 223, 258
fear 228, 240, 301
febrifuge 96, 119, 179, 182, 218, 239, 257
felwort 95
female sage 149
ferritin 89
fertilizer 68
fever 65, 69, 120, 206, 249, 257, 271, 282
fiber (or fibre): dietary 114, 207, 220, 223, 231, 267
fibroids
    uterine 70, 88
field horsetail 102
field mint 118
field sorrel 160
Fir: *See* **Subalpine Fir**
Fireweed 91–4, 99, 125–6, 142, 190, 287, 295, 300, 306, 322, 328, 336–7, 365, 368, 370, 375
    white fireweed 93
Fireweed Community Market 386
First Nations 54, 91, 93, 107, 151, 166, 168–9, 175, 193, 196, 210, 215, 226–7, 231, 235, 247–9, 252, 261, 267, 304
    coastal 86
    See also Algonquin; Blackfoot; Cree; Dena'ina; Dene (Slavey); Gwich'in; Northern Chipewyan; Nuxalk (Bella Coola); Ojibwa; Thompson; Tlingit ; Van Tat Gwich'in
flashes
    aids and remedies 119
    hot 70
flatulence 96
flower essence: *See* **essences: flower and plant**
Flower Essence Repertory 49, 71, 101
Flowers of the Yukon 90

flus: *See* respiratory system
folate: *See* vitamins: B-9
fomentations 116, 130, **319**
    See also **Recipes**
food and beverage
    ale 125, 198
    asparagus substitute 93
    baking 202
    bannock 240, 249
    bay leaf substitute 109
    beer 140, 211, 244, 262
    berry 188, 216
    berry (dried) 223
    berry substitute 166
    bitters (alcoholic) 97
    bread 71, 89, 113, 130, 190, 198, 211, 231, 240, 249, 255
    butter substitute 136
    cake 61, 99, 146, 169, 216, 219, 231
    cereal 113, 190
    chutney 198
    coffee substitute 37, 57, 80, 113, 285
    condiment 202, 211
    conserve 198, 211
    cookies 202, 216, 219
    crackers 89, 255
    dessert 207, 231, 244, 299
    dips 65, 291
    dressing (salad) 65, 80, 109, 130, 136, 159, 163, 181, 231, 245, 262, 296
    dumpling 267
    edible green 267
    egg dish 101
        omelette 113–4, 130
    filo pastry filling 113
    fish 210, 216
    fish (smoked) 237
    flavouring 97
    flour 113, 255, 267
    fowl 216
    fruit leather 188, 189, 192, 207, 220, 223, 231, 299, **376–7**
    game 210
    garnish 101
    gin 209, 212
    grains 71
    granola 246
    green drink 64–5
    hops substitute 97
    hummus 120
    ice cream 146, 216, 227
    ice cube (flavoured) 120
    "Indian ice cream" 227
    jam 145–6, 190, 192–3, 198, 206–7, 216, 219–20, 223, 231
    jelly 93, 120, 145 6, 166, 190, 192, 198, 200, 202, 206, 223, 227, 231, 262
    juice 53, 104, 126, 190, 192, 197, 200, 202, 206–7, 216
    lemonade 290
    liqueur 120, 193, 198, 207
    for livestock
        cattle 127
        goats 112
        poultry 65, 111–2, 127
        sheep 112
        swine 111–2, 127
    marinade 80, 211, 244
    mayonnaise 136, 296
    meat dish 61, 109, 207, 216

Part V: *For Reference*

muffin 61, 71, 89, 101, 113, 130, 190, 193, 198, 216, 231
oil 136
oil (salad) 159
onion (spring) 129
pancake 193–4, 198, 216, 230, 299
parsley replacement 113
pasta 65, 113, 136, 159, 231
pesto 65, 253
pickles 162
pie 198, 216, 223
popsicle 71, 120, 188, 262
porridge 190, 223, 230
poultry 231
preserves 201
quinoa substitute 113
rennet substitute 162
rice 231
salad 57, 61, 63–5, 71, 78, 93, 99, 101, 114, 116, 130, 136, 141, 143, 146, 154, 165, 219, 262
salad (fruit) 120
salt substitute 75, 262
sandwich 65, 71, 156, 159
sauce 71, 80, 89, 109, 113, 130, 136, 146, 197, 207, 216, 227, 231, 244, 291
sauce (green) 136
sauerkraut 210
sauerkraut substitute 159
seed 71, 89–90, 113
smoothie 61, 65, 113–4, 126, 140, 166, 188, 193, 216, 231, 283, 291
Soapberry Jelly Ice Cream 227
sorbet 146
soup 65, 71, 78, 80, 89–90, 101, 109, 113, 116, 126, 130, 136, 140, 159, 207, 211, 237, 240, 253, 258, 262, 271, 291
spanakopita (wild-weed) 65, 112
spice substitute 109, 113, 119, 130, 156, 181, 210
spinach substitute 87, 89–90, 101, 113, 126, 156
sprout 71
stew 65, 80, 89, 101, 109, 116, 211, 216, 237, 240, 253, 258, 262
stir-fry 71, 80, 89, 93, 101, 113, 126, 140, 142, 159, 271
stuffing (turkey) 153, 198
sweetener 267
syrup 146, 161, 192–3, 198, 206, 216, 219, 223, 242–4, 262, 346
tea (black) substitute 107
thickener 101
trail snack 140, 159, 209, 236, 240
vanilla substitute 169
vegetables 130, 160–2, 244
vegetables (leaf) 80, 89
vegetables (root) 80
vegetables (steamed) 93, 103–4, 116, 136, 142, 156
vinaigrette 219, 244
vinegar 79–80, 91, 125, 155, 158, 181, 203, 245
vodka 211
wild meat (smoked) 237
wine 198, 207, 216
yeast 211, 240
yogourt 190, 193, 219
See also **Recipes;** teas/infusion
Fragaria virginiana 229
free radicals: *See* radicals
Frisch, Sylvia 380

fumigant 212
fungistatic 79
fungus 53, 298
    candida 100
    chaga 245
    See also antifungal
galactagogue 60, 68, 79, 116, 119, 125, 172
Galium boreale 55, 57
Galium trifidum 55–6
Galium triflorum 55–7
gallbladder: *See* digestive system
garbling 33
gargle: *See* mouth
garlic 130
gastritis: *See* digestive system
Genest, Michele 227, 244, 262
genitourinary system: *See* urinary system
Gentian 95–7, 297
    Gentianella amarella 95
    Gentianella propinqua 95
Gentinaceae (gentian family) 95
Gerard, John 390
German chamomile 59–60
gingivitis: *See* mouth: gums
ginseng: *See* **Devil's Club**
GLA: *See* acids: fatty
Gladstar, Rosemary 39, 41
glandular system 125, 162
    adrenal glands 108, 139, 263
        infusion for 281
    prostate 69, 210
        benign prostatic hyperplasia/hypertrophy (BPH) 126, 240
        cancer 100, 219, 230
        inflammation 135, 216
        tincture for 295
    thyroid gland 97
glycerine 291, **292**
Golden Flower Body-and-Massage Oil 101
Goldenrod 98–101, 287, 320
"golden root" 137–8
goldenseal 138
goosefoot 111
goose grass 55
gout: *See* skeletal system: joints
government agencies, lists 385–6
gravel grass 55
Gray, Deanna 222
greater plantain 132
Greenland moss 106
Greenland tea 106
grief 76, 117, 143
Grieve, Margaret: *A Modern Herbal* 75, 97
Grossulariaceae (currant family) 204
ground birch 241
Growers of Organic Food Yukon (GOOFY) 386
gum: *See* resin (pitch)
Gurudas 255, 253
    The Spiritual Properties of Herbs 116
Gwich'in 10, 29, 262
    See also Van Tat Gwich'in
gyùutsanh 149
haemostatic 155, 226
hair
    aids and remedies 85, 101
    colour 177
    dye 227
    growth 57
    infusion for 284
    loss prevention

infusion for 281
oil 108
wash (shampoo and rinse) 103–4, 121, 127, 177, 211, 228, 246
Handley, Tanya 21
harmony 61, 177, 303
harvesting, timing of 23
harvesting equipment 24
harvesting tips 23
head
    headache 69, 96, 120, 177, 243, 245, 271
        tincture for 294
    migraine headache 176, 206
headache: *See* head
head lice: *See* skin: parasites
Healing Wise 65
Health Canada, guidelines for natural products 384
heart: *See* cardiovascular system
heartache 197
heartburn: *See* digestive system
Heart Chakra Vibrational Essence 147
heart-leaf arnica 47
heart-leaved arnica 47
hemostatic 125, 236, 243
hemp powder 340–1
hemp seeds 343, 358, 361, 370
hepatic 54, 79, 152, 161, 179
hepatonic 88
herbalists, history of 389–91
herbal medicine 25, 37
herbal usage chart 406–415
herbe militaris 178
herbs
    drying 33–5
    garbling 33
    labelling 35
    storing 35
herbs, usage of (chart) 406–415
heshkeghka'a 83
Hierochloë 167
Hierochloë odorata 170
Highbush Cranberry (Crampbark) 69, 100, **200–203**
Hippocrates 26
hoarseness 75–6
holy grass 167
homeopathic 49, 109, 301
hormones
    balance 69, 88, 146, 180
    estrogen 70, 180
    rooting 272
Horsetail 102–5, 124, 228, 320
    See also marsh horsetail
hot-water bath method 313
How Do We Make Use of Nature's Gifts? 357
How to Cook a Reindeer: The Reindeer Recipe Book 255, 357
Hoxsey Method 70
huckleberry 187, 189
Hudson Bay tea 106
humectant 147
hydration 138, 190
hydrosols 60, 99–100, 108–9, 119, 145, 147, 152, 168, 170, 179, 182, 209, 212, 243, 245, 248, 261, **321–3**
Hydrosols: The Next Aromatherapy 100, 109, 182, 212, 250, 263
hyperplasia: benign prostatic 93
hypnotic 60, 175

432 | The Boreal Herbal

Index

hypoglycemic 85, 179
hypotensive 130, 175, 179, 226
hysteria 71
immune system
    booster 197, 210, 245, 250, 263, 265, 267
    stimulant 109, 133, 139, 180,
        182, 209, 245, 266
incense 250
infection 85, 88, 134, 146, 250
    See also antibacterial; antiviral; bacteria; yeast
infertility: *See* reproductive system: fertility
inflammation 79, 257, 270
    See also body part, system or condition
influenza: *See* respiratory system: flus
infusions: *See* **teas/infusions**
Inonotus obliquus (chaga) 245
inositol 156
insect bite: *See* skin: inflammation
insomnia 139, 174, 176, 301
insulation 75
intuition 212
Inuit 141, 143, 159
    Inuvialuit 191
inulin 79
iodine 231
Irish daisy 77
irritability 70
isoflavones 69–70
    genistein 69
jàk zraii 187
Jason Winters Tea 70
jaundice 79, 88
Johnson, Steve 61, 81, 86, 93, 105, 109, 114, 116,
    127, 143, 147, 153, 170, 173, 177, 182, 193,
    216, 228, 237, 245, 258, 263, 268, 272
joints: *See* skeletal system
Journal of Ethnopharmacology 108
juice 64, 66, 88, 192, 197, 230
    See also food and beverage
juneberry 221
Juniper 99, 121, **208–13,** 236, 292, 322, 340, 354,
    357, **403–5,** 410
Juniperus communis 208–9
Juniperus communis ssp. *alpina* 208
Juniperus communis ssp. *depressa* 208
Juniperus horizontalis 208–9
Kakuzo, Okakura: *The Book of Tea* 289
kale 338
Kaminski, Patricia 49, 71, 101
kapporie (kapor tea) 93
Katz, Richard 49, 71, 101
Katzen, Mollie 367
Kaur, Sat Dharam 80, 308
k'ele t'lia 171
khehdi' 102
kidney: *See* urinary system (tract)
k'ii 241
"King of the Waters" 236
kinnikinnick 51, 54, 151
k'jeghi ch'da 75
Klondike gold rush 189
Kneipp, Sebastian 96
knitbone 116
k'oh 235
Kuch, Birch 245
Labiatae (mint family) 321
labour: *See* reproductive system: birthing
Labrador Tea 54, **106–10,** 151, 289, 322
Lamb's Quarters 111–14, 165, 319, 341, 359–61,
    363–70, 376, 383, 404, 410

Lamb's skin 47
Lamiaceae (mint family) 118, 150, 321
Lammers, Mickey 90
Land of My Ancestors: Trees and Forest 261, 263
Lappé, Frances Moore 369
Larch: *See* **Tamarack**
Larix laricina 265, 267–8
Larson, Lyndsey Berwyn 161, 383
larynx: infection (laryngitis) 75, 100, 134
laxative 56–7, 61, 64, 79, 85, 88, 91, 108, 125, 135,
    161, 205, 226, 248–9, 266
Ledum 109
Ledum groenlandicum 106
leopard's bane 47
lice: *See* skin: parasites
lichen 264
lidii masgit 106
Liliaceae (lily family) 129
limb (arm, hand, leg, foot)
    coldness 180
    foot
        odour 153
        soreness 153
        See also **baths: footbaths**
    leg
        restless-leg syndrome 79, 202
    soreness 173, 267
lingonberry 195–7
liniments 49, 60, 85, 201–2, 209, 239, 243, 248–9,
    252–4, 257, 261–2, 271, **317**
Linnaea americana 171
Linnaea borealis 171
Linnaeus, Carolus (Carl) 26, 67, 75, 137, 171, 230,
    391
lion's tooth 77
lips: *See* mouth
liver: *See* digestive system
locavore 333
Lodgepole Pine 251–5
loss 143
low-bush cranberry 195, 200
"low bush-cranberry" 200
lumps: *See* cysts
lungs: *See* respiratory system
lung tan (dragon's gall) 96
Lung-Tonic Infusion 283
Lungwort 115–17, 135, 328
lymphatic system
    aids and remedies 109
    circulation 308
    congestion 69
    inflammation 69
    swelling (in lymph nodes) 88
    tonic 57
maceration (extraction) 291
marc (medicinal properties) 291, 294
marsh horsetail 105
marshmallow root 210
marsh tea 106
marsh valerian 174
mastitis 69
Matthew Watson General Store 291
Matricaria discoidea 59
Matricaria matricariodes 59
Matricaria recutita 59–60
McCutcheon, Dr. Allison 108, 215
mealberry 51
measurement: simpler's measurement method 37
medicinal actions of plants 403–5
medicinal actions of plants chart 416-419

medicine chest 37
meditation 30, 116
    plant-spirit 81
memory
    aids and remedies 121, 139
    loss 188
menopause: *See* reproductive system
menorrhagia: *See* reproductive
    system: menstruation
menstruation: *See* reproductive system
menstruum (solvent) 291–2, 294, 304
Mentha arvensis 118
Mentha canadensis 118
Mertensia alaskana 115
Mertensia paniculata 115–6
methyl salicylate: *See* salicin
midwifery 219
100-Mile Diet 333
milfoil 178
Millar, Vanora 81
minerals 79, 140, 189
    aluminum 220
    calcium 54, 61, 66, 71, 79–80, 89, 104, 114, 121,
        127, 131, 147, 156, 182, 190, 193, 198, 203,
        207, 218–20, 223, 231, 245, 282, 296
    chromium 71, 127, 182, 203, 211, 220
    cobalt 211, 220
    copper 80, 114, 131, 162, 223, 231
    iron 54, 66, 71, 80, 89–90, 104, 114, 121, 126–7,
        131, 146–7, 156, 159, 162, 182, 190, 193,
        198, 203, 207, 211, 218–20, 223, 227, 230–1,
        299
    magnesium 54, 61, 66, 71, 80, 89, 104, 114,
        121, 131, 147, 182, 190, 203, 207, 211, 218–
        20, 223, 231, 245, 296
    manganese 54, 66, 71, 80, 90, 127, 162, 182,
        203, 207, 211, 220, 223, 231, 245
    phosphorus 66, 71, 80, 89, 104, 114, 127, 131,
        147, 156, 162, 182, 190, 198, 203, 207, 211,
        218–20, 223
    potassium 54, 61, 66, 71, 80, 89, 114, 127, 131,
        136, 147, 156, 162, 182, 190, 198, 203, 207,
        211, 220, 223, 231, 245, 296
    selenium 71, 80, 89, 114, 131, 147, 182, 203,
        211, 220
    sodium 80, 114, 147, 182, 190, 198, 211, 220
    sulphur (or sulfur) 104, 223
    tin 182, 203, 211, 220
    zinc 54, 66, 71, 80, 90, 97, 114, 121, 127, 131,
        147, 203, 220, 245
    See also chlorophyll; cobalt; silica
Mint 107, **118–22,** 125, 127, 228, 289, 292, 321–2
"Mitchell's Genuine Balsam" 257
mood enhancer 139
mood swings 70
Moorcroft, Lois 119
mooseberry 200, 225
The Moosewood Cookbook 367
Morton's neuroma 57
Mossberry 214–16, 340, 346, 354–5, 376, 403–4,
    411
motion sickness: *See* digestive system: nausea
motor skills 188
mountain alder 235
mountain bilberry 195
mountain cranberry 195
mountain sage 149
mountain sagewort 149
Mountain Sorrel 157–9
mountain tobacco 51

Part V: *For Reference*

mountain wormwood 149, 153
"mouse's rope" 171
mouth
    breath 153, 162, 249
        freshener 153
    gargle 100, 113, 152, 168, 180, 202, 206, 210, 219, 239, 249, 267
    gums 252
        bleeding 219
        cankers 161
        gingivitis 140, 197, 202
        infection 267
        inflammation 89, 180
        irritation 100
        pyorrhea 140
        soreness 161
        swollen 270
        ulcers 230
    infection 53, 180, 206
        thrush 100
    inflammation 76, 100
        canker 219
        ulcer 219
    irritation 75, 90
    lips
        balm 314
        sores 261
    teeth
        cavities 245
        cleanser 105
        decay 197
        infection 135, 180, 240
        inflammation 135
        irritation 100
        loose 202
        plaque 252
        soreness 270
        teething 60
        toothache 100, 135, 180, 219, 240
        toothpaste 245–6
        whitening 231
    ulcer 113, 165
    wash 53, 89, 112, 161, 165, 202, 219, 245, 249
mouthwash: *See* mouth: wash
mucilage 75
mucilaginous 133, 135
mucolytic 248
Mueller, Fritz 172
mugwort 149
muscles
    cramps 201–2
    cream for 315
    inflammation 49, 109, 254, 317
    pain 57, 61, 100, 151, 202, 216, 253, 262, 271
        baths for 325
        oil for 307
    relaxant 237
    soreness 100, 149, 152, 177, 209–10, 227, 236, 243, 249–50, 254, 309, 313, 317
    spasms (restless-leg syndrome) 79, 202
    spasms: oil for 307
    strain 65, 271
    tension 308
muskeg tea 106
nails
    infusion for 285
    wash 103
nakàl 191
narcotic 107
narrow dock (*Rumex crispus*) 87

narrow-leaved arnica 47
natl'at 195
Natural Health Products Regulations 384
nausea: *See* digestive system
neck: stiffness 100
neeyùu zraii 204
nephritis: *See* kidney: inflammation
nervine 49, 60, 68, 119, 175, 201–2, 294
nervous system 321
    aids and remedies 96, 108
    irritability 202
    nerve
        pain 100
        strengthener 176
        toner 176
    nerve endings
        inflammation 57
        pain 100
    sciatic nerve: pain (sciatica) 100, 197, 210, 271
    soothing 60, 69, 177, 253
    tincture for 294
    tonic 139, 230
neti pot: *See* respiratory system: nose
nets 268
Nettle: *See* **Stinging Nettle**
niacin: *See* vitamins: B-3
nichìh 144
"Nine Herbs Charm" 134
nitrogen 68, 127, 236
nivaqsiaq 142
non-timber forest products (NTFP) 381–2
northern bedstraw 55–6
northern black current 204
Northern Chipewyan 229
northern gentian 95
northern spinach 111, 113
nose: *See* olfactory system; respiratory system
numbness/paralysis 52
nursing: *See* breast: breastfeeding
Nutritional Herbology 127
Nuxalk (Bella Coola) 143
Ogham 235, 238, 241, 247, 256, 392
oils
    aromatic 107
    bath 86, 136, 250, 304
    body 86, 147, 250
    carrier 310–11
        almond (*Prunus dulcis*) 310
        avocado (*Persea americana*) 310
        castor (*Ricinus communis*) 310
        coconut (*Cocos nucifera*) 310
        grape seed (*Vitas vinifera*) 310
        hemp (*Cannabis sativia*) 311
        jojoba (*Simmondsia chinensis*) 311
        olive (*Olea europaea*) 311
        rosehip seed (*Rosa mosqueta*) 311
        sesame (*Sesamum indicum*) 311
        St. John's wort (*Hypericum perforatum*) 311
        sunflower (*Helianthus annus*) 311
    essential 99–100, 108–9, 119, 148, 175–6, 179, 182, 209, 212, 243, 245–6, 248, 250, 252–3, 261, 263, **321–3**, 327–8
    infused 108, 119, 125, 131, 133, 143, 179–80, 209, 239, 243, 257, 264, 304, **305–6**
        double-boiler method 306
        sun infused 306
    massage 58, 61, 86, 136, 147, 150, 152, 249–50, 271, 304, 308

medicinal 49, 52, 60, 64–6, 68, 74, 79–80, 85, 88, 91, 99–100, 108, 127, 135–6, 143, 145, 158, 168, 201, 206, 212, 239, 243, 248, 252–3, 257–8, 261–2, 271, 291, **304**
    shower 136
    volatile 118–9, 151
        removal 279
    See also **Recipes**
ointments 48, 50, 52, 64, 99, 101, 108, 148, 155, 250, 257, 304, 307, **309–14**
    double-boiler method 312
    hot-water bath or bain-marie method 313
    See also **Recipes**
Ojibwa 161
old man's beard 264
Olea europaea 311
olfactory system 175
    nose: bleeding 156, 179
Onagraceae (evening primrose family) 91, 141
Onion: *See* **Wild Onion**
Oplopanax horridus 83
osteoporosis 70, 104
Our Lady's bedstraw 55
Oxyria digyna 157
pain 49, 86, 108, 110, 180, 239, 243, 245, 249, 257, 270
    capsules for 301
    liniment for 254
    pharmaceutical remedies 271
    tincture for 294
    See also body part, system or condition
palmate-leaved coltsfoot 73, 75
paper birch 241, 245
Paracelsus 389
paralysis/numbness: *See* numbness/paralysis
parasites
    birch 245
    See also digestive system; fungus; skin
partridgeberry 195
pasture sage 149
paunnat 141
pectin 219
pectoral 74, 108, 116, 125, 133, 257
Pedersen, Nadine 366
peppermint 328
perception 237
perfume 61
Persea americana 310
perspiration: *See* skin
pesticides 81
    See also contaminants; poisons; toxins
Petasites frigidus var. *sagittatus* 73–4
Petasites var. *frigidus* 73
phlegm: *See* respiratory system: mucus; throat: mucus
The Physicians of Myddfai 53
phytochemicals 220
    flavonoids 79–80, 145, 147, 219, 230
        quercetin 98–100, 116, 143, 197, 215–6, 220
phytochemistry 401–2
phytoestrogens 162
phytonutrients 230
phytosterols 80
Picea glauca 259–60
Picea glauca var. *porsildii* 260
Picea mariana 259–60
pigment: *See* dye
pigweed 111–2
Pinacaeae (pine family) 247, 251, 259, 265

Index

Pine: See **Lodgepole Pine**
pineappleweed 59–61, 328
pine bark, inner 252, 356
pine needles 338
pink eye: See eye: infection: conjunctivitis
Pinus banksiana 252
Pinus contorta ssp. *latifolia* 251–2
Pinus contorta var. *yukonensis* 252
pioneer species 91
pissenlit 80
pitch: See resin
Planetary Herbology 65
Plantaginaceae (plantain family) 132
Plantago major 132
Plantago ovata 135
Plantain 124, **132–6**, 202, 210, 282–3, 292, 294, 297, 300–1, 307, 313–4, 319–20, 327–9, 338, 360, 366, 370
plant essence: See **essences: flower and plant**
plaster 68, 70
Pliny the Elder 95, 389
PMS: See reproductive system: premenstrual syndrome
Poaceae (grass family) 167
poisons 105, 134, 140
  to dogs 245
  for insects 156
  See also contaminants; pesticides; toxins
pollination 145
pollutants 114
  electromagnetic 121
Polygonacea (buckwheat family) 87, 157, 160
polyphenols 57, 188, 197
poor man's pepper 154
Poplar: See **Balsam Poplar**
Populus balsamifera 256
Populus tremuloides 238
poultices 49, 60–1, 64–5, 68–70, 74–5, 85–6, 88, 91, 93, 99, 104, 108, 116, 119, 133, 135, 152, 155, 158, 161, 179, 192, 197, 201, 206, 209–10, 219, 226, 230, 236–7, 239, 243, 248–9, 252–3, 257, 261, 267, **318–19**, 320
  eye 64
  spit 60, 100, 112–3, 124, 132–4, 155, 159, 179, 304, **319**
  See also Recipes
powders 99, 104–5, 121
power plants, 29
prairie sagewort 149
prebiotic 267
pregnancy: See reproductive system
prickly juniper 208
prickly wild rose 144
progesterone 180
prostaglandins 57
prostatitis: See urinary system: infection
protein 258
  collagen 104
  ferritin 89–90
  plant 66, 71, 121, 126, 147, 153, 182, 193, 198, 220, 223, 235–7
Prunus dulcis 310
psoriasis: See skin: inflammation
puberty: See reproductive system
purifier: See cleanser
puskwa 75
pussy willow 271
puzzle grass 102–3
pyorrhea: See mouth: gums
quaking aspen 238

quercetin: See phytochemicals
quinoa 113
radiation 93, 181
radicals, free 57, 161, 192, 215
ragweed 99
rashes: See skin: inflammation
Raspberry 155, 176, 192, **217–20**, 223, 271
raspberry leaf 69
Reading Yukon Forests 253
Recipes
  Alcoholic Drinks
    Dandelion Wine 81
  Boreal Baths and Steams
    Boreal Healing Bath 325
    Cold-and-Flu Bath 326
    Flower-Power-Peace Bath 325
    Oh-My-Aching-Bones Bath 326
    Rejuvenating Facial Steam 329
    Respiratory Steam 329
    Sweet-Balsamic Body Oil 250
    Tree-Medicine Purification Bath Oil 325
    Yarrow Oily Skin Treatment 183
  Breads
    Pine Bark Bread (Crackers) 255
  Capsules
    Allergy-Aid Capsule 301
    Pain-Aid Capsule 301
    Sleep-Aid Capsule 301
  Cereals
    Birch-Syrup Granola 246
  Cleaners
    Pine Household Cleaner 254
  Cleanses
    Juniper-Berry Cleanse 213
  Decoctions
    Rhodiola Decoction 140
    Rosehip Decoction 286
    Spring-Cleanse Dandelion Decoction 286
  Deodorants
    Bedstraw Deodorant 58
  Desserts and Sweet Treats
    Cooked Fruit Leather Process 377
    Creamy Cranberry-Vanilla Ice 373
    Dandelion-Petal Cake 371–2
    Flower-Delight Tempura 375
    Living Fruit Leather 377
    Raw Uncooked Fruit Leather Process 376
    Roasted Dandelion-Root Ice Cream 374
    Saskatoon-Berry Pie 224
    Wild-Blueberry Fruitsicles 373
  Ear Remedies
    Bedstraw Ear Oil 58
  Essential Oils and Hydrosols
    Basic Oil-Infusion Recipe 305
  Flower Essences
    Twinflower Vibrational Essence 173
  Fomentations
    Lungwort Fomentation 117
    Wild-Onion Fomentation 319
  Healing Herbal Creams
    Boreal Botanical Skin Cream 317
    Muscle-Ease Cream 317
    Sacred-Spirit Sweetgrass Cream 170
  Herbal Infusions
    Beautiful Hair, Skin, and Nails Infusion 285
    Dandelion-Root Coffee 285
    Detoxifying Infusion 283
    Get-Up-and-Go Infusion 283
    Good-for-Fever Infusion 282
    Good-for-Gout Infusion 284

  Good-for-Nausea Infusion 284
  Good-Sleep Infusion 283
  High-Calcium Infusion 282
  Moon-Time Infusion 284
  Tea-for-Two Pregnancy Infusion 284
  Herbal Tea or Tisane
    Afternoon Floral Tisane 281
    Basic Herbal-Tea Recipe 280
    Bearberry Tea 54
    Mountain Sorrel Iced Tea 159
    Red-Clover Tea Aid (Iced) 290
    Spirit-of-the-Boreal-Forest Tea 281
    Tamarack Tea 268
    Wild Chamomile Tea 62
  Jams, Jellies, Chutney, and Topping Syrups
    Basic Syrup Recipe 354
    Basic-Syrup Recipe 298
    Cranberry Chutney 198
    Dandelion Jelly 350
    Dandelion-Petal Syrup 355
    Fireweed Jelly 94
    Herb-and-Berry Sweet Summer Days Syrup 355
    Mossberry Syrup 216
    Raspberry Jam 220
    Rose-Petal Jelly 351
    Spruce Tip Jelly 352
    Wild-Blueberry Chutney 353
    Wild-Blueberry Jam 349
    Wildly Mint Jelly 122
  Liniments
    Juniper-Willow Liniment 317
    Pine Tree Ache-and-Pain Liniment 254
  Main Courses and Sides
    Healthful Weed Pie 368
    Lamb's Quarter Omelette 114
    Piquant Plantain Pie 366
    Wild-Greens Stir-Fry 370
    Wild-Weed Spanakopita 367
  Masks
    Blueberry Beauty RX (Blueberry Fruit Mask) 190
    Wild-Strawberry Exfoliating Face Mask 231
  Medicinal Oils
    Boreal Healing Oil 258
    Boreal Skin Oil 307
    Dandy-Lioness Oil 308
    Labrador Tea-Flower Oil 110
    Muscle-and-Pain Relief Oil 307
    Plantain Herbal Healing Oil 136
  Medicinal Syrups
    Coltsfoot Cough Syrup 76
    Cough-and-Cold Syrup 300
    High-Iron Syrup 299
  Muffins, Biscuits, and Pancakes
    Blueberry Muffins 342
    Cloudberry-Buckwheat Pancakes 194
    Cranberry-Mint Muffins 343
    Dandelion-and-Birch Cornbread 347
    Dandelion-Petal Pancakes 345
    Red-Clover Tea Biscuits 344
    Wild-Berry Buckwheat Pancakes 346
  Poultices
    Alder Poultice 237
    Plantain Poultice 319
    Spit Poultices 319
  Raw Juices and Smoothies
    Athletes' Blend Smoothie 340
    Basic Berry Juice Recipe 334
    Blueberry-Mossberry Juice 336

Part V: *For Reference*

Chickweed Shooter 339
Cool Cranberry Juice 335
Currant Juice 207
Herb-and-Berry Antioxidant Smoothie 340
Hot Tomato 338
Kalerific! 338
Pine-Forest Smoothie 338
Pink Drink: High-C 340
The Purple Drink 190
Revolutionary Raspberry Rosehip Juice 335
Spring-in-the-Boreal Smoothie 336
Strawberry Fields Forever Juice 335
Strawberry Spinach Smoothie 166
Sunshine Daydream Smoothie 337
Toon-Town Smoothie 337
Wild Rose and Mint Lassi 341
Wise-Weed Smoothie 337

Salves, Ointments, and Balms
All-Purpose Botanical Ointment 313
Arnica Ointment 50
Aromatic Herbal Balm 314
Basic-Salve Recipe 312
Boreal Balm 257, 314
Moisturizing Lip Balm 314
Spruce Winter Salve 264
Three-Tree Healing Salve 313
Wild-Rose Petal Healing Ointment 148

Sauces and Dressings
Borealis Green-Goddess Dressing 163
Chickweed-Garlic Green Dip 66
Dandelion Dressing 362
Extra-Green Pesto Sauce 356
Gratitude Gathering Cranberry Sauce 199
Green Sauce 360
Highbush Cranberry Applesauce 358
Juniper Butter 357
Nettle Pasta Sauce 358
Raspberry-and-Rosehip Vinaigrette 362
Wild-Onion Spread 131
Wild-Rose Petal Vinaigrette 361
Wild-Weed Dressing 361

Scrubs
Detoxifying Juniper Sea-Salt Scrub 213

Shampoos and Rinses
Soapberry Hair Shampoo 228

Soups and Salad
Cream of Wild-Weed Soup 364
Green Bouillon 363
River Beauty Summer Salad 143
Wild Dock Sorrel-Potato Soup 90
Wild-Weed Salad 365

Styptic Powder
Stop-the-Bleeding Styptic Powder 320

Tinctures, Vinegars, Elixirs, and Bitters
After-Dinner Elixir 297
Athlete's-Aid Tincture 294
Boreal Bitters 297
Clear-Skin Tincture 72
Cold-and-Flu Tincture 294
Gentian Tincture 97
Moon-Cycle Nourishing Vinegar 203
Nettle Tincture 128
Nourishing Vinegar Tonic 296
Pain-Aid Tincture 294
Prostate-Support Tincture 295
Shepherd's Purse Tincture 156
Skin-Disinfectant Spray 295
Urinary-Tract Tincture 294
Valerian-Root Tincture 177
Willow Inner-Bark Tincture 273

Tooth Remedies
Aspen Toothache Remedy 240
Horsetail Tooth Powder 105

Trail Herbs
The Hiker's Herb 153

red bearberry 51–2
Red Clover 67–72, 162, 218, 287, 289, 308, 340, 344, 364
  See also alsike clover; cow clover; wild clover
red osier 54
red sorrel 160
red willow 235
refrigerant 96, 119, 196, 218
remedies: *See* aids
repellent
  insect 109, 151–2, 179, 183, 237
  rodent 109, 119
reproductive system 217
  birth control 70–1
  birthing 53, 61, 69, 155–6, 162, 197, 202, 207, 219, 223, 266, 328
  cancer (abnormal cells) 69
  conception 218
  endometriosis 156
  episiotomy 181
  fallopian tubes
    scarring 69
  fertility 69, 218
  infection 53
    thrush 100
  menopause 70, 146, 156, 223
    excessive bleeding, clotting, cramping 202
  menstruation 53, 60–1, 69, 80, 88–9, 104, 140, 203, 207, 230
    cramps and pain (dysmenorrhea) 53, 60, 69, 88, 120, 146, 152, 176, 180–1, 192, 218, 223, 271
      tincture for 294
    delayed 120, 152, 156
    excessive bleeding (menorrhagia) 126, 155–6, 162, 180, 218, 223
    infusion for 284
    lack of a cycle (amenorrhea) 181
    nausea 218
    premenstrual water retention 181
    See also premenstrual syndrome (PMS)
  ovary: cancer 70
  pelvic congestion 180
  perineum 181
  placenta 266
  postpartum hemorrhaging 155–6, 218
  postpartum treatment 181
  pregnancy 54, 66, 69, 76, 97, 109, 121, 153, 156, 183, 212, 219, 254, 263, 272
    infusion for 284
    midwives 219
    miscarriage 202, 218, 223
    morning sickness 218, 230
  premenstrual syndrome (PMS) 80, 146, 206
    water retention 126
  puberty 89
  testes: pain 202
  uterus
    cancer 70
    cramps 197, 202
    pain (endometriosis) 202
    "sensitiveness" 202
    stimulant 209
    strengthening 218

vagina
  discharge (leucorrhea) 202
  douche (wash) 53
  infection 53
vaginitis 210
  ulcer 53
resin (pitch) 104, 247, 249, 253, 257, 259, 261–4, 267, 279
resin birch 241
respiratory system 125, 263
  aids and remedies 108, 116, 134, 212
  ailments 75, 85, 88
  breathing 73, 76
    wheezing 75
  bronchial
    aids and remedies 108
    bronchitis 53, 69, 75, 85, 88, 210
    congestion 134
    irritation 202
    steam 107
  colds 69, 75, 85, 88, 120, 130, 140, 151, 173, 180, 206, 210, 215, 230, 239, 249, 252, 267, 271
    baths for 326
    syrup for 300
    tincture for 294
  congestion 249, 257–8, 313, 319, 328–9
    syrup for 300
  coughs 65, 69, 73–5, 108, 115–6, 134, 173, 202, 239, 261
    coughing blood 227, 249
    dry hacking 75–6
    spasmodic 202
    syrup for 69, 74–6, 134, 300
    whooping 69, 75, 253, 262
  flus 85, 120, 140, 180, 206, 210, 215, 230, 239, 249, 252, 267, 271
    baths for 326
    tincture for 294
  infection 249
    croup 253, 262
    tuberculosis 75, 85, 103, 215, 227, 249, 253
  lungs 173
    cancer 75, 108
    cleanser 135
    congestion 253, 261
    infection 264
    infusion for 283
    irritation 108, 126, 133–4
    pneumonia 253, 262
    stimulant 75
  mucus 222, 249
  nose
    bleeding 156, 179
    neti pot wash 243
  sinus passages
    congestion 243
    sinusitis 210
    steam 107
  soothing 69
  stimulant 116
restorative 139
rheumatism: *See* skeletal system: joints
Rhodiola 137–40, 292, 294, 340
Rhodiola integrifolia 137–8
Rhodiola rosea 138
Rhodiola rosea L. ssp. *Integrifolia* 138
Rhododendron groenlandicum 106
Rhododendron tomentosum 106–7
rhubarb 70, 82, 161, 336
Ribes hudsonianum 204–5

Index

Ribes oxyacanthoides 204–5
Ribes oxyacanthoides ssp. *oxyacanthoides* 205
Ribes triste 204–5
riboflavin: *See* vitamins: B-2
Richters Herbs 71, 138
Ricinus communis 310
ringworm: *See* skin: infection
rinse: *See* wash
ripple grass 132
River Beauty 141–3
Rodia riza 138
Rogers, Robert 54, 57, 61, 68, 75–6, 85, 88, 96, 100, 105, 116, 126, 140, 143, 175, 180, 207, 212, 223, 227, 240, 263, 267
Rogers' Herbal Manual 267
Root Chakra Vibrational Essence 105, 212, 302
rope 268
Rosa acicularis 144, 147
rosacea 66
Rosaceae (rose family) 144, 191, 217, 221, 229
Rosa mosqueta 311
rosavins 139
Rose 99, **144–8**, 177, 287–8, 303, 321–2, 328
  hips 125, 289, 299
rosebay-willowherb 91, 94
rosehip 144–6, 148
Rosehip Syrup 299
roseroot 137
Royal Roads University, Centre for Non-Timber Resources 386
rubefacient 125, 209, 248–9
Rubiaceae (madder family) 55
Rubus chamemorus 191
Rubus idaeus 217
Rudge, Simone and Tom 383
Rumex acetosella 160
Rumex crispus 87
Rumex species, *Rumex arcticus* 87
sachets 59–61, 109, 153
sacred grass 167
"sacred herb" 134
sacred-spirit plant healing 28
SAD: *See* Seasonal Affective Disorder
sadness 143
Sage: *See* **Wild Sage**
Salicaceae (willow family) 238, 256, 269
salicin (methyl salicylate) 236, 239, 243, 257, 270, 273
salidrosides 139
Salix arctica spp. 269
Salix glauca 269
Salix species (Willow) 269
salmonberry 191
salves 49, 60–1, 64–6, 68–70, 75, 79, 85–6, 88, 91, 93, 99–100, 104, 116, 119, 121, 133, 135, 152, 168, 180, 201–2, 209–10, 236, 239, 243, 248, 250, 252–3, 257–8, 261–2, 264, 267, 271, 304, 307–8, **309–14**
  double-boiler method 312
  hot-water bath or bain-marie method 313
  See also **Recipes**
salvia boreale 150
Sami 162, 279, 357
saponins 65, 100, 114, 116, 226
Saskatoon Berry 221–4, 337
sauna herbs 108, 210, 249, 329
Saxifragaceae (saxifrage family) 204
scabies: *See* skin: infection
Schofield, Janice 76, 236–7, 323
Schroeder, Lori 253

sciatica: *See* nervous system
scouring rush 102
scour weed 102
scrape: *See* skin: wound
scrub 211, 213
scurvy 113, 159, 162, 197, 252, 261–2
scurvy-grass 157
Seasonal Affective Disorder (SAD) 139, 250
sebum: *See* skin: oily
sedative 68, 74–5, 175, 201, 239, 257, 294
  mild 60
  prescription 177
self-awareness 71
self-confidence 268
self-esteem 193
Service, Robert: "The Spell of the Yukon" 223
serviceberry 221
Sesamum indicum 311
shampoo: *See* hair: wash
sharp tree 208
shavegrass 102
Sheep Sorrel 160–63
Shepherd, John 225
Shepherdia canadensis 225
Shepherd's Purse 154–6, 291, 320
shock 49, 93, 143, 258
silica (silicon dioxide) 71, 80, 103–4, 127, 182, 211, 220
Simmondsia chinensis 311
simpler's measurement method 37, 294
sinus passages
  aids and remedies 107–8
  decongestant 162
Skeet-Addle 151, 179
skeletal system
  bones
    broken 49, 104, 116, 135
    density 104
    tonic 244
  joints 263
    arthritis 69, 85, 100, 206, 210, 216, 227, 236, 257
    inflammation 125
    pain 100, 109, 120
    rheumatoid 239
    baths for 326
    compress 100
    gout 69, 79, 105, 135, 206, 210, 243, 267, 271
    inflammation 125
    infusion for 284
    osteoarthritis 271
    pain 85, 151, 210, 253, 262
    pain (rheumatic) 49, 100, 176, 243
    repetitive strain 100
    rheumatism 69, 79, 85, 104, 206, 210
    See also anti-rheumatic
    soreness 69, 250
    sprains 49, 65, 104, 243, 271
skin
  bruises 49, 134
    oil for 307
  burns 91, 100, 104, 108, 239, 250, 258, 319
  cancer 65, 70, 219, 230
  cells 250
  chapped 168
  cleanser 81, 109, 114, 236
  congested 212
  cracked 108, 135, 257

  cream for 315
  detoxifier 81, 162
  dry 147, 213, 309
    oil for 307
  face
    aftershave 263
    astringent 211
    cleanser 57
    clogged pores 220, 245, 328
    creams 121, 147
    mask 162, 190, 231
    steam 56, 61, 66, 81, 101, 121, 147, 156, 182, 193, 220, 231, 236, 245, 253, 263, 328–9
    toner 231
    tonic 121, 147, 156, 182, 220
    wash 156
  infection 86, 100, 189, 264
    abscess 65, 135
    boil 64
    burn 135
    chickenpox 180, 236–7
    dermatitis 147, 212
    fungal 152
    leprosy 134
    measles 180
    spray for 295
    strep (*Streptococcus*) 202
  inflammation 65, 67, 104, 109, 182, 243, 257
    acne 66, 68, 93, 101, 121, 147, 152, 177, 206, 211–2, 227, 231, 245, 249, 253, 261
    acne (hormonal) 89
    blister 152, 313
    boil 65, 152, 227, 245, 249, 253, 261
    dandruff 108, 127, 177, 211, 245
    eczema 57, 68–9, 75, 79, 135, 212, 230, 245–6, 253, 257–8, 267
    insect bites 49, 65, 75, 100, 104, 108, 113, 132–3, 135, 146, 159, 180, 236–7, 249, 270, 313, 319
    pimple 89, 177, 236
    psoriasis 68, 135, 206, 230, 245–6, 257, 267
    rashes 66, 68, 89, 113, 120, 135, 239, 245, 250, 258, 264, 267, 319
      diaper 307
    redness 93
    scarring 231
    slivers 313
    sores 75, 89, 151, 227
    stings 60, 75, 100, 108, 113, 124–5, 127, 132–3, 135, 270
    sunburn 231
    warts 80, 89
  infusion for 285
  irritation 64–6, 68, 126, 257
    chapping 108, 309
  oil for 307
    eczema 93, 206
    frostbite 267
    itching 57, 108, 236–7, 239, 257
  oil for 307
    psoriasis 93
    tickling 57
  oily (sebum) 147, 156, 183, 193, 212, 231, 245, 263
  parasites
    head lice 85, 108
    ringworm 89, 134

Part V: *For Reference*

perspiration 85, 108, 120, 162, 206
pH balance 147, 231
pores 81, 147, 182
scabies 108, 126
shingles 134
sunscreen 238–9
tonic 193
toxins 318
ulcer 75, 108, 113
ulcer (leg) 135
wash 103, 113
wound 85, 100, 103, 113, 116, 135, 146, 152, 180, 197, 206, 212, 219, 227, 230, 239, 245, 249–50, 253, 257–8, 261, 270, 305, 313, 319
    bite 134
    bleeding 319–20
    infection 64, 267
    inflammation 61
    puncture 109
    scab 134
    splinter 253, 319
    spray for 295
    weeping 192, 236–7
sleep
    baths for 325
    infusions for 62, 89, 281, 283
    tinctures for 294
    See also insomnia
sleeplessness: *See* insomnia
slippery elm 70, 161, 210
small bedstraw *(Galium trifidum)* 55–6
Smith, Michael 79
smoking, 135
smoking mixture 74–5, 151
smudge 168–9, 212, 250
snakeweed 132
snowshoes 268
soak: *See* bath
Soapberry 225–8
soaps 121
sodium 69
Solar Plexus Chakra Vibrational Essence 49
Solar Plexus Vibrational Essence 302
Solidago canadensis 98–100
Solidago lepida 98–9
Solidago multiradiata 98
Solidago simplex 99
solvent: *See* menstruum
soopolallie 225
soothing 75, 133, 139
Sorrel: *See* **Mountain Sorrel; Sheep Sorrel**
sour dock 87
speckled alder 235
spelt 333
Spik, Laila 255, 357
spirit 210, 212, 248, 250, 301
    cleansing 167, 169
    healing 167, 169
    strengthening 156
The Spirit of Herbs 212
spirits of the woods meditation 30
spirulina 190
spleen 79
sprays
    body 263, **295**
    quit-smoking tincture 135
    throat 264
Spruce 166, 243, **259–64**, 287, 322, 328, 403
spruce tips 341, 348, 352
squashberry 200

stamina 138
starch 104, 253, 262
    removal of 279
Star lady 63
staunchweed 178
steams 52, 56, 61, 66, 81, 101, 104, 107, 121, 133, 147, 156, 182, 193, 220, 231, 236, 239, 243, 248, 252–3, 257, 261–2, 271, **328–9**
    sauna 329
    See also sauna herbs
    See also part of body, system or condition
Stellaria media 63
steroid compounds 143
sterols 104
stimulant 49, 119–20, 152, 175, 179–80, 209, 213, 218, 243, 248–9, 266
    appetite 79, 96, 130, 152
    uterine 155
sting: *See* skin: inflammation
Stinging Nettle 69, 112, **123–8**, 133, 155, 176, 190, 227, 236, 281, 289, 320
stinkgrass 149, 151
stinkweed 149
stomachic 60, 64, 79, 85, 96, 119, 226
stoneberry 51
storytelling tea 106
strain 49
Strawberry 222–3, 227, **229–31**
Strawberry Blite 112, **164–6**
strawberry spinach 164
stress 96, 126, 138–40, 146, 176, 301
    oxidative 192, 215, 222
styptic 52, 100, 104, 116, 133, 135, 155, 179
    powder **320**
    See also **Recipes**
Subalpine Fir 247–50, 306, 322, 328
sugars 253, 262
    fructose 244
    removal 279
    sucrose 244
sulfur 131
Susun Weed's Fertility-Enhancing Infusion 72
swamp red currant 204
sweating: *See* skin: perspiration
sweet coltsfoot 73
Sweetgrass 167–70, 322
sweet-scented or fragrant bedstraw 55
syrups 88, 96, 107, 116, 119, 125, 133, 145–6, 176, 192, 200–2, 205, 209, 215, 218, 222, 230, 243, 248, 261, **298**
    cough 69, 74–6, 134, 300
    high-iron 299
    See also food and beverage; **Recipes**
Tamarack (Larch) 265–8
Tang Center for Herbal Medicine Research 85
tannin 53–4, 57, 91, 104, 116, 145, 162, 168, 222, 258
    removal of 279
Taraxacum alaksanum 77–8
Taraxacum ceratophorum 77–8
Taraxacum lyratum 78
Taraxacum officinale 77–8
teas/infusions 52–4, 56–7, 60–2, 64–6, 68–9, 72, 74–5, 78–81, 85, 88, 91, 93, 96, 99–101, 103–4, 107–9, 112–3, 115–6, 119, 121, 125, 127, 130, 133–4, 139–40, 143, 145–6, 151–2, 155–6, 161, 165, 168–9, 173, 175–7, 181–2, 188–90, 192, 197, 201–2, 205–6, 209–10, 215, 218, 222–3, 226–7, 230, 239, 243, 245, 248–50, 252–4, 257, 261–3, 265, 267, 270–1, **279–91**, 281

cold 52–4, 120, 162
hot 53, 180
iced tea 71, 159, 216, **289**, 299–300
lunar infusion **289**
oil 108, 119, 125, 131, 133, 143, 179–80, 209, 236, 239, 243
oil 304
solar infusion (sun tea) 81, 146, **287**
tisane 40, 121, 145
See also **Recipes**
tea-tree oil 212
tendon
    inflammation (tendonitis) 100, 271
    tightening 57
tension 308
thiamine: *See* vitamins: B-1
Thiel, Lisa 29
Third-Eye Chakra Vibrational Essence 183
thirst 146
Thompson 166
Thomson, Dr. Samuel 27, 70
throat
    infection 249
        strep *(Streptococcus)* 202
    inflammation 76, 89
    irritation 65, 75, 90
    mucous membranes
        irritation 49
    mucus 75, 85, 88, 99, 120, 126, 134, 182, 210
    soreness 69, 75, 88, 100, 152, 168, 202, 210, 219, 239, 249, 261, 264, 267
    spray for 295
    ulcer 113, 165
thrush: *See* mouth: infection; reproductive system: infection
thujone 152
Tierra, Michael 65, 212
    The Spirit of Herbs: A Guide to Herbal Tarot 156
Tigner, Daniel 253
timing, of taking herbal medicines 38
tinctures 52–4, 56, 58, 64–6, 68–9, 74, 79, 81, 85, 88, 91, 96–7, 99, 104–5, 108, 119, 125–6, 130, 133–5, 139–40, 152, 155, 161, 175–7, 179–80, 192, 197, 201–2, 206, 209, 218, 230, 239, 245, 257, 264, 271, 282–5, **291–3**, 299
    See also **Recipes**
tiredness 61
tisane 280, 288
    See also **Recipes**; teas/infusions
tissue 308, 318–9
    aids and remedies 99, 116, 237
    compress 100
    firmness 168
    pain 139
Tlingit 85–6
Tlingit aspirin 83
tl'oodrik 129
tobacco 54
    See also mountain tobacco
tobacco root 174
tonic 52. 56, 68, 74, 79, 89, 96, 108, 116, 125, 146, 152, 155, 175, 179, 182, 196, 209, 218–20, 227, 230, 236, 239, 243–4, 248–9, 252, 257, 266, 270, 296, 299
    See also body part, system or condition
toothache: *See* mouth: teeth
toothpastes 121
topical treatments 304
toxins 85, 89, 108, 120, 180, 308, 318
    See also contaminants; pesticides; poisons

Index

trapper's tea 106
trauma 49, 93, 109, 143, 258
Trembling Aspen 238–40, 301
tremors 49
Trifolium hybridum 72
Trifolium pratense 67
tri'itthoh 87
tsa dzhi 164
ts'iiheenjoo 265
ts'iivii 259
ts'ivii nèechùu 259
tuberculosis: *See* respiratory system: infection
tumour: *See* cancer
turpentine 212, 249
Tussilago 74
Twinflower 171–3
ulcer 85, 97, 103, 120
    See also body part, system or condition
Uncle Berwyn's Yukon Birch Syrup 244, 383
upland cranberry 51
uranium 93
urethritis: *See* urinary system: infection
urinary system (tract) 125, 127
    aids and remedies 100, 135
    ailments 65, 243
    antiseptic 209
    bladder 100, 104, 210
        ailments 65
        cystitis 240
        infection 53–4, 104, 135, 181, 196
        infection (*E. coli*) 197
        irritation 146
    bleeding 135
    cramps 202
    incontinence 181
    infection (UTI) 53, 104, 189, 197
        cystitis 53, 57, 100, 135, 210, 240, 243
        prostatitis 53, 210
        urethritis 53, 100, 135, 210
    inflammation 57, 100, 270
    kidney 210, 263
        aids and remedies 80, 135
        ailments 65, 103, 155
        cleanser 155, 202
        efficiency 125
        infection 135
        inflammation (nephritis) 105, 126
        irritation 125, 146, 212, 264
        stones 80, 100, 104, 197, 243
        stones (calcium oxalate) 198
        tonic 100, 162
    odour 197
    renal dysfunction 90
    stones 53, 90
    tincture for 294
    tonic 162
    ureter: cystitis 240
    urine: test 197
urination
    aids and remedies 93, 104, 108
urine 80, 126
Urticaceae (nettle family) 123
Urtica dioica 123, 281
Urtica gracilis 124
Usnea 58, 100, 264, 292
Vaccinium ovalifolium 187
Vaccinium uliginosum 187
Vaccinium vitis-idaea 195
Valerian 20, 127, **174–7,** 292, 322
Valeriana capitata 174

Valerianaceae (valerian family) 174
Valeriana dioica 174–5
Valeriana officinalis 175, 177
Valeriana sitchensis 175
vanilla grass 167
Van Tat Gwich'in 51, 53, 87–8, 102–3, 106, 129, 144, 149, 178, 187, 191, 195, 204, 208, 210, 214, 235, 241, 259, 262, 265
vasoconstrictor 53
vasodilator 120
veins: varicose 146
vermifuge 60, 152, 161
Viburnum edule 200–2
Viburnum opulus 200
vinegars 218, 230, 291, **292**
    tonic 296
    See also food and beverage; **Recipes**
Virginia strawberry 229
visions 76
vitamins 296
    A (including beta carotene) 54, 61, 66, 71, 80, 89, 93, 97, 111, 113–4, 121, 127, 131, 136, 140, 143, 147, 153, 156, 159, 162, 182, 190, 198, 207, 211, 220, 231, 253, 262
    B-1 (thiamine) 71, 89, 104–5, 156, 182, 207, 211, 220, 245
    B-2 (riboflavin) 104, 156, 182, 190, 207, 211, 231, 245
    B-3 (niacin) 66, 71, 89, 104, 114, 127, 147, 156, 182, 198, 207, 211, 220
    B-5 207, 231
    B-6 231
    B-9 (folate or folic acid) 114, 190, 220, 231
    B-12 71
    B complex (including choline) 80, 105, 146, 156, 159, 162, 218–9
    C (ascorbic acid) 54, 61, 66, 71, 80, 89, 93, 97, 100, 104, 107, 109, 111, 113–4, 121, 127, 131, 136, 140, 143, 145–7, 156–9, 162, 182, 190, 192–3, 198, 200, 203, 207, 211, 216, 218–20, 223, 227, 230–1, 240, 245, 248–9, 253, 257–8, 262, 265, 268, 271
    D 78, 80, 127, 162
    E (tocopheral) 71, 114, 131, 147, 162, 190, 220, 304, 309
    K 71, 114, 121, 126, 136, 155–6, 162, 190, 203, 220, 231
    flavonoids 114
    See also inositol
vitamin chart 420-23
vomiting: *See* digestive system
von Bingen, Hildegard 181, 237, 239
vulnerary 56, 64, 104, 133, 152, 155, 181–2, 257, 270
Vuntut Gwitchin First Nation 11
warfarin 170
warming 130
wash 60–1, 65–6, 99–100, 104, 108, 155, 158, 179, 209, 211, 226–7, 236, 239, 243, 245, 248, 252, 261, 267, 320
    See also body part, system or condition
watercress 70
weakness 89
Weed, Susun 65, 69, 72
    Healing Wise 152
weight loss 57, 65, 80, 182
Welsh, Ruth 10, 11, 89, 29, 262
western coltsfoot 73
wheat-grass greens 336
Wheaton River Garden 383

white man's foot 132
white poplar 238
white sage 169
white spruce 259
"wild asparagus" 93
wild blueberry 187
Wild Chamomile 59–62, 107, 176–7, 236, 287, 289, 322, 328
wild chives 129
wild clover 67
wild endive 77
wild gooseberry 205, 207
wild heliotrope 174
wild mint 118–9
Wild Onion 129–31
wild red raspberry 217
Wild Rose College of Natural Healing 104
Wild Sage 54, 127, 149, **149–53,** 169, 228, 289, 322, 327
wild shamrock 67
wild spinach 111
wild strawberry 229
Willard, Terry 104
William, Steve 170
Willow 269–71
willowherb 91, 141
Wise Woman Herbal for the Childbearing Year 69
witch's broom 53
wolf's bane 47
wolverine's foot 75
Wood, Matthew 57, 89, 99, 181
Woodland Essences 240
wormwood 149, 151
woundwort 98, 100, 178
xenoestrogens 70
xylitol 245
Yarrow 26, 69, 100, 120, 151, 155, 178–83, 190, 202, 206, 289, 297, 306, 319–20, 327–8
yeast: *Candida albicans* 108, 206
yellowberry 191
yellow dock 87, 297
yellow root 87
Yukon Agricultural Association 386
Yukon Brewing Company 125, 198, 244, 383
Yukon white birch 241

Contributors

Contributing Authors

Gwich'in Elder **Ruth Welsh** (1931-2011), who wrote the foreword to this book, was a teacher of traditional plant medicine and the Gwich'in language. Ruth was born at Dootat Gwitshik (Husky River), Northwest Territories, near Fort McPherson. Her  mother, Elizabeth Blake, and her Aunt Esther began teaching her about medicinal plants from a young age. By the time she was old enough to make healing teas and salves, she knew that working with medicinal plants was going to be her life's work. Ruth taught thousands of people throughout the NWT and Yukon, as well as graduate-level ethnobotany students at the University of Victoria, about traditional plant medicine.

 Herbalist and botanist **Robert Dale Rogers**, who wrote "Understanding Chemical Constituents and Phytochemistry" for the reference section of this book, is the author of *Rogers' Herbal Manual* and *The Fungal Pharmacy: Medicinal Mushrooms and Lichens of North America*. He specializes in the plants of Canada's northern prairie and boreal forest regions. He is a professional member of the America Herbalist Guild, and the author of seven books on indigenous and introduced plants hardy in zones 1–3. He is on faculty at Grant MacEwan University and Northern Star College in Edmonton. Visit www.selfhealdistributing.com or www.northernstarcollege.com to learn more.

Main Contributing Photographers

Based in Whitehorse, Yukon, photographer **Cathie Archbould** has been capturing the essence of the land and people "North of 60" for twenty years. A former news photographer who photographed for the *Globe and Mail, National Post, Maclean's,* and the *Toronto Sun,* Cathie now makes her  living as a commercial and industrial photographer. Well known for capturing the people, places, and landscapes of the Yukon and beyond, Cathie photographs the North with a passion that comes from doing what she loves—living large in one of Canada's last frontiers. www.archbould.com

 **Fritz Mueller** has worked in northern Canada since the 1980s as a field researcher, government biologist, and, in recent years, as a wildlife and landscape photographer. Fritz's images have been published internationally (*National Wildlife, Nature's Best, Ranger Rick, Natural History, Canadian Geographic, Up Here, Defenders of Wildlife,* the *Globe and Mail*), and his photographs feature prominently in Yukon tourism campaigns. Fritz was awarded the Banff Mountain Photography Competition Grand Prize for his photograph of Tombstone Mountains. He lives in Whitehorse, Yukon, with his wife and two daughters. www.fritzmueller.com

Peter Long believes taking photographs helps him to see things. A love of the outdoors, a curiosity about the variety of plants in the woods near his house in Whitehorse, and an appreciation for the "macro" feature on digital cameras, led his photography to plants. Before long, the former Lost  Moose book publisher decided to make it a challenge to see how many Yukon plant species he could photograph. www.yukonviews.com

The following individuals also contributed photographs to this book:
Michelle Clusiau, Whitehorse, Yukon
Robert Frisch (1930–1985)
Randi Hausken, Telemark, Norway
Berwyn Larson, Dawson City, Yukon
Birch Kuch, Old Crow, Yukon
Marcella Nowatzki, Marsh Lake, Yukon
Robert Rogers, Edmonton, Alberta